KILLER LYMPHOCYTES

Killer Lymphocytes

by

GIDEON BERKE

Department of Immunology, Weizmann Institute of Science,
Rehovot, Israel

and

WILLIAM R. CLARK

University of California,
Los Angeles, California, U.S.A.

A C.I.P. Catalogue record for this book is available from the Library of Congress.

ISBN-10 1-4020-3269-2 (HB) Springer Dordrecht, Berlin, Heidelberg, New York
ISBN-10 1-4020-3270-6 (e-book) Springer Dordrecht, Berlin, Heidelberg, New York
ISBN-13 978-1-4020-3269-1 (HB) Springer Dordrecht, Berlin, Heidelberg, New York
ISBN-13 978-1-4020-3270-7 (e-book) Springer Dordrecht, Berlin, Heidelberg, New York

---

Published by Springer,
P.O. Box 17, 3300 AA Dordrecht, The Netherlands.

Printed on acid-free paper

Printed in the Netherlands.

TABLE OF CONTENTS

PREFACE

The existence of a unique kind of immune cell – the killer lymphocyte - which destroys other cells in a highly specific manner, has fascinated immunologists for almost half a century. How do these cells, whose precursors have lived in communal harmony with their host, decide that some of their cohabitants must die? And how do they kill them?

The definition of killer lymphocytes came from discovery of their roles in a wide range of in vivo phenomena such as transplant rejection, virus infection and its related immunopathologies, and anti-tumor responses. Yet for the most part almost everything we know about these cells has come from studying them in vitro. They have yielded their secrets slowly and reluctantly. To understand fully how they work, geneticists and immunologists had to unravel the major histocompatibility systems of vertebrates, a long and torturous road that provided some of the darkest hours of immunology. The search for antigen-sensing receptors on both T cells and NK cells was scarcely less frustrating. And the holy grail of cell-mediated cytotoxicity – defining the mechanism by which killer cells take down their adversaries – sorely tested the ingenuity, patience and mutual good will of laboratories around the world.

These questions have now largely been answered. But do we really understand these cells? We can tame them to a large degree in transplant rejection. It may yet turn out that we can harness their immunotherapeutic potential in treating viral and malignant disease. The pivotal role of CTL induction has become part and parcel of many vaccination schemes. But it has become less clear with time that the dramatic destruction of cells wrought by killer cells in vitro represents their true function in vivo.

In writing this book we had several goals in mind. First, we wanted to provide a definitive resource for the subject of cell-mediated cytotoxicity – killer lymphocytes. We felt it would be useful to have a single volume that traces the history of this field, telling its story in terms of key experiments and ideas that have shaped research into the function and biological meaning of cells that kill other cells. At the same time, we wanted to integrate, where possible, the major themes coursing through this subject in its fifty or so year history. And finally, we felt it is time to assess the evidence for and against a role of killer lymphocytes in vivo.

Having worked actively in the field for over thirty years, we were still surprised by how extensive it has become. We estimate – admittedly somewhat loosely – that well over 100,000 papers have been published on various aspects of cell-mediated cytotoxicity (CMC) since it began. Our goal could not possibly be to cover all of this information. We have tried to identify those papers that were key to the origin of each of the many themes in CMC, and then to identify key recent reviews that allow anyone interested in a given sub-topic to work their way back through the existing literature as suits their needs. We have done our best to keep the number of papers cited to a minimum, consistent with that goal. We apologize in advance to our many friends and colleagues whose many excellent and important papers have not been cited here.

All scientific fields are works in progress, and even as we bring this project to a conclusion the field of CMC is morphing in new directions. It may be worth coming back in five years or so to update both new developments and our interpretation of the history of this fascinating field. We hope readers of this book will find it useful, and we especially hope they will feel free to communicate to us their own thoughts on what we have presented, and what we have not.

Finally, we would like to acknowledge the valuable assistance of Drs. Dalia Rosen, Judith Gan and Orit Gal-Garber, and Mr. Steven Manch, all of the Weizmann Institute, for invaluable assistance in preparation of this book.

Gideon Berke
Rehovot
gideon.berke@weizmann.ac.il

Bill Clark
Los Angeles
wclark222@cs.com

January, 2005

Chapter 1

BASIC IMMUNOBIOLOGY: A PRIMER

This chapter is intended for those readers whose primary field of interest is not immunology, and for whom some degree of orientation in the basic structure and function of the immune system would be useful. A comprehensive discussion of the immune system is beyond the scope of this book; for that purpose the reader is referred to any of a number of current excellent textbooks and reviews in immunology, such as Paul's *Fundamental Immunology*[1] or various volumes of *Annual Reviews of Immunology*. However, for purposes of orientation to those aspects of immunity most relevant to understanding the immunobiology of killer lymphocytes, we offer the following greatly simplified summary. Many of the topics that follow will be dealt with in more detail throughout this book.

1 THE IMMUNE SYSTEM

The immune system is composed of a number of interacting organs, tissues and cell types, whose principal function is to protect the host organism from uncontrolled growth of potentially pathogenic microorganisms. The immune system, including killer lymphocytes, may also be involved in suppressing the development of malignancy, but from an evolutionary viewpoint this is likely to be a secondary function. The immune system can also cause harm, through such hypersensitivity reactions as allergy and asthma, through "overshoot" reactions to intracellular parasites (whether the parasites are harmful or not), and through autoimmune disease. These are presumed to be evolutionarily acceptable costs of an otherwise efficiently protective microbial defense system. And the immune system – in

[1] *Fundamental Immunology*, 5th ed.. Lippincott, Williams & Wilkins, 2003.

particular killer lymphocytes – constitutes the major barrier to potentially life-saving organ transplants.

From a functional point of view, it is useful to think of immune reactions as falling into two major categories: innate immune responses, and adaptive immune responses. Innate immune reactions are evolutionarily older, occurring in one form or another in virtually all animals and most if not all plant species. Innate responses are part of the germline inheritance of each organism, and require no further modification to be maximally effective. In general, innate immune reactions in vertebrates rely on recognition of relatively invariant molecular motifs that are, in turn, part of the genetic heritage of potentially harmful microorganisms (Janeway and Medzhitov, 2002). In vertebrate immune systems, various cells have this capacity, in particular dendritic cells (DC), but also macrophages and granulocytes. Among killer lymphocytes, natural killer (NK) cells are considered part of the innate system of defense (Chapter 6). Innate immune responses are highly effective and absolutely essential for survival. Many diseases based on an absence of innate immunity are lethal. The possibility of self-harm from innate immune responses is for the most part neutralized phylogenetically through natural selection.

The major shortcoming of innate immune responses is that they cannot respond quickly to a changing pathogenic environment. One wonders how many established plant or animal species, dependent entirely on innate immune defenses (all living organisms other than vertebrates), have disappeared from the planet because of the sudden emergence of a new and lethal (for the affected species) pathogen. The ability to change quickly in response to a changing antigenic environment is *the* hallmark of adaptive immunity. Adaptive immune responses are based on the principle of ontogenetic generation of essentially unlimited numbers of different receptors with which to probe the antigenic universe. This highly polymorphic receptor repertoire is generated completely randomly, by recombination of a limited number of germline gene fragments. The survival of any particular genetic recombination is dependent on selection by antigen, and a system for negative selection of self-reactive receptors to avoid autoimmunity. Although defects in the adaptive immune system can be lethal, often they are not.

1.1 Anatomy of the immune system

The mediators of adaptive immunity in vertebrates consist of the B lymphocytes, which produce antibodies, and T lymphocytes, which include both T helper cells and killer T cell - the cytotoxic T lymphocytes, or CTL. Antibodies are produced in response to the presence, in the extracellular fluids of the body, of non-self or altered-self biological materials, and are released into the bloodstream, where they trigger a series of reactions culminating in inactivation and/or removal of the non-self material (antigen) from the system. CTL participate in the destruction of self cells compromised by intracellular infection or oncological transformation. Both B-cell and CTL responses are aided by a subset of T lymphocytes called helper T cells, which in addition to their help function are also potent mediators of inflammation. We explore the basic parameters of these reactions in the remainder of this chapter.

The major organs of the vertebrate immune system are the bone marrow, the thymus, the spleen and various lymph nodes scattered throughout the body, which comprise both adaptive and innate immune elements. The bone marrow is the locus of hematopoietic stem cells characterized by, among other things, self renewal, and the presence of a surface marker called CD34. (A list of CD designations used to define immune system cells is given in Table 1.1).

These stem cells give rise to the cells - called generically white cells or leukocytes - involved in immune responses (Figure 1.1), as well as red blood cells. As such, CD34 cells are critical in bone marrow transplantation, and are used for various strategies in gene therapy for diseases such as AIDS and SCID (severe combined immune deficiency). CD34 cells may also serve as pluripotent adult stem cells for generation of non-lymphoid cells and tissues.

Table 1-1. CD designation

CD antigen	Previous Designation(s)	Function(s)
CD1	T6; Leu 6	Presentation of non-peptide (lipid) antigens to T cells ($\alpha\beta$, γ/δ, NKT)
CD2	LFA-2	Intercellular adhesion (with CD58/LFA-3) on NK cells, T cells; activation
CD3	OKT3; T3; Leu 4	T-cell receptor signal transduction
CD4	OKT4; T4; Leu 3a; L3T4	T-cell coreceptor for MHC class II
CD8	OKT8; T8; Lyt2,3; Leu 2	T-cell coreceptor for MHC class I
CD11a	LFA-1	Intercellular adhesion (with CD54/ICAM-1)
CD16	Leu 11a; Fcγ RIIIA	Low affinity Fc receptor on NK cells
CD28	T44	T-cell receptor for B7 (CD80, 86)
CD30	Ber-H2; Ki-1	TNF-family receptor
CD34	gp 105	Adhesion molecule; hematopoietic stem cell marker
CD40	Bp50	Growth factor receptor
CD54	ICAM-1	Intercellular adhesion (with CD11a)
CD56		Intercellular adhesion (immune system ligand unknown
CD57	NKH-1; Leu 7	Intercellular adhesion (?)
CD58	LFA-3	Intercellular adhesion (with CD2)
CD64	FcγRI	Mediates ADCC

CD antigen	Previous Designation(s)	Function(s)
CD80	B7.1	Ligand for CD28, CTLA-4
CD86	B7.2	Ligand for CD 28,
CD95	Fas; APO-1	Mediation of apoptosis
CD152	CTLA-4	Negative regulation of T-cell function
CD159a	NKG2A	NK cell signaling (inhibition)
CD161	NKR-P1A	Regulation of human NK function
CD178	Fas ligand CD95L	Mediation of apoptosis

There are now over two hundred cell-surface markers used to define cell populations within the immune system. These markers were defined by antibodies developed in laboratories around the world, and the initial names for these markers were those chosen by the producing laboratory. It was not unusual for the same marker to have more than one name. In an attempt to bring uniformity to the nomenclature for such markers, an ongoing series of international workshops and conferences on Human Leukocyte Differentiation Antigens are periodically held, in which candidate antiens are tested by multiple laboratories and assigned a CD number (www.hlda.org). While these workshops focus on human CD antigens, where identity is clear the same designation is used for mouse antigens as well. A current list of CD designations can also be obtained at www.ncbi.nlm.nih.gov/prow/.

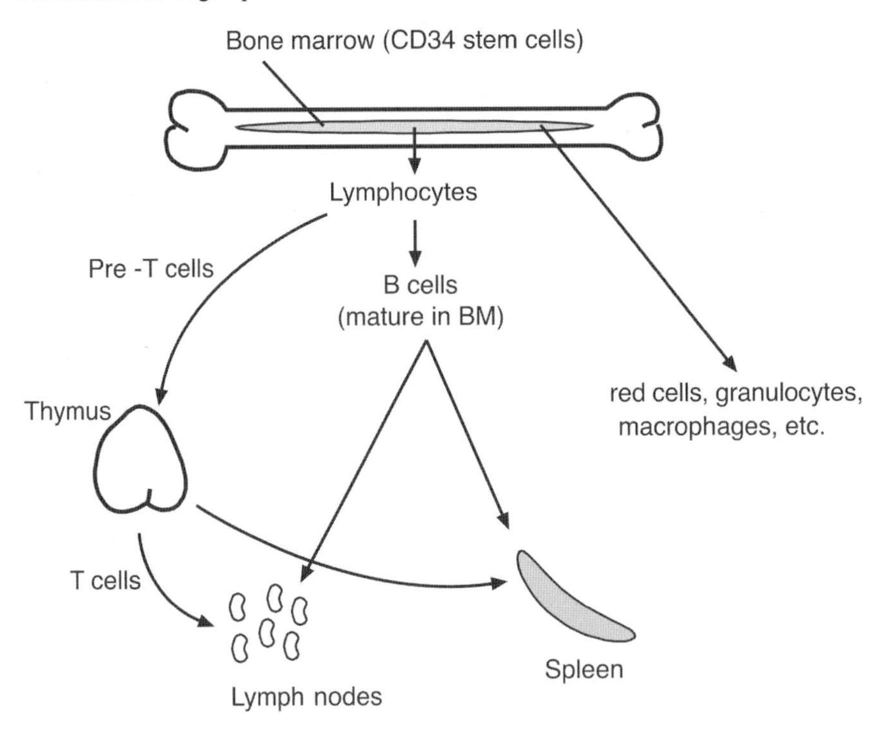

Figure 1-1. The origin of B and T lymphocytes. B-cells complete their maturation in bone marrow and seed directly into B cell compartments of lymph nodes and spleen. T-cells complete their maturation in the thymus.

Bone marrow is also the site of B-cell maturation. After maturation, the B cells seed into specific regions of the lymph nodes and spleen, where they encounter antigen. Once activated to become plasma cells (the activated, antibody-producing form of B cells), many of them migrate back to the bone marrow, which provides space for expanded plasma cell replication and antibody production. And finally, bone marrow can also be a site of T cell encounter with antigen (Feurer et al., 2003), although this is probably a minor venue in comparison with lymph nodes and spleen.

The thymus is the site of T-cell maturation, the 'T' in fact denoting the role of the thymus in their developmental history. Like B cells, T cells originate in the bone marrow, but early in their life history they translocate to the thymus for their final stages of maturation. In the thymus, the T cells differentiate into killer cells (CD8) and helper cells (CD4) (Germain, 2002), and undergo a series of positive and negative selection steps to fine-tune their ability to distinguish self from non-self.

From an immunological point of view, "self" refers to biological materials derived from normal healthy tissues and cells of the host organism. "Non-self" may refer to biological material of extra-self origin (whether pathogenic or not); to altered self (self materials that have been mutated, damaged or in some other way altered); or self materials that appear only secondary to processes such as tumorigenesis. This ability to distinguish between self and non-self is largely a property of T lymphocytes. B cells with potential self-reactivity are often present in adult animals, but their ability to react with self components is circumvented by various strategies, such as functional deprivation of T-cell help.

1.2 Lymphocyte circulation

The lymph nodes serve as filters for antigenic materials brought to them in lymph fluid, which drains virtually every location in the vertebrate body. The spleen filters antigens only from arterial blood. Cells such as macrophages, follicular dendritic cells and B lymphocytes resident in lymph nodes can entrap these materials. More importantly, particularly for killer cell responses, highly specialized antigen-entrapping cells such as dendritic cells can also capture antigen in various tissues of the body (e.g. at the site of a skin allograft), and transport the antigen via the lymphatic circulation to a regional draining lymph node (Figure 1.2). In either case, T cells then scan the entrapping cells for anything interpreted as non-self and thus a potential threat to the host organism's existence. We will discuss the mechanics of this process, and the ensuing activation of T cells, in greater detail below in "Antigen processing and presentation". The spleen plays a role similar to that of the lymph nodes in the entrapment of antigen and as a site for lymphocyte activation. Once activated by antigen, T cells leave the lymph nodes or spleen via efferent lymphatic vessels and circulate throughout the body, looking for the source of the activating antigen.

The circulation of lymphocytes and many of the cells with which they interact takes place in both the blood and lymphatic fluid. Lymph fluid is created when plasma leaves capillaries to deliver oxygen and nutrients to body tissues, and to gather waste products excreted by cells. This fluid collects into lymphatic sinuses, which give rise to lymphatic vessels, which in turn coalesce into larger lymphatic trunks that ultimately empty into the bloodstream at the great veins of the neck. Lymphatic vessels drain every region of the body served by blood vessels; failure of the lymphatic drainage system results in local pooling of lymph fluid in tissue spaces (lymph edema).

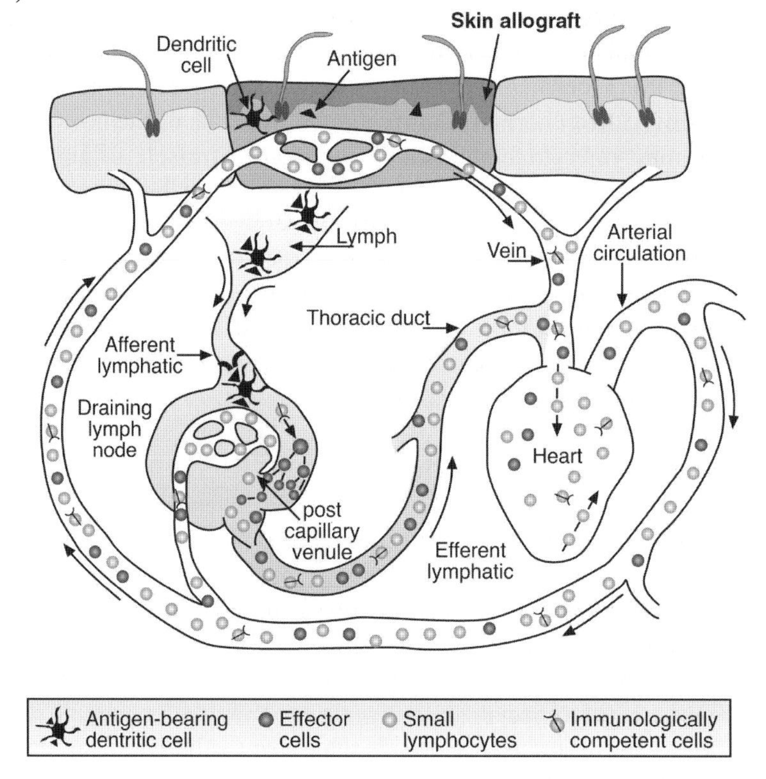

Figure 1-2. Lymphocyte circulation. Antigenic material from a skin allograft is carried by dendritic cells into a draining lymph node, where it interacts with T lymphocytes. Lymphocytes recognizing graft antigens leave the node via the efferent lymphatic and eventually enter the bloodstream and circulate throughout the body until they encounter the graft site.

At different points in their life cycle, T cells may travel in either lymph or blood. Newly generated lymphocytes arrive at lymph nodes from the thymus or marrow via arterial blood, and with the aid of surface receptors home to T- or B-cell-specific regions of the nodes. They leave the blood circulation by squeezing between endothelial cells lining post-capillary venules in the middle and deep zones of the lymph node cortex, which brings them into the lymphatic circulation. Upon activation by antigen, they exit the lymph nodes via the efferent lymphatic vessels, and after passing through varying numbers of "downstream" lymph nodes, re-enter the blood circulation at the neck.

In the antigen-activated state, T cells do not express the surface markers required to interact with post-capillary venules in lymphoid tissues, and thus remain in the blood stream. However, they can interact with chemokine receptors on post-capillary venules at sites of inflammatory reactions, which allows them to leave the blood and enter tissue spaces, where they are once again in the lymphatic circulation. As will be discussed in detail later, some of the activated T cells differentiate into a memory state, and once again express receptors that allow them to take up long-term residence in lymphoid tissue until they encounter cognate antigen again, resulting in their reactivation.

As mentioned above, dendritic cells scattered throughout the body are very important in bringing antigen into the lymph nodes, and converting it to a form recognized by T cells. Dendritic cells that have acquired antigen become mobile, and slip away from their tissue-resident sites into the surrounding lymphatic fluid. From there they travel to the nearest downstream lymph node, where they enter T-cell-rich areas and can be examined by T cells. Dendritic cells spend very little time in the blood circulation.

1.3 T cells and their functions

T cells are divided into two main functional types: helper T cells and killer T cells (CTL). Helper T cells, distinguished by the presence of the CD4 surface marker, are mainly amplifier cells (Figure 1.3). They release chemical signals (cytokines) used by cells of both the innate and adaptive immune systems to carry out their functions (Table 1.2). B cells, for example, require certain T-helper cell cytokines to mature into antibody-producing plasma cells. There are also receptors for many of these cytokines on non-lymphoid tissues, including even the brain, suggesting the possibility of two-way communication between the nervous system and the immune system. T-helper cells are often further divided into T_h-1 and T_h-2 subsets based on the pattern of cytokine production. It is likely that the division into these subsets occurs at the time of initial activation of CD4 cells by antigen.

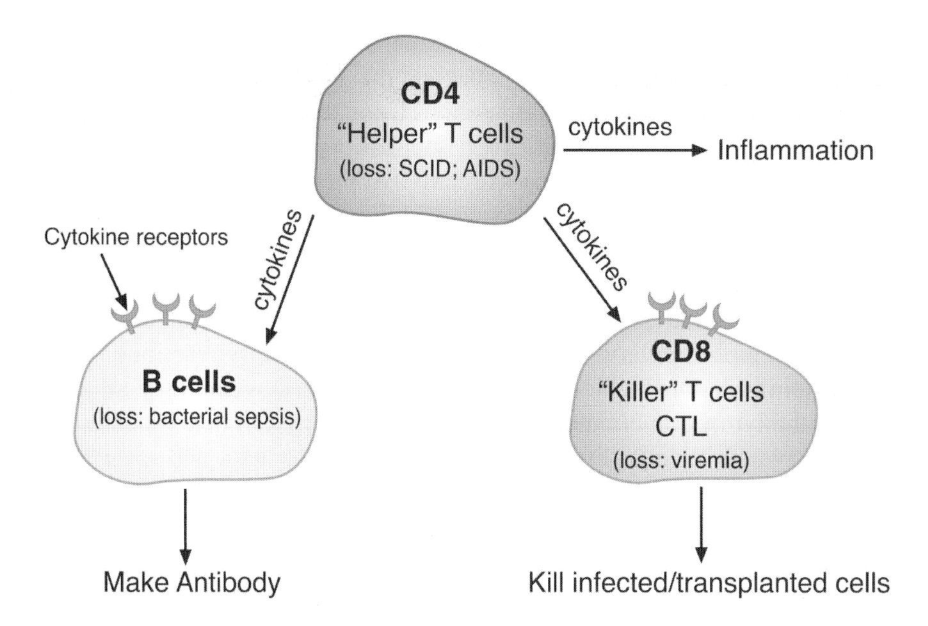

Figure 1-3. T helper (CD4) and killer (CD8) cells. The figure depicts the centrality of CD4 T cells in inflammation, induction of CD8 killer cells as well as in helping antibody production by B lymphocytes.

Table 1-2. Some major cytokines involved in immune system function.

Cytokine	Secreted by	Function	Target cell(s)
IL-2	CD4 T cells	Growth, proliferation	Antigen-primed T, B and NK cells
IL-4	CD4 cells. NK and NKT cells	Growth, proliferation	Most immune system cell
IFN-γ	CD4, CD8, NK cells	Activation or inhibition, depending on target	CD4 cells, B cells, macrophages
TNF-α	Macrophages, NK cells	Depends on target; can be cytotoxic	

In addition to their role as amplifier cells, CD4 T cells can in certain situations kill other cells in vitro, i.e., they can be cytotoxic. The relevance of this killing in vivo is difficult to discern; it would seem counterproductive to kill cells destined to be helped. By far the majority of cell-mediated cytotoxicity resides in the subset of T cells distinguished by the CD8 surface marker (CTL). CTL are mostly involved in elimination of cells within the body compromised by microbial infection, or possibly by oncogenic transformation. In vitro, CTL induce such cells to undergo self-destruction via apoptosis. Like CD4 cells, antigen-stimulated CTL release a spectrum of cytokines that exert varying effects on other cells, particularly those involved in inflammatory responses. Inflammatory reactions promoted by CTL may also be involved in the elimination of foreign cells, or parasite-infected self cells. Coincidentally, from nature's point of view, CTL are the major barrier to clinical transplantation of tissues and organs (see Chapter 7).

In addition to the major populations of helper and killer T cells, there are two other minor T-cell subsets that we will encounter in later chapters. γ/δ T cells have an antigen receptor that is similar to, but molecularly distinct from, that used by CD4 and CD8 cells (described below); these cells are involved largely in microbial defense and in stress responses. They themselves display neither CD4 nor CD8. The so-called NK/T cells (Chapter 6), as the name implies, share properties with both T cells and NK cells. NK/T cells use the same T-cell receptors as CD4 and CD8 cells; they are involved in tumor surveillance and response to stress. Both γ/δ T cells and NK/T cells differ from CD4 and CD8 cells in that they display antigen receptors of greatly restricted variability.

2 COGNITIVE PROCESSES IN T-CELL IMMUNITY

Although the needs of immune cells for interaction with their environment are more varied and complex than for most cell types, the cellular and molecular processes involved are similar to those used by all other cells. Because T cells, in particular, must interact with such a wide range of other cell types throughout the body, both immune and nonimmune, a universally distributed system of cell-associated molecular markers recognizable by T cells is needed. These markers are the cell-surface proteins encoded by histocompatibility genes. All antigen recognition by T cells involves interaction of the T-cell antigen receptor with fragments of antigen complexed to histocompatibility proteins.

2.1 The T-cell receptor for antigen

The molecular nature of the T-cell antigen receptor (TCR) was for many years a complete mystery. The obvious candidate was cell surface immunoglobulin (Ig), which since the late 1960s had been accepted as the antigen receptor for B cells. Indeed, there were numerous reports of immunoglobulin components detectable at the surface of T lymphocytes, mostly detected using antibodies to B-cell receptors (e.g., Binz et al., 1976; DeLuca et al., 1979; Marchalonis, 1980). But this was clearly a case of wanting to see what everyone thought must be there, when in fact it was not. Despite vigorous investigation over ten years or more, no convincing evidence could be obtained for expression of Ig genes within T cells.

Contributing to the confusion were reports (Wigzell, 1969) that although B cells specific for a given antigen could be trapped on a column matrix to which the cognate antigen had been bound, T helper cells, equally specific for the same antigen, could not be retained on the same column. However, CTL specific for a given set of transplantation antigens could be selectively removed from a suspension of lymphocytes by incubation on a monolayer of cells displaying those same antigens (Brondz et al., 1968; Golstein et al., 1971; Berke and Levey 1972; Kimura and Clark, 1974). These two seemingly contradictory results would be resolved once the nature of the T-cell receptor and its interaction with antigen were worked out. Finally, once genomic probes for Ig genes were available, it became obvious that Ig gene components in T cells are in an un-rearranged state, proving that this locus is inactive in mature T cells (Kronenberg et al., 1980).

The molecular nature of the T-cell receptor was not resolved until the receptor itself was physically isolated in the early 1980s. This came about primarily from the work of two laboratories, using different but complementary approaches. In the early 1980s, several laboratories began producing monoclonal antibodies to surface components of mouse monoclonal T-hybridoma cells. They reasoned that among the antibodies thus produced there must be antibodies recognizing the clonotypic T-cell receptor expressed on each hybridoma cell. By testing the ability of various monoclonal antibodies to interfere specifically with T-helper cell function, the Marrack and Kappler lab was the first to identify candidate antibodies against the T-cell receptor. These antibodies were used to immunoprecipitate molecules from membrane fractions of T-hybridoma cells, which led to the detection of a dimeric molecule composed of 40-44 kD disulfide-linked subunits, which they called α and β (Haskins et al., 1983).

The second approach, used by Davis and Hedrick, looked for genes uniquely expressed in T cells, reasoning that among such genes must be those encoding the T-cell receptor or its components. They also speculated that, like immunoglobulin genes, the genes for the T-cell receptor would undergo rearrangement during ontogeny. Using a library of T-cell-derived cDNA that had been "subtracted" by hybridization with B-cell mRNA, they found T-cell cDNA sequences that were absent in other cell types, and that underwent gene rearrangement between fetal and adult DNA. The sequence information derived for these cDNA matched well the initial protein sequence information obtained by Marrack and Kappler (Hedrick et al., 1984a, b).

These seminal studies at the gene and protein level were followed by a very productive period of several years, involving many different laboratories, culminating in the model for the T-cell receptor shown in Figure 1.4. It quickly became apparent that the T-cell receptor (TCR) is actually a complex of molecules, of which the α- and β-chains are only one part. The α- and β-chains are, as predicted by Marrack and Kappler, the polymorphic portions of the complex interacting with antigen. The second part of the TCR complex consists of five polypeptide chains, referred to collectively as CD3. These five chains (γ,δ,ε,ζ,η) associate to form three different dimers: γε, δε, and either ζζ or ζη (ζ and η are made from alternately spliced forms of the same gene).

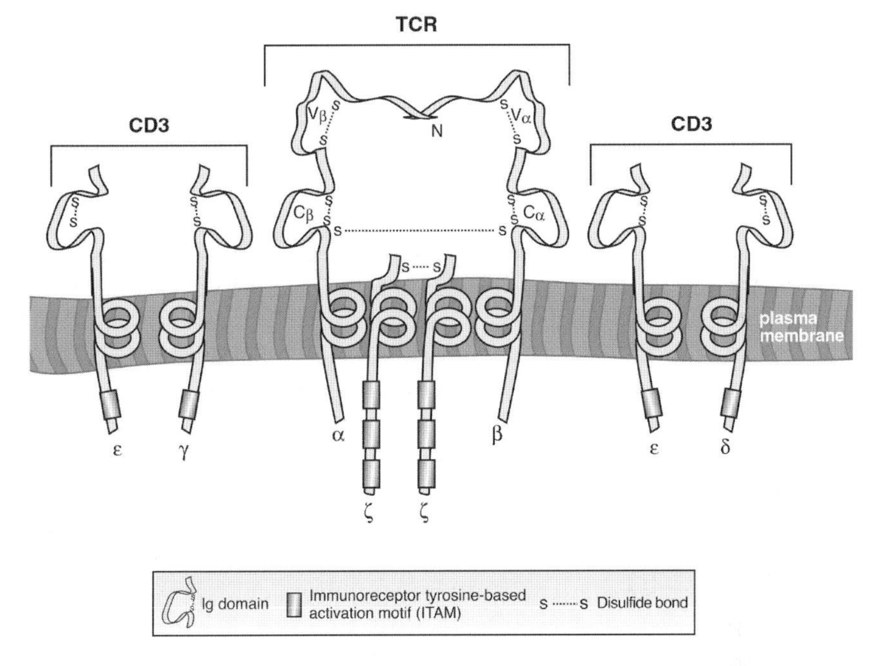

Figure 1-4. The T cell receptor complex. The TCR complex of MHC-restricted T cells consists of the αβ chains of the TCR non-covalently linked to the CD3 proteins and the ζ chain.

CD3 structures are closely, although not covalently, associated with the TCR-α and -β in the surface membrane. CD3 forms this association with TCR-α and -β chains in the endoplasmic reticulum, and this association is absolutely required to bring α and β to the surface. The CD3 chains play a key role in transducing the signal generated by the engagement of TCR α/β with an appropriate antigenic structure during T cell activation. The TCR-α and -β chains have large and complex extracellular domains, but very short intracellular domains, much too short for signal transduction. The CD3 chains have the opposite structure: very short extracellular regions, but extensive intracellular domains. The cytoplasmic portion of the ζ chain contains three repeats of the ITAM (immunoreceptor tyrosine-based activation motif) sequence (Tyr-x-x-Leu/Ile-x; "x" being any amino acid), whereas the γ,δ and ε chains have one ITAM sequence each. ITAM sequences are rapidly phosphorylated upon TCR cross-linking by cognate antigen. This is the crucial first step in bringing the dormant T cell to its fully active state (see below).

Shortly after discovery of the α/β T-cell receptor, a second TCR was defined and named the γ/δ T-cell receptor. The γ chain of the CD3 complex had already been named when this second receptor was discovered, but it was decided to name the second TCR γ/δ anyway, to emphasize its close relationship to the α/β receptor. The vast majority of T cells display the α/β receptor; γ/δ T cells play a largely regulatory role, but can also be cytotoxic. From a structural and molecular functional point of view, the α/β and γ/δ receptors are highly similar.

The genomic elements encoding the T-cell receptors reveal how variability of these receptors is generated. The N-terminal variable regions of each component chain of both the α/β and γ/δ receptors are not encoded as such in the germ line, but rather are assembled by combining either two or three distinct genetic elements (V, J and D segments) in essentially random fashion. In addition to this combinatorial diversity, variability can also be generated by adding or subtracting nucleotides at the joining sites for these various elements (junctional diversity). The great advantage of this scheme is the ability to generate enormous numbers of different T-cell receptors from a modest stock of germline elements. This in turn is what underlies the adaptive nature of T- (and B-) cell immune responses: enormous numbers of different T cells are generated without reference to the antigenic universe, and T cells "adapt" to that universe through selection by antigen of T cells bearing cognitive receptors.

2.2 Histocompatibility genes and their products

Histocompatibility systems of vertebrates, as the name implies, are systems of genes and proteins that determine whether or not cells and tissues transplanted between different individuals are compatible – whether they will survive or be rejected by the immune system. While there may be up to two dozen histocompatibility genes in any given vertebrate species capable of triggering graft rejection between genetically disparate (allogeneic) individuals (Little, 1956), each species has a set of genes referred to as the *major histocompatibility complex* (MHC), the products of which provide the strongest barrier to transplantation, in terms of the vigor and pace of rejection. Although discovered in connection with transplant rejection, it was obvious from the beginning that causing rejection could not be their biological function. Why would there be a system of genes and gene products dedicated to regulating tissue transplantation, a medical procedure that has no counterpart in nature? We now know that the protein products of MHC genes facilitate recognition by T lymphocytes of potentially antigenic materials, either foreign or altered self, present inside of cells throughout the body.

The discovery of cell-associated products of the major histocompatibility gene systems came about as the result of work by Peter Gorer in the 1930s (Gorer, 1936) on the serology of blood cell antigens in the mouse. One of the red-cell antigens, which he initially called "antigen II," turned out to be present not only on red cells, but on all cells in the body. Subsequent work of Peter Medawar, George Snell, and many others established antigen II, or H-2 as it came to be known, as the major gene system controlling strong graft rejection among different inbred strains of mice. Sorting out the various genes comprising MHC systems by serological means, and determining the functions of the associated gene products, has had a long and often confusing history. This story has been admirably set forth in an excellent book by Jan Klein (1986) and a review by Hugh McDevitt (2000), and will not be discussed here.

The structures of the human and mouse MHC systems, called HLA and H-2, respectively, are shown in Figure 1.5. This arrangement of MHC genes is typical of most vertebrate species. There are two major types of MHC genes, referred to as class I and class II, associated with all MHC systems. Class I gene products (HLA-A, -B, and -C transmembrane proteins) carry fragments of peptides synthesized within the cell to the cell surface, where CD8 T cells monitor them. Class II gene products (HLA-DP, -DQ, -DR proteins) carry to the cell surface fragments of proteins endocytosed by the cell from its environment. Class II gene products are examined at the surface by CD4 cells. The true function of MHC systems is thus revealed as a means of informing CD4 and CD8 T cells about the proteins present within cells and in their immediate environment.

The MHC gene locus also contains varying numbers of other genes related to immune function, such as complement genes and cytokine genes, and the so-called "non-classical" class I MHC genes whose structure is similar to the "classical" MHC genes just discussed, but whose distribution and mode of antigen presentation to CTL is quite different. MHC complexes also contain several genes unrelated to immune function, such as the hydroxylase genes, and varying numbers of unexpressed pseudogenes structurally related to both class I and class II genes.

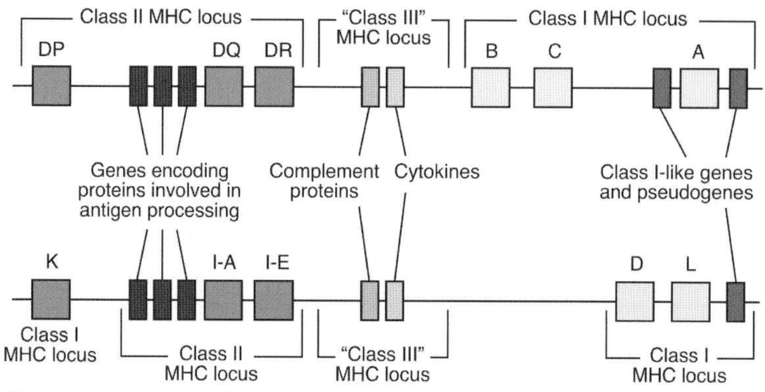

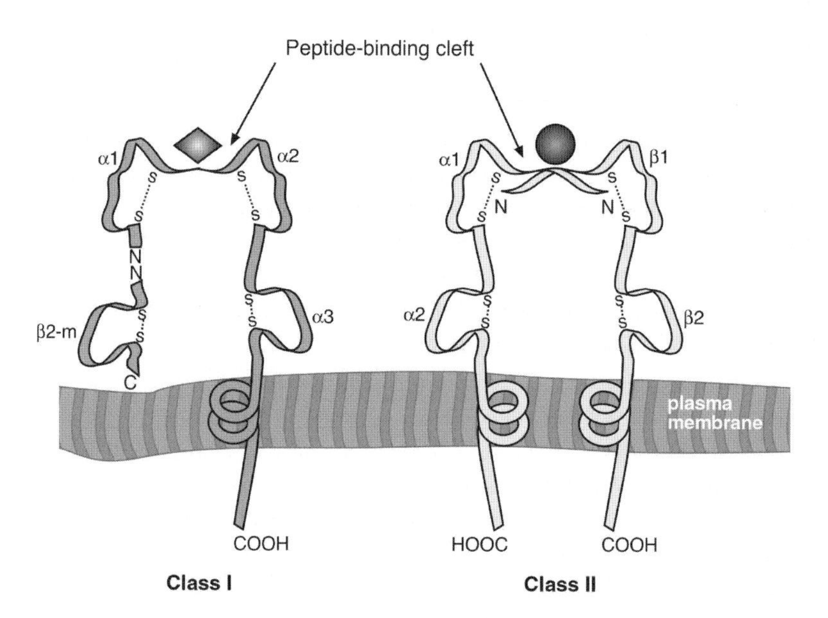

Figure 1-5. MHC gene loci and proteins in human (HLA) and mouse (H-2).

The overall structure of the mouse MHC gene complex, called H-2, is also shown in Figure 1.5. The number and types of class I and II genes is similar to the human system, the main difference being that one of the murine class I genes (H-2K) has been transposed from the telomeric to the centromeric end of the complex. This arrangement is found in only a few other species, such as the closely related rat MHC system (Rt-1). The non-classical MHC genes associated with H-2 are also similar to those found in HLA. Class I and class II MHC products of mice function in exactly the same way as the corresponding proteins in humans.

The structure of the MHC molecules encoded by class I and class II genes varies little among vertebrate species. MHC proteins are part of the structurally defined immunoglobulin super-family of genes and proteins. This family includes heavy and light chains of immunoglobulin; the T-cell receptor α and β chains, as well as the γ, δ and ε chains of the CD3 complex; class I and II MHC molecules; CD4 and CD8; and intercellular adhesion molecules such as ICAM and VCAM. Proteins encoded by this family are characterized by compact globular subdomains of approximately 110 amino acids, with alternating stretches of α-helix and β-pleated sheet that dictate a unique and recognizable three-dimensional folding pattern.

Class I genes encode an approximately 350-amino acid peptide chain called the α-chain, which has three distinct globular domains called α-1, α-2 and α-3, a transmembrane domain, and a short intracellular domain. The α-chain associates in the endoplasmic reticulum with a non-MHC-encoded molecule called β-2 microglobulin (β-2m), which appears to be necessary only for proper folding and transport of the α-chain to the cell surface. β-2m is in essence a single Ig-family domain, and is thought by some possibly to represent an evolutionary ancestor of the Ig superfamily. Class I molecules are N-glycosylated at two or three sites, depending on the species. Each moiety contains about 15 sugar residues, for a total carbohydrate mass of about 3300 daltons. Blockage of the addition of these carbohydrate groups has little or no effect on class I expression or function.

Class I molecules are found on nearly all cells of the body, the exceptions including germ cells and pre-implantation embryos. They are found in the highest concentration on lymphocytes (ca. 10^5 molecules per cell), and in moderate concentrations (ca. 10^4 molecules per cell) on other somatic cell types. Some cells, such as neurons, monocytes and hepatocytes, display fewer than 10^3 molecules per cell. It is interesting that Gorer originally detected his "antigen-II" on red blood cells, since these cells display among the lowest density of class I molecules ($< 10^2$ per cell). Class I molecules bind 8-10 amino acid peptide fragments of proteins synthesized within the cell (we will discuss antigen processing and presentation in the following sections). A precise model for class I-peptide interaction derives from X-ray crystallographic studies, which reveal a peptide-binding cleft consisting of two alpha-helices, one derived from part of each of the two amino terminal (α1 and α2) domains, supported by a floor of 7 pleated sheets (Bjorkman et al., 1987a, 1987b; Garrett et al., 1989; Saper et al., 1991).

Class II molecules are composed of two separately encoded polypeptide chains called α (ca. 33 kDa) and β (ca. 28 kDa), which associate non-covalently in the surface membrane. Each of these chains consists of two globular extracellular domains, a transmembrane domain and a short intracellular domain. The overall three-dimensional structure of class II proteins in the membrane is quite similar to the structure formed by the class I α-chain plus β-2m. Peptide fragments of 20 or so amino acids derived from proteins processed through endocytic pathways are presented in a cleft formed by helical regions of the α_1 and β_1 domains. Most vertebrate species have three kinds of class I molecules, and between two and four major types of class II molecules.

Class II MHC molecules are much more restricted in cellular distribution than are class I molecules. They are normally found only on B-cells, macrophages, and dendritic cells. These are the cells that gather antigen from their environment and present it to CD4 T cells. Thus, information about the extracellular antigenic universe is gathered and presented to the immune system only by a specialized subset of cells within the body, referred to collectively as antigen-presenting cells (see following sections). Information about the production of intracellular proteins, however, is presented to the immune system by each cell in the body.

One of the most striking properties of both classes I and II genes, in virtually every species examined, is their tremendous polymorphism. The number of intra-specific alleles at these various loci runs from as few as a half-dozen to many dozens or even a hundred or more. The evolutionary precursors of histocompatibility genes are unknown, but are presumed by many to be gene systems controlling social interactions of single-cell organisms, especially mating. In MHC proteins, polymorphism is restricted to the α_1 and α_2 domains of class I molecules, and to the α_1 and β_1 domains of class II molecules. The significance of this polymorphism is a subject of ongoing debate.

From a biological point of view, the polymorphism of MHC proteins does not benefit the individual. Individuals, depending on their inherited MHC alleles, differ considerably in their ability to respond to foreign proteins. It is often assumed that MHC polymorphism is driven by the need for a species as a whole to deal with a wide range of pathogenic challenges – to bind as many different peptides to MHC molecules as possible. Although it is clear that MHC polymorphism is advantageous for a species, it is not obvious how polymorphism (or any other genetic trait) can be selected and maintained at a species level. At a practical level, this enormous polymorphism in MHC genes is at the heart of the problem of matching donors and recipients for organ transplantation.

Major histocompatibility loci are the most common basis for defining genetic relationships among individual immune systems. Two members of the same species who share the same MHC gene alleles (in practice, identical twins in humans, and members of the same inbred strains of laboratory animals) are referred to as MHC-syngeneic. Members of the same species with one or more differences in MHC genes are referred to as MHC-allogeneic. MHC differences between species are referred to as xenogeneic. When immunologists use the terms syngeneic and allogeneic without further definition or qualification, they are referring to MHC genes. With respect to cell-mediated cytotoxicity, immunologists often speak of "alloimmune" or "alloreactive" CTL. This refers to CTL recognizing allogeneic MHC products. In terms of cell, tissue or organ grafts, exchanges between genetically distinct members of the same species are referred to as allografts; between two different species, xenografts.

Because both class I and class II MHC protein products were originally defined using specific antibodies, they are often referred to as MHC antigens. We will use the term "MHC molecules", "MHC antigens", and "MHC structures" interchangeably throughout this book.

2.3 MHC restriction and the recognition of self

A breakthrough in understanding the central role of MHC molecules in antigen presentation came with the discovery of MHC restriction. One of the first observations of this phenomenon was made by Katz and his colleagues (Katz et al., 1976), who found that efficient cooperation between T and B lymphocytes in an antibody response required identity of the cooperating cells in the MHC class II region. The significance of this was not immediately obvious, but one explanation was that efficient cooperation depended on the positive recognition of self-MHC by the interacting cells. Rosenthal and Shevach, using strain 2 and strain 13 guinea pigs, in even earlier experiments, had contributed another key observation. They found that antigen-primed strain 2 T-helper cells would proliferate in vitro when presented with antigen on strain 2 macrophages, but not when antigen was presented on strain 13 macrophages. The same was true in the opposite strain combination (Rosenthal and Shevach, 1973). Subsequent experiments using MHC-recombinant mice showed it was identity in the class II region of MHC that was critical for an optimal response. These experiments also demonstrated that T cells do not interact directly with nominal antigen, but rather with antigen presented on the surface of what came to be called antigen-presenting cells (APC).

At about the same time, Zinkernagel and Doherty (1975, 1997) were analyzing virus-specific CTL activity in infected mice. They showed that, similar to the interaction of T and B cells in antibody production, virus-specific CTL would lyse only MHC-syngeneic, but not MHC-allogeneic, virally infected target cells (Figure 1.6). In this case, however, the requirement for MHC identity was in the class I region. Thus, the target cell killed by a CTL appeared to be analogous to the APC activating T-helper cells. It was not immediately apparent whether CTL recognized class I molecules and some portion of antigen, e.g. virus, as separate entities on the target-cell surface, or as a single combinatorial entity. The same dilemma applied to T helper cell-B cell interactions. The first model requires two T-cell receptors, one for MHC and one for antigen, whereas the latter requires only a single receptor. A single receptor in the case of CTL was anticipated by the fact that virus- or hapten-specific CTL were not inhibited by soluble forms of either MHC proteins or viral antigens alone (Matzinger, 1981). This controversy was definitively resolved in favor of a single receptor when the structure of the T-cell receptor was finally deciphered. We now know that it is the combined structure of MHC plus peptide that is recognized by a single T-cell receptor, rather than either structure alone.

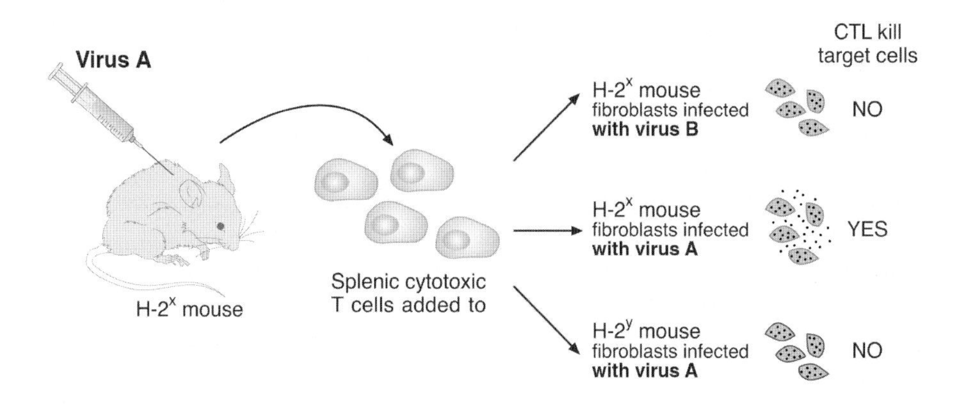

Figure 1-6. MHC restriction of CTL action against virus-infected cells.

Thus, in order for both helper and killer T cells to interact with potentially antigenic foreign peptides, they must first be able to interact with syngeneic MHC molecules, which display these peptides at a cell surface. The ability of T cells to interact with syngeneic MHC molecules is determined during the process of thymic maturation. In their earliest stages of development, just after entering the thymus from the bone marrow, all T cells are "double-positives", displaying both CD4 and CD8 molecules. By a process that is presently incompletely understood, these double-positives almost immediately sort themselves into single-positives, either CD4 or CD8.

The maturing thymocytes are then subjected to a critical selection test: the ability to interact "appropriately" with self-MHC molecules. This positive selection is a critical step in T-cell maturation. The MHC molecules present in the thymus are the guideposts to what is self. T cells must be able to interact with them with sufficient affinity to allow sampling of the peptide embedded in the peptide-binding site, yet they must not interact with them with excessive affinity, which would lead to T-cell activation – against self. If T cells are unable to interact at all with an MHC molecule in the thymus, they fail to display a receptor for a key thymic cytokine necessary for their survival, and they die by apoptosis. Those T cells that interact too strongly with an MHC molecule, whether because of direct interaction with the MHC molecule itself, or because of the peptide fragment embedded in the MHC molecule, are activated. Activation in the thymus is lethal, and these T cells are negatively selected; they also die by apoptosis. Between 90 and 95 percent of thymocytes passing through the thymus die because of an inability to interact appropriately with the MHC molecules displayed there. An understanding of the self-nonself selection process in the thymus will be critical when we discuss the phenomenon of alloreactivity.

What about T cells that are appropriately selected for MHC recognition in the thymus, but which did not encounter in the thymus self-peptides that occur only elsewhere in the body? It is clear that such potentially self-reactive T cells, both CD4 and CD8, do in fact leave the thymus and enter the periphery. The mechanisms by which these T cells are controlled are not entirely clear. Some T cells seem to be eliminated in the periphery; others remain, but are rendered permanently inactive. A fuller discussion of this problem can be found in most comprehensive immunology textbooks under the heading "peripheral tolerance."

2.4 Antigen processing and presentation

The concepts of antigen presentation and antigen-presenting cells arose in part from the experiments of Mosier (1967), who showed that T and B cells cannot respond efficiently to antigen in the absence of adherent cells, which were presumed to be macrophages, but which we now recognize must have included dendritic cells as well. The precise role played by adherent cells was unclear until the experiments of Emil Unanue in guinea pigs. Unanue and his coworkers showed in an elegant series of experiments that for many and perhaps most protein antigens, adherent cells (which he, too, assumed were macrophages) must first process the antigen – ingest it, break it down proteolytically, and display it on the presenting cell surface (Allen and Unanue, 1987).

This clarified a number of puzzling observations, for example, studies showing that T-helper (unlike antibody) recognition did not discriminate denatured from native protein (since processing cells denature it anyway), and the fact that T-helper cells often responded equally well to native antigen and to proteolytically derived protein fragments thereof (which processing cells can present directly, by absorption of cognate peptide fragments to surface MHC structures). Together with the overall data on MHC restriction, these observations suggested a model in which class II MHC molecules formed molecular complexes with peptides derived from macromolecular protein antigens, normally after some sort of digestion and processing by the APC.

For CD4 T cells, antigen-presenting cells include, as we have seen, dendritic cells, macrophages, and B cells. All three cell types are class II positive, consistent with the requirement for class II MHC restriction. Macrophages can be important in the processing of complex antigens such as bacteria or other cells, particulate antigens, and large protein complexes. However, the dendritic cells (DC) scattered throughout the body are perhaps the most potent antigen-presenting cells. They are nearly as effective as macrophages in breaking down antigens, although for very large antigens some degree of pre-processing by macrophages may be required. DC are highly sensitive to many products associated with microbes, and interaction with them, even independently of ingestion, can drive DC into an activated state. Moreover, DC are equipped both with surface ligands and secretable cytokines needed for the stimulation of both T cells and NK cells (Janeway and Medzhitov, 2002; Saas and Tiberghien, 2002). In their immature state, the level of these ligands, and the ability of the DC to release cytokines are low, but all of these activities are up-regulated by microbial infection, or simply by contact with microbial agonists (Henderson et al., 1997). As will be discussed below, DC can be further activated by interaction with CD4 T cells.

After activation by antigen in the periphery (e.g. in the skin), DC migrate to T-cell areas in regional lymph nodes, where they present acquired antigen to T cells (see Figure 1.2). Resting virgin B cells in lymph nodes are less effective as phagocytes than either macrophages or DC, yet in order to receive T-cell help, it is absolutely essential that B cells present processed antigen in association with class II antigens. B cells can internalize smaller antigens using their surface antibody receptors for receptor-mediated endocytosis, but it is likely that for larger antigens, B cells depend on the processing and release of smaller antigenic fragments by macrophages or dendritic cells.

The pathway for endocytic processing of antigen for presentation to CD4 T cells is shown in Figure 1.7. APC ingest antigen by phagocytosis, macropinocytosis, receptor-mediated endocytosis, or some combination thereof. In a process that normally takes less than two hours, antigen passes through a series of endosomal compartments, and arrives in lysosomes, where digestion is completed. Class II molecules are shunted from the Golgi complex in vesicles that intersect with this endosomal/lysosomal pathway. Fragments of peptides in the 10-30 amino acid size range bind to the class II molecules and are carried out to the APC surface, where CD4 T cells probe them.

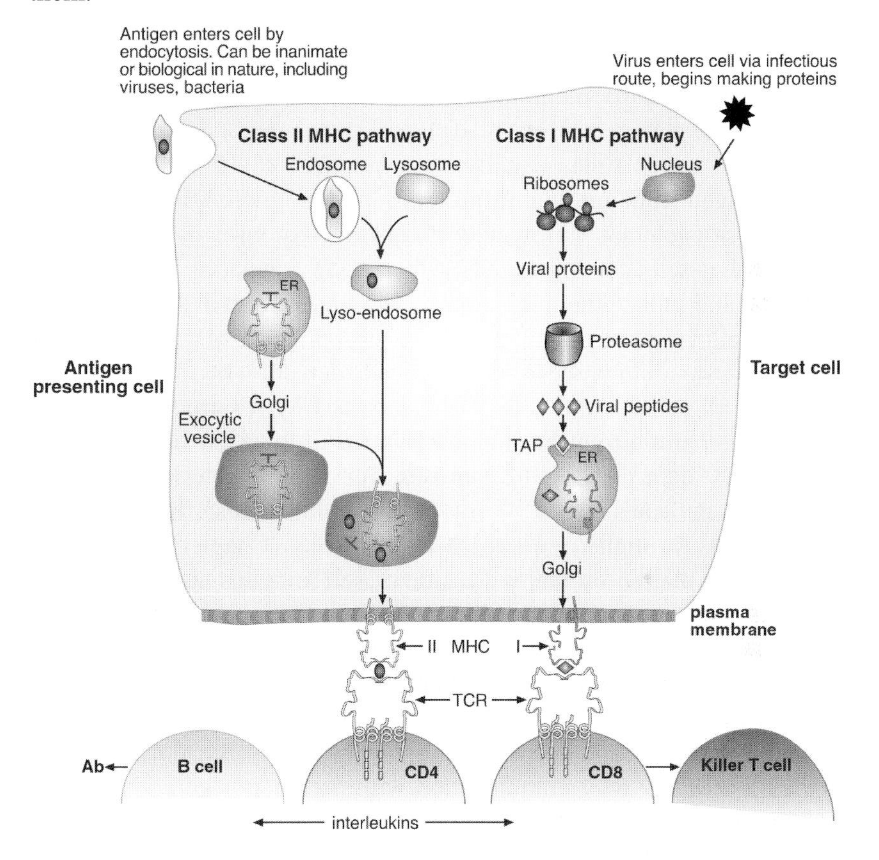

Figure 1-7. The two MHC pathways of antigen processing and presentation to CD4 and CD8 T cells.

It was less obvious initially how antigen might be processed and presented in association with class I MHC molecules for inspection by CD8 T cells. After some early observations, Townsend suggested some sort of antigen digestion process would likely be involved here as well, since CTL generated in response to viruses often reacted with internal proteins of the virus, rather than the more accessible coat proteins (reviewed in Townsend and Bodmer, 1989). Moreover, it became clear from these studies that CTL were in fact recognizing short peptides derived from these proteins.

Several different lines of evidence converged to provide the class I processing pathway also shown in Figure 1.7. Proteins in the cytosol are hydrolyzed by large multicatalytic structures called proteasomes. Proteasomes regulate the half-lives of proteins in the cytosol, and are also responsible for clearing denatured or misfolded proteins from the cell. Some of the peptide products thus generated are transferred into the endoplasmic reticulum (ER) by the "transporter associated with antigen processing" (TAP) and, after binding to newly assembled MHC class I molecules, are transported to the plasma membrane. Since each APC displays 10^5 – 10^6 class I molecules, a large variety of different endogenous peptides will be displayed at any given time. Usually these peptides (8-10 amino acids in length) are derived from autologous proteins and are ignored by the immune system due to self-tolerance. However, if cells display foreign peptides (e.g. viral or mutated-self gene products), then CTL consider the APC as compromised, and induce them to commit suicide. As we will see in Chapter 8, some intracellular pathogens, particularly viruses like CMV, can interfere with this peptide-loading process and thus evade detection by CTL.

Essentially all cells of the body express class I MHC structures, and as such are capable of alerting the immune system to invasion by genetically viable parasites. Often these will be viruses, but certain bacteria and single-cell eukaryotic parasites can also invade vertebrate cells; these too will reveal themselves through the class I MHC pathway (Chapter 8). But CD8 cells may also be alerted by the expression of self antigens that the immune system has not seen before, or to which it has not been made tolerant, such as altered self molecules, tumor-associated antigens, or cryptic fetal antigens not normally encountered in the adult, and this can result in autoimmunity (Chapter 10).

The nature of the antigen-processing and presentation system for class I proteins makes it clear why immunization with proteins or killed pathogens does not normally prime CTL responses in vivo. Macrophages, B cells or dendritic cells will take up antigens of this type, and derivative peptides will be presented via the endocytic class II but not the class I pathway, activating T-helper cells but not CTL. The class I pathway in most cells only presents proteins made within the cell, not those taken in from outside. This understanding has led to an exciting prospect for immunization in the future – immunization with DNA coding for an antigen of interest. For viruses in particular, immunization with live or attenuated viruses (such as HIV) are often considered too risky, and immunization with killed viruses induces only an antibody response that may be partially or even totally ineffective in clearing an infection. Introduction of selected viral genes into host antigen-presenting cells such as DCs results in presentation of the corresponding peptides at the cell surface, triggering a CD8 response. We will discuss these strategies in more detail in Chapter 11.

On occasion, one does see CTL activation to proteins that could not have been synthesized within the cell, yet appear on cell-surface MHC structures. There are several ways this can happen. One was mentioned above: cells can absorb pre-digested peptides directly onto MHC structures able to bind them, because MHC molecules are in dynamic equilibrium with their peptide ligands, and the latter are subject to displacement. One can take advantage of this to "pulse" selected peptides onto APC, which can be a useful strategy for vaccines (Chapter 11). Another explanation for the processing of material not synthesized within the cell into the class I pathway is through something called "cross-priming" (reviewed in Zinkernagel, 2002). Basically, any process that introduces a protein into the cytosol of a cell will probably result in that protein ending up in the class I pathway, because it will almost certainly at some point be picked up by proteasomes. Sometimes this may reflect a unique ability of an endocytosed antigen to escape from the lysoendosomal compartment: some bacteria produce toxins that disrupt endosomal membranes, allowing them to escape into the cytoplasm; certain fusogenic viruses can disrupt internal cell membranes. Dendritic cells have a unique ability to send endocytosed material directly into the cytosol, giving exogenous materials access to the class I pathway (Albert et al., 1998; Rodriguez et al., 1999). This imparts a great deal more flexibility to DC in generating activating signals for both CD4 and CD8 T cells, and is one of the reasons underlying the special APC role of dendritic cells.

And finally, the interaction of CTL with antigen-presenting cells such as DC raises an interesting question – do the CTL kill these cells in the same way they kill other cell types, or are APC somehow protected from cytolysis? In fact increasing evidence suggests that CTL do kill their APC (Wong and Pamer, 2003). This, too, will ultimately have important implications for vaccine design.

3 THE MOLECULAR BASIS OF CTL ACTIVATION

The process driving virgin CD8 T cells to the cytokine-secreting and cytotoxic phenotype encompasses a complex sequence of events, mediators, and signals that culminate in the production of cytocidal cells capable of specifically recognizing and lysing target cells. As postulated many years ago, CTL activation (like helper cell activation) is a two-step process (Bretscher and Cohn, 1970). Contact of virgin pre-CTL with an appropriate peptide-modified self or allogeneic MHC class I structure is the first step in the activation sequence (signal 1). Signal 1 can also be provided by cross-linking the antibody to the T-cell receptor, if (and only if) the antibody is itself cross-linked, e.g. via Fc receptors on dendritic cells (Weber et al., 1985; Jung et al., 1986, 1987). Signal 2 is provided by the DC after it has itself been activated by interaction with a CD4 helper cell (see below), or by certain microbial products that directly activate dendritic cells (Bernard et al., 2002). We will discuss these signals in more detail in Chapter 2.

Initially, it was thought that CTL-APC interactions would be mediated simply by interaction of the T-cell receptor with class I MHC structures, stabilized by CD4 or CD8 interactions with their respective MHC ligands. However, it was soon apparent that the α- and β-chains of the TCR, with their truncated cytoplasmic tails, are unlikely signal transduction molecules. The interaction between the T cell and its APC is now viewed as occurring across a molecularly complex structure called an "immunological synapse," our understanding of which is still evolving (Grakoui et al., 1999; Bromley et al., 2001; van der Merwe, 2002; Kupfer and Kupfer, 2003; Jacobel et al., 2004). Formation of this synapse, which is triggered by contact of T cells with antigen, involves the rearrangement of structures in both the T cell and the APC plasma membranes. A key element of signal 1 is the aggregation of the γ, δ, ε and ζ chains of the CD3 complex with the TCR α- and β-chains (Figure 1.4). These are so-called "adaptor molecules", bridging the TCR-α/β chains to the intracellular signal transduction machinery. The intracellular tails of the CD3 chains contain ITAM sequences that interact with src-family tyrosine kinases. ITAM sequences are not present on the short intracellular tails of TCR-α and -β chains. Subsequent phosphorylation events initiate a signal transduction cascade resulting in new gene expression required for cytotoxic and cytokine secretory functions. Binding of the src-family kinases also provides a docking site for SH2-type phosphatases (SHP-1 and SHIP), which can attenuate the activation sequence. Engagement of this TCR complex with MHC class I molecules on the APC, and activation of the signal transduction machinery constitute signal 1.

But the aggregation of TCR-α/β chains and CD3 chains takes place within an even larger context that defines the full immunological synapse (Figure 1.8). Short-range molecular interactions, such as those involving the TCR-α/β and CD8 molecules with APC class I molecules, form the center of the synapse. APC such as dendritic cells are more effective in antigen presentation to CD8 cells if they have first interacted with a CD4 cell. One of the results of CD4 cell interaction with class II MHC molecules on DC is up-regulation of a number of costimulatory molecules and adhesion molecules on the DC surface. Co-stimulatory signals provided by interaction of the CD28 molecule on the responding T cell with its ligand, B7, present on the surface of the DC, are important for CD8 cells, particularly in cases where the antigen concentration on the APC is relatively low.

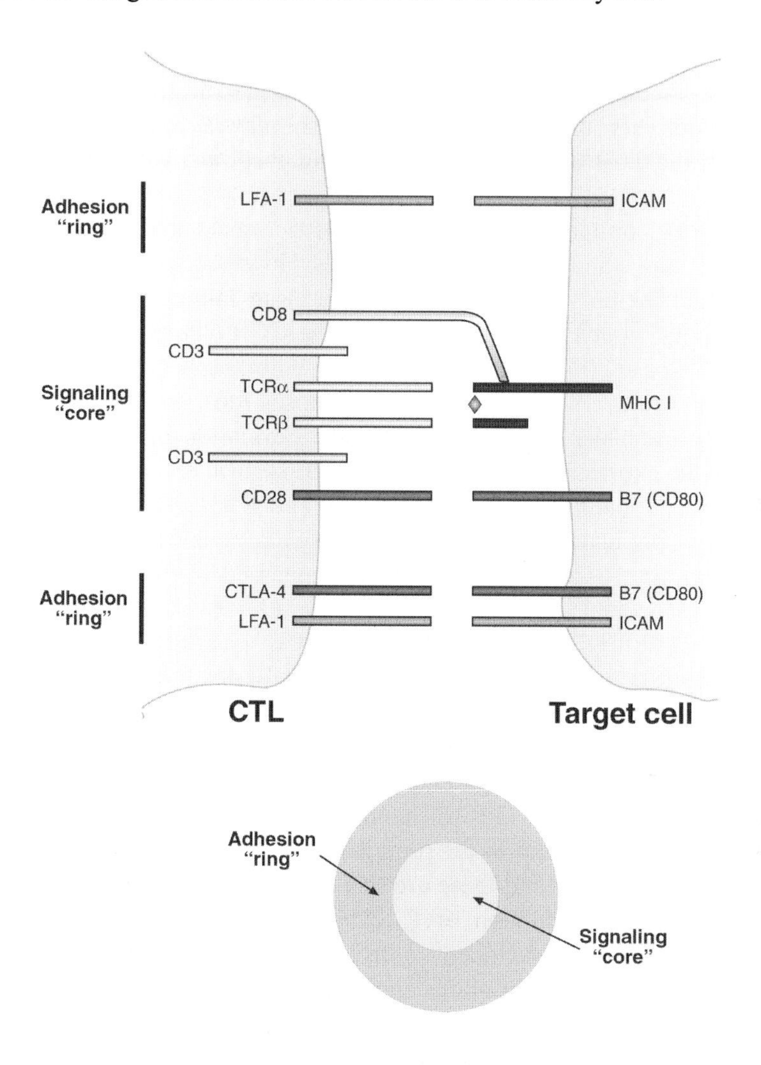

Figure 1-8. The immunological synapse of T cells.

The failure of CD28 to engage with B7 in such cases may result in T-cell anergy and tolerance (Lenschow et al., 1996). CD28 and other co-stimulatory molecules such as CD2 and CD8, together with the TCR/CD3 complex, occupy the center of the synapse. Surrounding this complex are stabilizing receptors such as LFA-1 and VLA-3, 4, and 5, which may engage their respective DC ligands even prior to engagement of the TCR (see discussion in Chapter 3). Engagement of these costimulatory and adhesion-stabilizing molecules on the CTL constitutes signal 2.

The inner core of the CD8 CTL synapse is thought to have space for extrusion of cytolytic granules during the killing process (see Chapter 5 for a fuller discussion). Shortly after engagement with APC class I, the T-cell receptor is rapidly down-regulated. Although the synapse is primarily involved in T-cell activation (Davis and van der Merwe, 2001), and coordinates delivery of signal 2, it may also play a role in the down-regulation of the TCR as a means of tempering T-cell activation (Lee et al., 2002).

Initial work on immunological synapses was carried out almost exclusively in naïve and activated CD4 T cells. However, it is now clear that both naïve and memory CD8 T cells also form an immunological synapse with cells with which they interact via the TCR (Potter et al., 2001; Stinchcombe et al., 2000). The CD8 synapse with target cells differs from CD4 synapses with APC in that the CD8 synapses form and break up much more quickly (Van der Merwe et al., 2002). Formation of the synapse is followed by the expression of high-affinity CTL surface receptors for several key cytokines, including interleukin-2 (IL-2) and IL-12. If IL-2 is present, we know from population studies that DNA synthesis and cell proliferation commence within 16-24 hours. This is accompanied by the expression of new genes associated with effector function (Agarwal and Rao, 1998; Grayson et al., 2001). However, development of mature, functional CD8 CTL, even from naïve pre-CTL, is almost certainly much less dependent on prolonged activation through the immunological synapse than are CD4 T cells. We will discuss this point in more detail in Chapter 7. The role of the immunological synapse during CTL destruction of target cells has been explored in considerable detail (Stinchcombe et al., 2001).

In most reactions leading to the production of CTL, CD4 helper cells are simultaneously triggered, and these are one source of needed cytokines such as IL-2. Autocrinic production of IL-2 by activated CTL has also been documented, and is often sufficient to drive pre-CTL to maturity (Paliard et al., 1988). Under the continuous influence of IL-2, CTL mature to proliferating blasts and develop cytoplasmic granules containing the lytic protein perforin, serine proteases, and other components. IL-2 is also critical for the development of CTL memory (Chapter 11). CTL also express the Fas ligand as a result of activation. CTL begin to express cytocidal activity 2-3 days after their activation. These still rapidly dividing lymphoblasts (now 10-20μ in diameter) differentiate into 'secondary', small-to-medium sized (7-10μ) effector CTL that continue to express a potent cytotoxic function. After one to two weeks, cytolytic function is no longer detected, since the CTL either die or become memory CTL.

As will be discussed in Chapter 4, fully cytotoxic CTL can engage a target cell, kill it, and then detach and recycle to a new target cell and restart the killing process. Previously activated but minimally cytotoxic memory CTL can be brought to a fully cytotoxic state by re-engagement with a target cell or APC. To our knowledge, the activating mechanisms in these cases are essentially the same as those used to activate a virgin pre-CTL for the first time, except that the requirements for a full signal 1 appear to be reduced. This may reflect the fact that structures composing the immunological synapse (at least within the CTL) are already assembled, and that the CTL already express the needed interleukin receptors.

As with all immune cells, the activation state of CTL depends on a balance between positive (activating) and negative (inhibiting) signals. A major inhibitory signal for CTL arises through the interaction of CTLA-4 (a CD28 homolog) on the CTL with APC-associated B7 (Walunas et al., 1994; Ostrov et al., 2000). CD28 is expressed on resting CTL, but CTLA-4 does not appear until some hours after the initial activation of CTL, when it is found in the plasma membrane as a homodimer. Whereas engagement of CD28 with B7 is a critical step in developing full T-cell activation, several lines of evidence point to a negative regulatory role for the engagement of CTLA-4 with B7. For example, mice in which the gene for CTLA-4 has been deleted die at 3-4 weeks of age of rampant proliferative disease triggered by T-cell expansion (Tivol et al., 1995). The mechanism by which CTLA-4 inhibits CTL activation is not entirely clear, but may involve the release of the inhibitory cytokine TGF-β (Chen et al., 1998), and/or inhibition of progression through the cell cycle (Krummel and Allison, 1996). Blocking CTLA-4 with soluble antibodies enhances T-cell responses; cross-linking CTLA-4 inhibits TCR-triggered proliferation and cytokine production (Chambers and Allison, 1999).

As we will see when discussing NK cells (Chapter 6), many transmembrane negative signaling molecules involved in inhibiting immune cell functions have ITIM (immunoreceptor tyrosine-based inhibitory motif) sequences in their cytoplasmic portions, which interact with and perturb phosphorylation reactions involved in activating signal transduction. CTLA-4 has such a sequence in its cytoplasmic tail (Sinclair, 2000), but the exact role of this regulatory pathway in CTL has yet to be delineated. The affinity of CTLA-4 for B7 on APC is much greater than that of CD28, so the tendency of CTLA-4 to block access of CD28 to B7 could be a major factor in determining a balance between the influences of these two receptors. Moreover, CTLA-4 coprecipitates with the CD3-ζ chain and SHP phosphorylases, suggesting that CTLA-4 can influence dephosphorylation of the T-cell activating pathway.

A second potential negative regulatory mechanism in CTL involves some of the same negative regulators operating in NK cells, such as Ly49A in mice and KIRs in humans, and CD94/NKG2 heterodimers in both species present (Young et al., 2001; Uhrberg et al., 2001; Vely et al., 2001). We will discuss these receptors in detail in Chapter 6. It was unexpected, but certainly intriguing, to find these molecules associated with CTL. They are present on relatively few CTL, mostly on more mature cells that have been in prolonged contact with antigen in vivo (McMahon and Raulet, 2001; Moser et al., 2002;). However, they are not found on resting T cells, or even on recently primed CTL. The exact interplay of these inhibitory receptors with the overall biology of CTL is difficult to assess at present.

Chapter 2

CYTOTOXIC T LYMPHOCYTES

Generation and Cellular Properties

While the involvement of CTL in the destruction of cells compromised by intracellular parasites has driven much of the recent research into the biological role of CTL, in fact their initial discovery emerged from a long history of trying to understand the basis for tumor immunity, hypersensitivity reactions, and especially allogeneic transplant (allograft) rejection (Chapter 7). All of these phenomena had the hallmarks of immune reactions, such as the involvement of lymphoid cells, specificity, and memory, but they could not be fully accounted for by antibody – the only immune mediator with these properties known in the first half of the twentieth century.

The first experimental evidence suggesting direct killing of specifically recognized allogeneic target cells by in vivo-sensitized lymphocytes was provided by Weaver et al. (1955). ("Target cells" are defined as cells bearing the same allogeneic class I MHC molecules as the cells used to induce CTL activity in vivo or in vitro; in 1955 these were referred to simply as "transplantation antigens"). Tissues transplanted into a mouse previously sensitized to MHC antigens displayed on the transplant were rapidly destroyed. But Weaver found that when these secondary grafts were placed inside a diffusion chamber permeable to antibodies but not to cells, they were not rejected. This was consistent with previous evidence that antibodies were not sufficient for graft rejection (Mitchison, 1953). However, if spleen cells from a mouse that had previously rejected a graft of the same type were placed into the same diffusion chamber as the secondary transplant, the transplant within the chamber was rapidly destroyed. The authors clearly stated they believed transplant rejection was caused by contact-mediated destruction of graft cells by immune cells, rather than soluble molecules. Interestingly, when mouse tissues from a rat that had previously rejected mouse tissues were placed into a diffusion chamber, they were quickly destroyed. This destruction of transplanted xenogeneic, as opposed to allogeneic, tissues by soluble factors is a characteristic of hyperacute rejection, which is mediated largely by preformed antibodies, and by species differences in complement molecules (Platt, 2001).

A major transition in the study of cell-mediated cytotoxicity came about in 1960, with the report by Andre Govaerts that thoracic duct lymphocytes isolated from dogs after rejection of a renal allograft were capable of destroying, in vitro, monolayer cultures of kidney cells obtained from the original donor's contralateral kidney, but not kidney cells from an unrelated donor. The cytocidal potential of lymphocytes harvested from allografted animals was quickly confirmed in several laboratories, notably that of Werner Rosenau (Rosenau and Moon, 1961). These investigators described an antigen-specific clustering of in vivo-sensitized killer cells around monolayer cells if and only if the monolayer cells were from the same strain of animal as the graft donor (see Figure 2.1). This clustering was followed by cytopathological changes in the targeted cells, culminating in their death, with no evidence of involvement of either target cell-specific antibodies or complement.

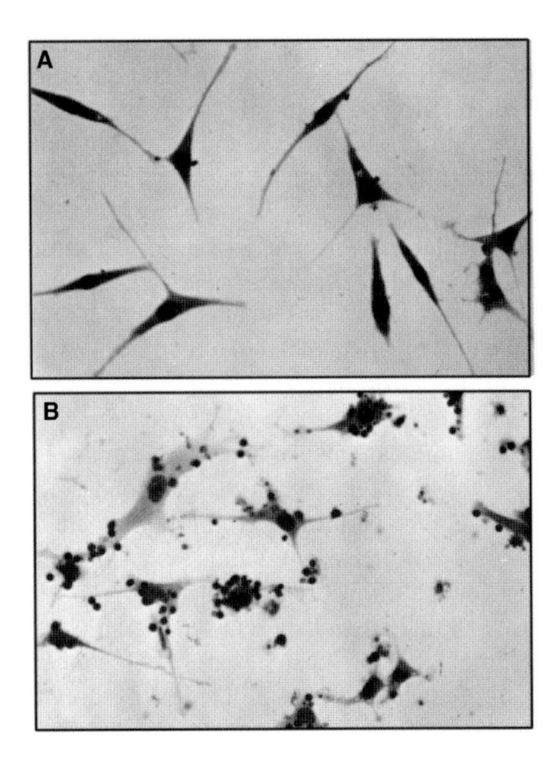

Figure 2-1. Contactual agglutination. Immune lymphocytes from (A) non-sensitized and (B) sensitized donors seeded on specific target cells (From Rosenau and Moon, J. Nat. Cancer Inst. 27: 471-86 (1961)).

The finding that cytocidally active cells could be obtained from the thoracic duct of animals that had rejected allografts showed that the attacking cells were part of a recirculating population of lymphocytes. In the 1960s, the existence of distinct T and B lymphocyte subpopulations was not yet appreciated. Once the existence of these subsets was demonstrated, and a marker found for T cells - the so-called theta (θ) antigen – the identity of killer cells with T cells was established (Cerottini and Brunner, 1974). Later, subsets of T lymphocytes were recognized, based on the CD4 and CD8 surface markers (Table 1.1), and the killer lymphocytes, now called CTL (cytotoxic T lymphocytes) were found to belong to the CD8 subset. This correlation is not absolute, as we have noted, since cytotoxic CD4 lymphocytes have also been described. Moreover, CD8 CTL are involved in functions other than just killing.

1 THE GENERATION OF CYTOTOXICITY IN VIVO

The initial reports by Govaerts and by Rosenau and Moon initiated a flurry of studies that quickly established the major parameters of the generation of killer lymphocytes in vivo. One way of generating CTL continued to be through transplantation of allogeneic tissues into a recipient animal, either free cells (usually allogeneic tumor cells) or skin or other normal tissue transplants. The animal models of choice rapidly became mice and rats, because of the availability of inbred strains; the genetic homogeneity of such strains greatly facilitated comparisons of results within and between laboratories.

Many investigators favored skin transplants because they are relatively simple to do, induce a strong CTL response, and allow easy monitoring of the rejection process. Skin transplanted from one inbred strain to another quickly heals and becomes vascularized, the skin acquiring a healthy pinkish color within 24 hours. After about seven days, lymphocytes capable of destroying target cells in vitro become detectable in regional draining lymph nodes. At about the same time, the graft becomes infiltrated by lymphoid cells, is devascularized, and within a few days dries up and detaches from the host dermis. CTL can be found in lymph tissue other than draining nodes, such as the spleen and other lymph nodes, but in general cytotoxicity is less vigorous, most likely reflecting a preferential homing and lodging (at least in primary reactions) of activated killer cells in regional nodes draining the graft site.

The injection of live allogeneic tumor cells into the peritoneal cavity of mice results in a more rapid appearance of cytotoxicity, most likely because of immediate contact of host lymphocytes with the allogeneic cells. Grafts of skin or other solid tissues not vascularized surgically require one to two days to become fully vascularized through the outgrowth of regional host blood vessels, and to initiate lymphatic drainage. This is necessary not only for graft viability but also for initiating contact between host immune cells and graft cells. Contact may take place at the graft site itself, or may require the migration of graft cells, such as dendritic cells, to draining lymph nodes; either way, vascularization is a critical first step.

After intraperitoneal injection of allogeneic tumor cells, low but detectable cytotoxicity is found in the spleen by day three, peaking at day 10-11, and falling off slowly thereafter. Some level of cytotoxicity is evident up to two to three weeks after immunization (Brunner et al., 1968; 1970). At the peak of the reaction, cytotoxic lymphocytes can also be found in the blood, thoracic duct and lymph nodes, as well as in the peritoneal cavity itself (see below) (Sprent & Miller, 1971). At peak reaction times, cytotoxicity is associated with large (blast) lymphoid cells in all tissue compartments. As the cytolytic response begins to subside, residual cytotoxicity shifts to medium and small lymphocytes.

Another system used to analyze in vivo generation of cytotoxicity is the graft-vs-host (GVH) reaction. This reaction occurs when immunoincompetent hosts are injected with a source of viable lymphocytes containing allogeneic T cells. The host may be naturally immunoincompetent (e.g., an F_1 animal with respect to either parental strain), or rendered incompetent by radiation. The donor T cells home in to host lymphoid tissues, recognize as foreign host MHC antigens on relatively radiation-resistant cells such as dendritic cells or macrophages, and mount a cytotoxic response.

The development of cytotoxicity in GVH reactions is very rapid: in fact cytotoxicity is apparent as early as two days in laboratory animals, usually peaking by day four. (Irradiated hosts normally die by day 6-7, from a combination of radiation and immune damage.) Again, the speed of the reaction is probably accounted for in part by the immediate contact of donor and host cells. In addition, most host lymphoid cells and most non-reactive donor cells die rather quickly, leaving an enriched source of CTL. The fact that donor cells expand by proliferation during the course of the reaction has been shown by systemic administration of mitotic inhibitors, which greatly decrease the resulting cytotoxicity. As with other allograft responses, anti-host cytotoxicity is found in the spleen, blood, lymph nodes, and thoracic duct, plus the thymus and, more rarely, bone marrow. (The latter may be due to circulating T cells contaminating any harvest of bone marrow cells). Cytotoxicity can be abrogated by treating either the cells used to induce GVH, or the cells resulting from GVH, with anti-Tcell antibodies plus complement.

One of the most potent in vivo systems for generating CTL involves the production of immune peritoneal exudate lymphocytes (PEL). In addition to being a highly enriched source of CTL, PEL are interesting because, apparently unique among primary CTL generated in vivo or in vitro, they appear to lack the perforin/granule exocytosis mechanism of target cell lysis (Chapter 5). Brunner and colleagues in the 1960s showed that mice injected intraperitoneally with allogeneic tumor cells would develop spleen and lymph node cells with measurable cytotoxicity (Brunner et al., 1968). Earlier works by Amos (1962) and by Baker (Baker et al., 1962) suggested that the tumor cells were rejected within the peritoneum by macrophages. At the peak of allogeneic ascites tumor growth, peritoneal macrophage populations are monocyte-like, consisting primarily of small cells with compact cytoplasm and rounded nuclei, and exhibit only slight phagocytic activity for the tumor cells. As the number of tumor cells decreases, larger, more active macrophages containing irregular lipid granules and tumor cell fragments, and which adhere to the tumor cells in vivo, are observed. After growing progressively for 5-7 days, rejection of an ascitic tumor allograft occurs rapidly; a 25 g mouse harboring up to $5x10^8$ tumor cells in the peritoneal cavity (corresponding to 2-3 kg in a 75-kg human adult) can clear this huge tumor mass almost within 48 hours (Figure 2.2), and a large proportion (up to 50%) of the cells recovered from the peritoneal cavity shortly after rejection are macrophages.

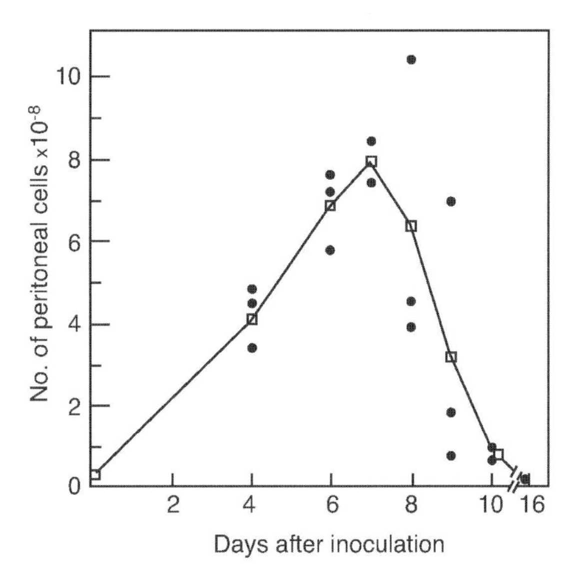

Figure 2-2. Graft rejection in the peritoneal cavity. The growth and rejection of leukemia cells injected intraperitoneally into allogeneic mice. Individual scores and averages of tumor cell members are shown. Rejection is complete on day 11 (from Berke & Amos, 1973).

However, when macrophages and other adherent cells are depleted from the immune peritoneal cells, for example by adsorption on nylon wool columns, the cytotoxicity is found to reside in the non-adherent cells, mainly lymphoid. Most if not all of the cytocidal activity of the unfractionated peritoneal cells can be attributed to small-to-medium sized (7-10 μ) PEL. These cells exhibit superior cytocidal activity against the sensitizing tumor target cells, far exceeding that of spleen cells from the same mice (Berke et al., 1972 a,b, c) (Figure 2.3). The PEL system soon became a useful source of highly potent CTL in in vivo settings.

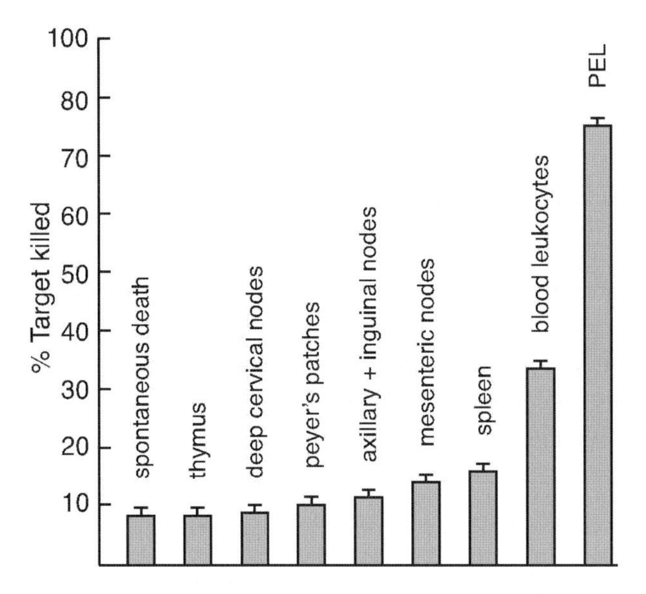

Figure 2-3. CTL action. The lytic activity of allogeneic CTL obtained from various sources 11 days after primary immunization (from Berke et al, 1972a).

The response of PEL to allogeneic tumors, as with the generation of cytotoxicity in other systems, is accompanied by cell division. When ^{3}H-thymidine is administered during the response in vivo, virtually all PEL able to bind to target cells in vitro are found to be ^{3}H-labeled upon autoradiography (Denizot et al., 1986). Upon restimulation in vitro in the presence of T-cell growth factors (TCGF) or recombinant interleukin-2 (rIL-2), the small, non-dividing PEL transform into large, IL-2-dependent, dividing PEL blasts (PEB) that in short-term (2-4 hr) assays express specific lytic activity similar to that of the original in vivo-primed PEL. PEB, in contrast to PEL, possess massive quantities of typical intracytoplasmic granules with readily detectable granzymes and perforin. (Chapter 5).

We will discuss the generation of cytotoxicity in response to allografts in vivo in more detail in Chapter 7, when we examine the role of cell-mediated cytotoxicity in graft rejection.

2 THE GENERATION OF CYTOTOXICITY IN VITRO

While the generation of cytotoxic lymphocytes in vivo soon became routine, and provided reliable sources of activated CTL for analyzing the killing process, the mechanism underlying cellular and molecular events involved in the activation process itself remained obscure. Thus the discovery by Ginsburg (Ginsburg, 1968; Berke et al., 1969a), that killer lymphocytes could also be generated in vitro, was a major step forward in the investigation of cell-mediated cytotoxicity. It was found that co-cultivation of unprimed rat or mouse lymphocytes with murine embryonic fibroblast monolayers for 4-5 days resulted in the transformation of the lymphocytes into blast cells, concurrent with the development of cell-mediated cytotoxicity specific for the MHC haplotype of the stimulating monolayer.

Thus, rat killer cells generated on monolayers of one mouse strain, and transferred to test monolayers of other mouse strains (differing in the MHC locus) killed only target cells with the same MHC as the stimulating cell monolayer (Berke, 1969a). This was surprising, because in tissue transplantation in vivo, rats in secondary reactions cannot distinguish among individual mouse strains. We now realize this is likely due to antibody responses, which play a much stronger role in xenogeneic reactions and are not restricted to MHC differences. However, Ginsburg's results strongly implicated MHC as the cognate target structures on mouse target cells, recognized by either mouse or rat killer cells, but precise mapping of the MHC loci involved, requiring MHC recombinant strains, had not been done at that time.

In parallel, other investigators explored the mixed leukocyte culture (MLC) reaction as a possible system for generating cytotoxicity in vitro. Bain and her colleagues (1963) had demonstrated that when peripheral blood leukocytes from two unrelated humans are mixed and cultured for a number of days, the cells undergo blastoid transformation, incorporate tritiated thymidine, and divide. This proliferative response, which appeared to involve less than one percent of the responding lymphocytes, occurred in the absence of prior immunization; as with the Ginsburg system, it was truly a primary in vitro reaction. Mixed leukocyte cultures between monozygotic human twins did not undergo transformation, and only weak responses were obtained with cells from closely related donors. Maximal proliferative responses were obtained when cells from HLA fully unrelated donors were mixed.

These observations suggested that the proliferation seen with cell mixtures from unrelated donors was related to their immunogenetic disparity, and indeed, it was quickly shown that the degree of MHC difference between individuals determined the vigorousness of the proliferative response in vitro. This in turn suggested that MLC reactivity might form the basis of a useful histocompatibility matching procedure. For many years, mixed leukocyte cultures were used to assess histocompatibility in transplant procedures, and they are still used on occasion to monitor potential graft compatibility between individuals.

As originally described, the MLC involved the reaction of each donor lymphocyte population against the MHC molecules of the other (two-way MLC). A more precise assessment of the involvement of MHC in guiding this reaction became possible when one of the two donor populations was treated with drugs (e.g. mitomycin C) or radiation to render it unable to proliferate (Bach and Voynow, 1966). In such cases, the reactivity of the untreated donor lymphocytes against the treated stimulating cell population could be studied to determine precisely which donor MHC antigens were involved in triggering proliferation (one way MLC). We now know that the stimulation of responding lymphocytes to proliferate in an MLC is mostly induced in CD4 T cells by allogeneic class II MHC antigens of the stimulating cells (Yunis and Amos, 1971). CD8 T cells responding to class I-associated antigens contribute a positive but relatively minor part of the overall proliferative response, while developing essentially all of the cytotoxicity.

The first reports of generation of killer lymphocytes in one-way MLC came in 1970 from four different laboratories (Hayry and Defendi, 1970; Hardy et al., 1970; Hodes and Svedmyr, 1970; Solliday and Bach, 1970). These labs showed that killer cell populations generated in MLC were more potent than anything that could be generated in vivo, with the possible exception of GVH sensitization. In the MLC, similar to the GVH reaction in vivo, the stimulating population dies off as a result of radiation, the responding cells selectively expand, and non-responding lymphocytes die off. In fact, GVH reactions in irradiated hosts were often referred to as "MLC reactions in vivo." As had been shown in GVH reactions, the inclusion of mitotic inhibitors during the MLC essentially abrogated the development of cytotoxicity, showing the importance of the proliferative phase of the reaction in the development of cytotoxicity. As with killer cells generated in vivo, and in Ginsburg's system, there was a high degree of target cell discrimination based on class I MHC. The MLC reaction rapidly became the system of choice for exploring allograft reactions in vitro.

3 CYTOKINES INVOLVED IN GENERATION OF CTL FUNCTION

Cytokines are the principal means by which all cells in the immune system communicate with one another and with non-immune-system cells (Table 1.2). The cytokines used by the immune system are low molecular weight proteins and glycoproteins secreted by various white blood cells as well as other cells of the body, including nerve cells. Receptors for immune system cytokines are present on other immune cells, and on non-immune cells such as vascular cells and nerve cells. Cytokines are involved in the initiation, maintenance, execution, and regulation of immune responses. The end product of cytokine engagement with a specific receptor is invariably the modulation of gene expression in the targeted cell. However, the precise form of this modulation is dictated by the receptor, and the intracellular pathways to which it is linked, rather than by the cytokine per se; cytokine effects within the immune system are thus pleiotropic. Cytokines may also act autocrinically, affecting the same cells that secrete them.

The development of in vitro systems for generation of cytotoxicity allowed the definition of a number of cytokines that promote the generation of CTL from CTL precursors during interaction with cognate antigen. The heterodimeric interleukin 12 (IL-12) is crucial to CD8 development, and is produced by dendritic cells (Trinchieri, 1994) in the initial activation steps described in the preceding section. IL-2 and IL-4 are produced primarily by CD4 T-cells, and influence the maturation of B cells as well as CTL. IL-7 is produced by bone marrow and thymic stromal cells (Hickman et al., 1990), and is a potent amplifier of CTL production (Sin et al., 2000). There is currently a great deal of interest in the use of cytokines as biological response modifiers to influence the outcome of encounters with antigen, for example during vaccine administration (e.g., Gherardi et al., 2001), or as part of standard therapy (e.g., Kumagai et al., 1997) or gene therapy (e.g., Möller et al., 1998) strategies for treating cancer. We will discuss these uses of cytokines in later chapters. The cytokine IL-15, also produced by DC, influences the migratory patterns of CTL at various stages in the activation process (Weninger et al., 2001), and promotes the transition of activated CTL into memory CTL (Zhang et al., 1998; Schluns and Lefrançois, 2003).

In addition to mediating cytotoxicity, CD8 CTL also release a spectrum of cytokines that are important in the host response to infection, and in triggering immune-based diseases such as hypersensitivity and autoimmunity. During the primary generation of CTL in response to modified self or allogeneic antigens, few cytokines are produced. However, fully activated CTL can produce a wide range of cytokines when re-stimulated with cognate antigen, including many, such as IL-4, which are thought of as "classical" CD4 cytokines (Kelso et al., 1991; Seder et al., 1992; Erard et al., 1993). CD8 cells can also produce IL-2 (von Boehmer et al., 1984; Sad and Mosmann, 1995), and can even provide help to B cells (Maggi et al., 1994). From an effector function point of view, the most significant cytokines produced by CD8 cells are probably IFNγ, TNFα and TNFβ. IFNγ inhibits viral replication, increases macrophage activity, and up-regulates surface expression of class I proteins (Harty and Bevan, 1999). TNFβ (lymphotoxin) can be directly cytotoxic to certain cell types (Ware et al., 1996). Activated CTL also express a surface form of TNFα, which can induce apoptotic death in cells displaying an appropriate receptor (Chapter 5).

The type of cytokines released by activated CTL is to some extent determined by the cytokine environment in which the CTL are originally generated. CTL generated in the presence of copious quantities of IL-2 will secrete predominantly IFNγ when subsequently challenged with antigen. The amount of IFNγ secreted will be greatly increased if IL-12 was also present during the sensitization period. CTL generated in the presence of IL-4 release less IFNγ upon reactivation, but produce substantial quantities of IL-4 and IL-5 (Croft et al., 1994; Sad et al., 1995). These CTL also release IL-6 and IL-10. These differing patterns of cytokine production are often used to distinguish two distinct subsets of CD8 CTL (Fong and Mosmann, 1990; Mosmann et al., 1997). However, there is no evidence at present to suggest that such subsets are developmentally determined in the same way that the CD4 and CD8 subsets are, e.g. during thymic development and with predetermined functions prior to antigen stimulation.

During their generation from precursor CD8 cells, CTL require cytokines to complete their maturation, and are thus dependent on CD4 cells to provide the cytokines. But if CD8 cells can produce most of the same cytokines as CD4 cells, why are they not able to mature into cytotoxic effectors on their own? Indeed, there have been reports that under some circumstances they do (von Boehmer et al., 1984; Sprent and Schaefer, 1985). However, it must be remembered that CTL, once activated, likely kill their APC (target cells), thus removing the stimulus that drives cytokine production and secretion. CTL may thus be suboptimal for providing themselves with help.

4 CTL LINES, CLONES AND HYBRIDOMAS

By the late 1970s, the generation and expression of CTL-mediated cytotoxicity were fairly well characterized within the limits of the populations available for study, that is, CTL generated from whole lymphoid populations either in vivo or in vitro. But even the most highly enriched CTL sources, such as CTL generated in GVH reactions or in MLC, were still contaminated with radiation-resistant stimulating cells and varying numbers of non-responding cells, impeding accurate biochemical analysis. Moreover, it was difficult to generate the large populations of activated CTL needed for detailed biochemical and molecular studies.

One way around these problems was suggested in 1976 in a report from Gallo's laboratory (Morgan et al., 1976). They found that culture supernates from human lymphocytes stimulated with the mitogenic lectin phytohemagglutinin (PHA) contained factors promoting growth of T lymphocytes in vitro. These factors, which were referred to collectively as T-cell growth factor (TCGF) for many years, would turn out to include a number of cytokines produced largely by CD4 cells, including most importantly the cytokine IL-2. Two years earlier, two laboratories had described repeated stimulations of CTL in mixed leukocyte culture, providing a useful system for expanding CTL and for studying the generation and reactivation of CTL memory (MacDonald et al., 1974; Hayry and Anderson, 1974). In 1977, Steve Gillis and Ken Smith combined these two approaches (repeated restimulations in vitro, with the addition of exogenous TCGF) to develop, at first, long-term CTL lines, and then T-cell clones specific for class I alloantigen (Gillis and Smith, 1977; Nabholz et al., 1978; von Boehmer et al., 1979). A short while later, a similar approach was applied to derive the first T-helper cell clones (Watson, 1979.)

In some instances investigators attempted to maintain CTL lines without periodic restimulation by antigen, driving them solely with crude preparations of TCGF. While numerous lines were indeed established that depended solely on TCGF for continuous proliferation, it soon became apparent that these lines contained numerous karyotypic and other abnormalities. They also showed a strong tendency to lose their MHC specificity and kill target cells nonspecifically.

CTL clones maintained by repeated antigen stimulation and moderate amounts of IL-2 showed fewer chromosomal abnormalities, but there has been some concern, largely unsubstantiated, that what really drives these lines may be transformation by either adventitious or endogenous viruses. CTL clones may also on occasion lose their specificity, or acquire NK-like killing activity. However, a large number of stable, antigen-specific clones soon became widely available, and they were used for the study of many different aspects of CTL biology (see e.g., MacDonald et al., 1980; Zinkernagel and Doherty, 1979). An important capability that emerged from the limiting dilution technique used to establish T-cell clones was the ability to estimate precursor-CTL frequency in various lymphoid cell populations (Ryser and MacDonald, 1979).

Another approach for deriving large-scale populations of CTL for study was the establishment of CTL hybridomas. The ground-breaking work of Köhler and Milstein (1975) in generating the first B-cell hybridoma naturally stimulated attempts to reproduce this feat with functional T cells. The first T-cell hybridoma produced utilized an enriched source of T-suppressor cells fused to a T-lymphoma cell line using polyethylene glycol (PEG). A functional hybridoma cell line was obtained, but uncertainty about exactly what T-suppressors are, and what they recognize antigenically, limited their widespread adaptation. A fully functional T-helper cell hybridoma was produced in 1980 (Harwett et al., 1980), which ultimately was used to isolate the T-cell receptor (Chapter 1).

Initial attempts to produce hybridomas using CD8 CTL fused to tumor cells were unsuccessful (reviewed in Gravekamp et al., 1987; Rock et al., 1990). This failure was initially thought to be due to the fact that the CTL, brought into close contact with the tumor partner cell by PEG, probably killed the tumor partner, thus abrogating hybridoma formation. Direct testing of this assumption found it baseless (G. Berke, unpublished). Indeed, successful production of CTL hybridomas was finally reported by two groups working simultaneously but independently (Nabholz et al., 1980; Kaufmann et al., 1981). CTL hybridomas generated through somatic fusion of in vivo primed CTL (PEL) and thymoma tumor cells express antigen-specific lytic activity (Kaufmann et al, 1981). They display TCR-α and -β chains, but, interestingly, neither CD4 nor CD8, and grow without an external source of IL-2. Although expressing constitutive lytic activity, both cytotoxicity and IL-2 production by these CTL hybridomas are augmented upon stimulation by mitogenic lectins such as Con A, or by specifically recognized stimulating cells. Subsequent experiments indicated that the hybridomas most closely mimic memory CTL (Kaufmann and Berke, 1983). The brief period (2 to 3 hr) required for stimulation by antigen or mitogen suggests that the expression of potentiated effector function(s) after stimulation does not require cell replication.

The great advantage of CTL hybridomas is that they grow continuously in culture without the need for either antigenic restimulation or exogenously supplied growth factors (Kaufmann and Berke, 1983). A potential downside is the unknown contribution of the tumor partner to overall hybridoma function. Nevertheless, they provide an important tool for analyzing CTL biology and response. Unfortunately, the initial successes in producing CTL hybridomas by the Nabholz and Berke laboratories have not led to a burst of new CTL hybridoma production in other CTL systems. The mouse thymoma most commonly used as a fusion partner, BW5147, may be relatively non-permissive for the expression of some essential CTL functions and/or receptors. BW5147 may not allow the expression of cell-surface CD8 molecules, which promote cell-cell interaction. The expression of CD8 is particularly critical when the CTL partner has a lower affinity for the desired target. The utilization of CD8-transfected BW5147 cells enabled the generation of a higher proportion of MHC class I alloreactive hybridomas, consistent with the auxiliary function(s) of CD8 molecules in T-cell reactivity (Rock et al., 1990). Nevertheless, CTL clones, rather than hybridomas, have remained the long-term CTL lines of choice for most studies.

5 γ/δ T CELLS

The enigmatic γ/δ T-cell subset was discovered only through its T-cell receptor. The γ chain was actually the second rearranging TCR gene discovered, after the β chain, and was at first assumed to be the anticipated α chain gene. When it proved to lack the required α-chain glycosylation sites, it was set aside as a rearranged T-cell receptor gene of unknown function. The true α chain was quickly found, and shortly after that a partner δ chain was discovered for the γ chain (Isobe et al., 1988). Thus, researchers now had two T-cell receptors to account for.

The significance and function of γ/δ T cells remained a mystery for nearly a decade, and are still only partially understood. They constitute a major resident population early in thymic ontogeny, but once the α/β receptor begins to rearrange, they become a minor population in the thymus, and remain a life-long minor population throughout most of the adult body (Fenton et al., 1988). The exception is epithelial surfaces such as the gut and skin, where they may constitute up to a third of resident T cells. While rarely comprising more than a few percent of circulating T cells in healthy human adults, during infection with a range of microorganisms, they may reach 30-50 percent of T cells in the blood (Balbi et al., 1993; Bertotto et al., 1993; Caldwell et al., 1995; Jouen-Beades et al., 1997). There is also evidence that γ/δ T cells are involved in autoimmune disease (Peng et al., 1996; Chapter 10), and in tumor surveillance (Giradi et al., 2001; Carding and Egan, 2002; Chapter 9).

Although the genetic elements and combinatorial mechanisms used to generate γ and δ chains are as large and complex as those used for α/β receptors, the repertoire of most γ/δ receptors is limited, and to some extent tissue-specific. Tissue-resident γ/δ cells in particular are highly restricted; circulating γ/δ cells show somewhat more heterogeneity. For reasons that are not yet understood, in both mice and humans the entire repertoire of resident γ/δ T cells may be based on just a few V region segments paired with a limited number of J segments, and no junctional diversity. In humans, for example, most circulating γ/δ T cells express TCR using Vγ9, often paired with Vδ2 (Bendelac et al., 2001). A minor subset utilizing Vδ1 is found mostly in epithelial tissues (Hayday et al., 2000). It is presumed that these TCR repertoires are selected in response to specific antigens, including stress-related antigens and microbial antigens (Carding and Egan, 2002).

The γ/δ receptor is associated in the plasma membrane with the same CD3 complex as α/β T cells, and is capable of signal transduction (Wu et al., 1988) that is enhanced by CD28/B7 co-stimulation (Sperling et al., 1993). However, γ/δ T cells are negative for both CD4 and CD8. Unlike α/β T cells, γ/δ T cells are normally not restricted by classical class I MHC structures. Rather, they appear to interact with antigen presented in association with non-classical class I products such as CD1 (Spada et al., 2000), MICA and MICB in humans (Groh et al., 1998, 1999) and T10 and T22 in mice (Crowley et al., 2000; Wingren et al., 2000). Whether these structures interact directly with the γ/δ T-cell receptor, however, is not always clear. The antigens presented, if any, by these structures are also not generally known, although as just noted, there is evidence for their association with stress-related and microbial antigens. MICA and MICB may also signal oncological transformation of cells (Groh et al., 1999). Antigens associating with CD1 structures are often lipid in nature, and probably of microbial origin; CD1-bearing APC are known to present lipid antigens to α/β receptor-bearing T cells (Beckman et al., 1994; Rosat et al., 1999). We will discuss this again in connection with NKT cells (Chapter 6), which also interact with CD1 structures.

Human γ/δ T cells release IFNγ and IL-2 when stimulated by CD1-positive dendritic cells (Spada et al., 2000). They express perforin constitutively (Nakata et al., 1990), and under some circumstances use both perforin and Fas to effect cytolysis (Spada et al., 2000). (These lytic pathways are described in detail in Chapter 5.) Cytotoxic granules of γ/δ T cells also contain the bacteriolytic molecule granulysin. Nothing at present would suggest that cytotoxicity mediated by γ/δ T cells is different in any way from the cytotoxicity mediated by CTL or NK cells.

From a functional point of view, the types of reactions in which γ/δ cells have been implicated are more reminiscent of innate rather than adaptive immunity; a discussion of them could as well be placed in Chapter 6 as here. The occasional involvement of γ/δ T cells in adaptive immune responses has recently been reviewed (Chen and Letvin, 2003). A full appreciation of the importance of the cytolytic function of these cells in the overall biology of the host has yet to be developed, but it is unlikely to be unimportant.

6 CD4 CYTOTOXIC T CELLS

Although originally defined on the basis of providing help to B cells and CD8 CTL, CD4 T cells have also been found under certain conditions to kill target cells in vitro. Probably the first description of this phenomenon involved the generation of cytotoxicity in an MLC between two strains differing only in the class II region of the MHC (Wagner et al., 1975). One has to say "probably", because the surface phenotype of the resulting CTL was not determined, and there have been occasional reports of CD8 CTL that recognize target-cell class II antigens (e.g., Golding and Singer, 1985). Investigators were by and large too busy exploring the more readily understood helper functions of CD4 T cells to follow up this unusual observation. But additional reports in 1981 described the same phenomenon, and provided more details (Swain et al., 1981; Dennert et al., 1981). Class II-specific CTL were cloned and shown to bear the Lyt-1 (CD4) surface marker. Increasing reports of CD4-positive, class II-specific CTL emerged over the next few years, although it was clear this was a minor subset in comparison with CD8 CTL. There was also concern that such cytotoxicity might be an artifact of maintaining the cells in long-term culture, especially in the presence of autocrinic IL-2. In viral responses, where CD4 CTL are most often detected, the frequency of CD8 CTL produced far outweighs that of CD4 CTL (Bourgault et al., 1989). In the case of allogeneic responses, the frequency of CD4 pre-CTL responding to class II MHC has been estimated to be ten percent or less that of the response of CD8 cells to class I (Golding and Singer, 1985).

A detailed analysis was made of clones of CD4 CTL lines, comparing them with CD8 CTL clones across a number of points (Lancki et al., 1991). To rule out differences in the efficiency of cell binding among different clones, lysis was examined in a redirected lytic assay, in which target cells are first coated with an out-facing CD3 antibody, which engages any CTL clone and induces lysis. All CD8 clones lysed both red blood cells and nucleated target cells. Some CD4 CTL lysed both target types as well, although some were able to lyse only one or the other. All of the CD4 clones able to lyse RBC expressed the gene for perforin, and released serine esterase from granules. CD4 CTL also induced DNA fragmentation in nucleated target cells.

Further studies have made it clear that CD4 CTL use both the degranulation pathway (Williams and Engelhard, 1996; Yasukawa et al., 2000) and the Fas pathway (Stalder et al., 1994; Zajac et al., 1996; Chirmule et al., 1999) in mediating target cell lysis. Other lytic mechanisms, such as TRAIL, have also been implicated in CD4-mediated lysis (Canaday et al., 2001; Thomas and Hersey, 1998; Bagot et al., 2000; Dorothee et al., 2002).

(These various lytic pathways are described in detail in Chapter 5.) The case has been made that the degranulation pathway may be more important in cases where target cells, such as certain tumor cells in vivo, become resistant to Fas killing, and that normally the Fas pathway is more important for CD4 CTL (Rivoltini et al., 1998). It has been suggested that these cells may be involved more in immune regulation in vivo than in target cell destruction per se (Hahn et al., 1995). Interestingly, cytotoxic CD4 T cells can still function to provide help to B cells displaying the appropriate antigen (Yasukawa et al., 1988; Van Binnendijk et al., 1989). How they do this without killing the B cell they are helping is not entirely clear; perhaps they do.

7 GENERATION OF KILLER CELLS IN THE ABSENCE OF ANTIGEN

7.1 Generation of cytotoxicity using antibody to the CTL antigen receptor

In the late 1970s, a group of researchers set out to develop a collection of monoclonal antibodies to human T-cell surface molecules that might prove useful in probing T-cell functions. Doubtless they had in mind uncovering the T-cell receptor for antigen, among other things. One of the monoclonals they developed, OKT3, proved useful as a general T-cell marker, since it was present on 100 percent of human T cells (Kung et al., 1979). The molecule with which OKT mAbs interacted indeed turned out to be part of the T-cell receptor, but one of the CD3 chains rather than the antigen-binding α- and β-chains. Within a year it had been established that the OKT3 antibody, in the presence of Fc-receptor-positive accessory cells (macrophages or dendritic cells), was a potent mitogen for human T cells (Van Wauwe et al., 1980, Van Wauwe and Goossens, 1981).

CD3 antibodies were found to block target cell lysis by CTL (Chang and Gingras, 1981), but also, surprisingly, to "activate" CTL clones to kill non-specifically (Hoffman et al., 1985; Leeuwenberg et al., 1985; Mentzer et al., 1985). This latter effect was later shown to be true only against Fc receptor-positive target cells (van Seventer et al., 1987; see also Chapter 4, "Antibody redirected lysis"). CD3 antibodies were also found to induce primary activation of pre-CTL, if accessory cells and needed cytokines were provided (Weber et al., 1985; Jung et al., 1986, 1987)). CD4 as well as CD8 CTL could be activated by CD3 antibody (Nishimura et al., 1992). Primary activation could also be achieved with antibodies to the α/β portion of the TCR (Leo et al., 1987; Schlitt et al., 1990).

The signal provided by MHC-associated antigen can thus be mimicked by cross-linking various components of the CTL receptor with antibody, which will result in the generation of cytotoxicity as long as the same factors required for CTL development in response to antigen are provided. The most important factor will be the presence of accessory cells, which not only present a cross-linked (via Fc receptors) form of the antibody, but also provide ligands for proper formation of an activating immunological synapse. If the antibody used for activation is clonotypic, i.e. if it recognizes only those TCR determinants involved in interacting with a specific MHC/peptide complex, then the response will be clonotypically restricted, as if that peptide complex itself had triggered the response. If the antibody used is directed to other determinants of the TCR, then the degree of activation will be variable (Staerz and Bevan, 1986). If the antibody is against CD3 determinants associated with the TCR, then essentially 100 percent of T cells – both CD4 and CD8 – will be activated, and any cytotoxicity induced will be polyclonal and directed toward a wide range of MHC-peptide targets.

7.2 Activation of cytotoxic function by mitogenic lectins

Lectins are sugar-binding proteins of plant and animal origin. Some are powerful yet selective agglutinins of mammalian cells, mostly of red blood cells. The ability of lymphocytes to proliferate in response to certain plant-derived lectins, such as phytohemagglutinin (PHA) or Concanavalin A (Con A) (mitogenic lectins) and not to others such as soy bean agglutinin (SBA) or wheat germ agglutinin (WGA) (non-mitogenic lectins), was noted already in the early 1960s (see Janossy and Greaves, 1972, for a discussion of the early work). This lectin-driven proliferation was studied intensely as a model for early events in lymphocyte activation, but the relationship of this proliferation to that induced by specific antigen was unclear, particularly in terms of the acquisition of specific immune functions. In the early 1970s several labs reported that lymphocytes stimulated for several days with mitogens such as PHA and ConA could generate cell-mediated cytotoxicity, but this cytotoxicity could only be seen if the same mitogens were present during the cytotoxicity assay (Möller et al., 1972; Sharon and Lis, 1972). Moreover, the cytotoxicity was non-MHC-specific; even target cells bearing self-MHC antigens were killed. So the question remained, what is the relation of this non-specific lysis to the MHC-specific and highly restricted killing generated against alloantigen?

This question was settled in several papers showing activation of cytotoxic function in lymphocytes by Con A (Clark, 1975; Waterfield et al.,

1975; Heininger et al., 1976). The cytotoxicity generated in response to this lectin was shown to consist of multiple clones of CTL, each mediating a different antigen-specific cytotoxicity. Thus all potential CTL clones in a given population are activated by lectin, making the cytotoxic activity of any single emerging clone much less potent than would be seen after activation in response to a single set of MHC antigens. In the earlier experiments, low levels of activity against individual targets in the absence of lectin in the assay was in fact seen, but was overlooked in face of the much greater killing seen in the presence of lectin.

Within twenty-four hours of exposure to lectin, the initially small lymphocytes aggregate, undergo blast transformation and divide. Proliferation is vigorous, and cytotoxicity is evident in stimulated cultures within three days. Since any given CTL clone is represented at a low frequency, polyclonal activation results in a relatively low level of antigen-specific cytotoxicity against randomly selected target cells. However, if lectin is carried over from the activation cultures or is freshly added into the cytotoxicity assay, the activated CTL (whether stimulated by antigen or lectin) will lyse virtually any target cell type via lectin-dependent cellular cytotoxicity (LDCC) (Chapter 4). Thus, in order to see the receptor-mediated, antigen-specific lysis mediated by individual subclones of lectin-stimulated CTL, it is necessary to include soluble carbohydrate blocking agents (alpha-methyl mannoside in the case of ConA) in the cytotoxicity assay to neutralize any carry-over lectin.

Although the nature of the initial triggering event induced by mitogenic lectins is still unclear, the sequence of events leading to the activation of cytotoxic function appears to be the same as in triggering by specifically recognized antigen. The most likely explanation for activation is that the lectins, all of which interact with carbohydrate groups, react with carbohydrate structures on various components of the T-cell receptor complex, thus cross-linking receptors in the same way as the receptor antibody. From that point of view, there would be no difference in the initial activation of CTL function by antigen, by the receptor antibody, or by mitogenic lectins; all deliver the same signal 1. Activation of CTL function by lectins, as with TCR antibody, is dependent on the presence of accessory cells (Rosenstreich et al., 1976), although their precise function is unclear. It is known that the same mitogens that trigger CD8 cells also activate CD4 cells, which are present in the initial cell mixture, causing them to release their full spectrum of amplifying cytokines, which would provide adequate signal 2. The same lectins that stimulate resting CD8 cells to proliferate and express primary cytotoxicity will preferentially reactivate cytotoxic function in memory alloreactive CTL populations (Chakravarty and Clark, 1977), or in memory virus-specific CTL (Tsotsiashvili, 1998) as long as a source of critical cytokines (usually supplied by accompanying memory CD4 cells) is provided.

7.3 Activation of cytotoxicity by oxidation of the T-cell surface

While chemically probing the surface of lymphocytes in an attempt to identify molecules interacting with mitogenic lectins, Novogrodsky observed that oxidation with $NaIO_4$ caused extensive DNA synthesis and proliferation of spleen cells (Novogrodsky, 1971). He later found that a combination of neuraminidase and galactose oxidase (NAGO) did the same thing, suggesting that $NaIO_4$ acted by oxidizing terminal sialic acid residues (Novogrodsky and Katchalski, 1973). Further probing suggested that the responding cells were T lymphocytes, and that accessory cells were required for the response (Novogrodsky, 1974; Biniaminov et al., 1975). As with mitogenic lectin stimulation, Novogrodsky was able to show that oxidation of cell surfaces resulted in not just proliferation, but the generation of cytotoxicity. The cytotoxicity generated was polyclonal, and so cytotoxicity against any given target cell was low, unless the target cells were themselves oxidized, in which case cytotoxicity was vigorous and effective against any target including itself. This latter phenomenon, oxidation-dependent cellular cytotoxicity (ODCC) (O'Brien et al., 1974; Kuppers and Henney, 1977; Grimm and Bonavida, 1979; Keren and Berke, 1986) will be discussed in Chapter 4.

7.4 Activation of cytotoxic function by mitogenic cytokines

Cytotoxic function can also be generated polyclonally by incubating lymphocytes in high concentrations (500-1000 U/ml) of mitogenic cytokines such as IL-2. The resulting effector cells are called LAK (lymphokine-activated killer) cells. Initially, LAK cells were thought to be a unique cell population, capable of lysing, among other things, freshly excised tumor cells, but not normal cells (Grimm et al., 1982; Lefor and Rosenberg, 1991). Later, LAK activity was attributed primarily to IL-2-activated natural killer (NK) cells. It is now evident that LAK cells can be generated from both NK cells (see Chapter 6) and T cells (Ortaldo et al., 1986). Although IL-2 is a widely used activator of NK cells, other factors, such as interferon-γ (IFNγ) and tumor necrosis factor-α (TNFα), may also be involved. LAK cells are large granulated lymphoblasts rich in perforin- and granzyme-containing cytolytic granules, which will be discussed in Chapter 5. In CD8 cells, this mode of activation bypasses the need for signal 1, which among other things up-regulates the expression of surface IL-2 receptors on T cells. But even resting, unstimulated T cells express low levels of IL-2 receptors, and can be triggered by sufficiently high concentrations of IL-2.

7.5 PMA and ionomycin activation of cytotoxic function

One of the earliest events to follow the triggering of T cells, whether by antigen, TCR antibody or mitogenic lectins, is a rapid influx of extracellular Ca^{2+} (O'Flynn et al., 1984). Thus, calcium ionophores were tested to see if this step alone could bypass activation via the TCR, but by themselves ionophores could not trigger T cells to proliferate (Hesketh et al., 1977, 1983; Truneh et al., 1985a). Upon closer examination of early T-cell triggering events, it was noticed that Ca^{2+} influx was always preceded by the breakdown of phosphatidylinositol to produce diacylglycerol, which activates protein kinase C. Reasoning that the latter might be critical in intracellular signaling pathways, a group of French investigators tried a combination of Ca^{2+} ionophore and a phorbol ester (PMA) known to induce PKC. This combination induced both T-cell proliferation and cytotoxicity in a previously primed population of T cells, as long as a source of IL-2 was present (Truneh et al., 1985a,b).

In both CD8 and CD4 CTL clones, PMA and ionomycin induce substantial non-antigen-specific target cell lysis, through up-regulation of the Fas ligand (Glass et al., 1996; Kojima et al., 1997; Thilenius et al., 1999). The Fas lytic pathway is discussed in Chapter 5.

All of these antigen-independent methods for generating cytotoxicity were studied in an attempt to understand the molecular basis of CTL triggering, in particular the structures on the stimulating cell required to provoke cytotoxic (and proliferative) responses. Once these questions had been addressed by studying the direct response to antigen, research into the basis of activation by antigen-independent means came to a halt. Many intriguing questions about activation in antigen-independent systems were never answered. For example, an interesting question relating to all such antigen-independent systems for T-cell activation is how or whether the need for the carefully orchestrated interactions taking place within the immunological synapse (Chapter 1) is bypassed, in particular those interactions, such as the interaction of CD28 with B7, which are dependent on contact with an APC/target cell. To date, no one has looked at the organization of T-cell receptor and activation structures during activation by antibody or lectin. In the case of direct cytokine activation, the T-cell receptor complex would appear to be bypassed altogether, and the assumption is that pathways downstream of T-cell receptor engagement are directly activated. And in the case of lectin and oxidation-generated cytotoxicity, how exactly do lectin and oxidation modify the stimulating cell surface to render it activating? While interesting, it is unlikely that resolving these ancillary questions would add anything to our knowledge of T-cell activation, and so they were not pursued. Nevertheless, lectin activation and activation by PMA and ionomycin in particular are still often used to generate large pools of polyclonally activated CTL for general study, since most researchers are convinced the resulting CTL are not different from those generated in response to antigen.

8 CELLULAR INTERACTIONS IN THE GENERATION OF CD8 CTL

The development of the MLC system greatly extended the possibilities for studying cellular and molecular requirements for inducing cytotoxicity in lymphocytes. Early events detectable in the responding population include RNA synthesis (mostly ribosomal), followed by a round of early protein synthesis. RNA and protein synthesis start in again later as the cells begin to divide. Enlarged (blast) cells begin to appear at about 14 hours, followed by the onset of DNA synthesis and cell division. These early events involve mostly CD4 T cells responding to class II alloantigens. Pre-CTL, in the meantime, are processing signals 1 and 2 (allogeneic class I plus co-stimulatory factors; see below), and must await the provision of IL-2 and other cytokines either from themselves or from maturing CD4 cells before beginning to express cytotoxicity. In primary cultures the first cytotoxicity is detectable at about 48 hours, although it is likely that cytotoxicity appears earlier when the frequency of CTL is too low to exert a measurable effect. CD4 cells decrease their proliferative response after two-three days. There is some continued expansion of CD8 cells for another day or so, but by five-six days the cultures have become markedly less active. Cytotoxicity also begins to fall off after day four.

One of the first things that became clear from the study of MLC reactions was the requirement for an adherent "accessory cell" in the stimulating cell population (Davidson, 1977). Accessory cells are in fact antigen-presenting cells. In MLC reactions, accessory cells present peptide-modified class I and II allogeneic MHC molecules to CD8 and CD4 cells, and provide needed cell-bound and soluble co-stimulation signals to each cell type. Accessory cells are normally some form of dendritic cell (DC), but they may also be macrophages or even (for CD4 T cells) B cells. Contact of naïve CD8 cells with an appropriate APC for as little as two hours can provide enough of a signal to drive full expansion and differentiation of CTL in vitro (van Stipdonk et al., 2001). But this may be misleading. In vitro, key cytokines produced autocrinically by this relatively brief exposure accumulate in the immediate vicinity, and quickly reach (and sustain) levels sufficient to drive maturation forward. In vivo, these cytokines diffuse away from the activation site, and a longer period – up to twenty hours of contact between the APC and CD8 cell – is required to achieve complete maturation (van Stipdonk et al., 2003).

The logistics governing the delivery of CD4 help to CD8 cells via a dendritic cell was for many years something of a puzzle. The frequency of CD4 and CD8 cells recognizing a given set of allogeneic or modified syngeneic MHC-associated signals must be very low, particularly in a primary response. In the initial view of CD4 help as primarily delivery of cytokines (Keene and Forman, 1982), one had to imagine a CD4 cell and a CD8 cell simultaneously and by chance encountering a single DC, which could present MHC structures both cells could interact with. Only then would the two reacting cells be in sufficient proximity to one another to allow meaningful delivery and uptake of soluble helper factors. A means of getting all three cell types - two of which are circulating, and one of which is semi-cessile – in the same place at the same time, was difficult to imagine.

In the newer view of 'help', the primary function of the CD4 cell is to modify (activate) the DC so that if and when a CD8 cell, able to recognize the dendritic cell's class I MHC signal, encounters that DC, the CD8 cell can receive a complete set of activating signals (1 and 2) from the DC and proceed on its own down the road to full maturation and development of cytotoxic function independently of CD4 cells (Ridge et al., 1998; Bennett et al., 1998; Schoenberger et al., 1998). This activation of the DC takes place at least in part through the interaction of the CD4 T-cell CD40 ligand (CD40L) and dendritic cell CD40, during a process which also leads to activation of the CD4 cell. The activated DC up-regulates co-stimulatory ligands needed for both CD4 and, subsequently, CD8 activation. The activated DC also produces cytokines of its own, such as IL-12, which promotes CD8 maturation in some (Trinchieri, 1994; Sad et al., 1995) but perhaps not all cases. This role for CD40/CD40L has been validated by studies showing that the need for CD4 cells in such cases can be bypassed with agonistic CD40 antibody, and blocked by CD40L antibody (Yang et al., 1996). A small subset of CD8 cells also express CD40L, but in greatly reduced surface concentration (Lane et al., 1992; Hermann et al., 1993). CD8 cells by themselves generally cannot activate dendritic cells through CD40 (Wu and Liu, 1994).

This scheme greatly simplifies matters, in that the CD4, CD8, and DC do not all need to be in the same place at the same time. The CD8 cell will still need helper factors to complete its maturation, but CD8 cells can provide many of these on their own (Paliard et al., 1988), and they can also utilize the abundance of such factors released by various cell types in the general vicinity of an allograft or other inflammatory reaction. As we will see in Chapter 9, DC can also be activated, in the absence of CD4 cells, to the CD8-helping state by various microbial products. Thus, in an animal in which an antimicrobial inflammatory reaction is underway or has recently occurred, DC may already be in a heightened state of reactivity. This view of CD4-CD8-DC interaction may explain why some CD8 responses can occur in the apparent absence of CD4 help (Buller et al., 1987; Hou et al., 1995). We will discuss this scheme for CTL activation in somewhat more detail in Chapters 7 and 8, when we look at details of CTL activation in vivo.

The proliferative phase of the ensuing reaction results largely from the response of CD4 T cells to allogeneic class II antigens on DC. The end result of the differentiation step involving CD4 cells, in addition to proliferation, is the production of numerous cytokines (Table 1.2). These cytokines are not all needed for the development of cytotoxicity; the same spectrum of cytokines is also produced by CD4 T cells activated in the course of a humoral immune response, where some of the cytokines are used by B lymphocytes as they mature to antibody-producing cells. Many of them are involved in the initiation of generalized inflammatory responses. There is no selective activation of CD4 cytokine production determined by the type of response the CD4 cells are involved in; they are either on, and discharge their full spectrum of cytokines, or they are off.

While the interaction of responder CD4 and CD8 T cells with allogeneic MHC products on dendritic cells results in maximal generation of cytotoxicity in vitro, there is evidence that alternate CTL activation pathways must exist. For example, in the early 1970s a number of MHC-mutant mouse strains were developed that differed only by a single mutation in a single class I gene. When cells from one such strain, called bm1, were injected into the wild-type parental strain (C57BL/6), or vice-versa, no antibody was generated in response to the slightly altered class I MHC proteins. Yet substantial cytotoxicity could be generated both in vivo and in MLC reactions between two such strains, yielding CTL specific for the minor difference in class I MHC structure (Berke and Amos 1973; Bach et al., 1973). Moreover, such strains reciprocally reject each others skin grafts, and are mutually sensitive in GVH reactions. Subsequent studies showed that it was the CD8 cells in host strains that responded to the mutant class I MHC molecules; host CD4 cells by themselves were incapable of mounting a cytotoxic response (Rosenberg et al., 1986).

These data present a challenge for the ideal MLC reactions just described. As we will discuss in more detail in Chapter 8, one interpretation is that responder CD8 T cells get signals 1 and 2 directly from mutant allogeneic class I-bearing stimulator DC. But CD4 cells will not have activated these DC, so they should not function effectively in the activation of CD8 cells. The cytokines necessary for full development of a cytotoxic response could come from an indirect reaction: responder CD4 T cells could be stimulated by self dendritic cells that have taken up and digested fragments of mutated stimulator cells, subsequently expressing mutant class I peptides on self class II molecules. The IL-2 thus released could be sufficient to support CD8 cell differentiation. But the problem of the absence of CD4 activation of the DC used by CD8 cells remains unresolved.

Chapter 3

CYTOTOXIC T LYMPHOCYTES

Target cell recognition and binding

1 CELL-CELL INTERACTION: A CONTEXT

The interaction of all cells with elements of their environment involves an initial recognition event in which a genetically encoded cell surface receptor binds to a complementary environmental molecule (the ligand). The mechanisms underlying informational cell-environment interactions show an astonishing universality among all living cells, from free-living single-cell organisms through the most complex multicellular life forms where each cell is, in effect, part of a larger social structure.

The cognate ligand may exist as a free molecular entity, or it can be associated with the surface of another cell. In the latter case, the purpose of the recognition event may be simply to orient the receptor-bearing cell within its environment – for example, to guide migrating cells to specific sites during embryonic development, to localize and stabilize cellular aggregates within tissues or organs, or to guide circulating cells into and out of blood vessels. However, in many cases, the interaction of a cell-surface receptor with a free or cell-bound ligand may result in an alteration - either activation or inhibition - of the state of the receptor-bearing cell. In those cases where the perceived ligand is an integral part of the membrane of another cell, the receptor-ligand interaction may alter the activation status of the ligand-bearing cell as well – receptor and ligand are relative terms in such a situation. Changes in the activation state of either cell in a bicellular interaction, or of any cell with a soluble ligand, occurs through the signal transduction machinery present in all cells. Following the establishment of a stable association between interacting cells, the receptor (or the ligand) generates an intracellular signal through spatial reorganization or stereochemical distortion of one or more submembranous signal-transduction molecules.

More than almost any other system in the body, the proper functioning of the immune system depends on a broad and highly complex set of interactions of immune cells with other cells of the immune and other systems, and of immune cells with soluble ligands. Immune cells spend part of their lives in various immune tissues, where they interact mostly with a broad range of other immune cell types. But many of them, especially T cells, spend a good portion of their lives circulating throughout the body, where they must interact with an even wider range of cell types – virtually every cell type in the body – as well as soluble regulatory chemical signals (cytokines) and foreign antigen. As discussed in Chapter 1, T-cell interactions with other cell types are regulated by class I or class II histocompatibility molecules.

In the following sections, we examine at the cellular level the first step involved in the process of target cell death at the hands of killer T cells – the MHC-guided recognition of the target cell by the killer cell, and the consequent formation of stable, bicellular or multicellular aggregates ("conjugates") that allow the killing reaction to proceed. We explore the nature of the killing reaction itself in Chapters 4 and 5.

2 THE BINDING OF CTL TO TARGET CELLS

2.1 CTL binding to cell monolayers

In the early 1960s, exploring further the seminal observations of Andre Govaerts, several groups studied microscopically the initial physical interactions of what were presumed to be killer lymphocytes, harvested after allogeneic sensitization in vivo, with target cells. In these early experiments, following the procedures established by Govaerts, the target cells were monolayers of cultured cells such as fibroblasts or macrophages. A specific clustering behavior involving immune lymphoid cells was immediately noted around target cells against which they had been sensitized, but not against other target cells (Figure 2.1). This cell clustering, initially termed "contactual agglutination," was generally observed to be complete within a short time in culture and always preceded target cell death. Although non-immune lymphoid cells, or cells sensitized against "third-party" cells, exhibited a superficially similar clustering behavior, such associations were transient and the adhering lymphoid cells could be easily dislodged by gentle agitation. True killer cells, however, formed stable bonds with their target cells, which could not be disrupted even by relatively vigorous pipetting (Rosenau and Moon, 1961; Kowproski and Fernandes, 1962; Taylor and Culling, 1963; Wilson, 1963).

Brondz and co-workers (Brondz, 1968; Brondz et al., 1975) were among the first to establish the immunological specificity of killer-target binding. His group depleted the cytolytic activity of immune lymphoid cell populations by incubation on macrophage monolayers displaying "transplantation antigens" (class I MHC molecules) identical to the immunizing cells. Lymphoid cells recovered from such monolayers ("non-adherent cells") showed greatly reduced cytotoxicity toward specific target cells. However, incubation on allogeneic monolayers of cells not displaying the sensitizing MHC products did not deplete the killing activity. This clearly established that a firm and antigen-specific adhesion of the killer cell to its target preceded killing. That the functional depletion observed was not due to inactivation or annihilation of effector cells, induced by the adsorbing monolayers, was demonstrated independently by Golstein et al. (1971) and by Berke and Levey (1972), who showed that CTL specifically adhering to such monolayers could be recovered and in fact exhibit increased cytocidal activity against the specific target cell phenotype employed for immunization and subsequent absorption.

Since most species have two or more genes encoding class I MHC antigens, and each cell in the body can express up to two alleles of each of these genes, the question arose as to whether an individual CTL recognizes a single species of class I molecules on an allogeneic cell, or more than one. Brondz et al. (1968, 1975) were the first to demonstrate in the mouse the generation of at least two distinct subpopulations of alloimmune CTL in terms of binding to target cells – one specific for the H-2K class I molecule, and the other for the H-2D molecule. Experiments with PEL showed clearly that the number of responding CTL increased with the number of MHC differences between recipient and donor (Berke, 1975). Moreover, CTL from animals simultaneously immunized with two antigenically distinct tumor allografts virtually never form heteroconjugates, that is, a CTL bound simultaneously to two allogeneically differing target cells. Taken together, these results suggested that individual CTL interact with only one isotype of a class I molecule on target cells (Berke, 1980).

The monitoring of killer cell binding to non-adherent cells such as leukemia cells was initially made possible by the use of poly-L-lysine (PLL)-fixed leukemia cell monolayers. Because the charged bonds holding the target monolayer to the plate are different from those binding CTL to their target, it was possible to establish Mg^{+2}, pH, and temperature dependence of CTL-target binding (Stulting and Berke, 1973).

Nearly all current assays of CTL-target interaction involve target cells in suspension (see following section). However, in vivo, the interaction between T cells and specifically recognized cells in organized tissue is biologically very important, and the interaction of CTL with cells growing in a monolayer may be a better in-vitro model for this type of interaction. CTL-induced disruption of cell monolayers by detachment of live target cells from their substrate is an early event in CTL interaction with cells bound to a physical substratum and interacting with neighboring cells. This most likely reflects detachment of damaged targets from neighboring tissue cells in vivo. Target cell detachment is an MHC antigen-specific event. Detachment is a distinct event in the sequence of killer cell-target interaction, and can occur in the absence of cell lysis (Abrams and Russell, 1991).

2.2 Conjugate formation in suspension

Brunner et al. (1968) were the first to develop an assay for cytotoxicity in which both killer cells and target cells were mixed in suspension, rather than seeding CTL onto target cell monolayers. This not only greatly streamlined cytotoxicity assays, but also provided an opportunity to examine more closely the interaction of CTL and their targets. Initial studies of CTL lytic function had been carried out with populations of lymphocytes sensitized either in vivo (during graft rejection) or, later, in vitro (mixed leukocyte cultures) to allogeneic class I MHC antigens. The resulting immune populations were indeed cytolytic but of widely varying lytic potency. This was presumed by most investigators to be due to varying proportions within the immune population of actual CTL able to bind to and kill target cells, but proof of this assumption came only after the development of suspension assays to assess the binding of CTL and target cells (Berke 1975; Martz, 1975). In suspension assays, an immune population of CTL is mixed with target cells, usually hematopoeitic tumor cells. The mixture is centrifuged at room temperature to promote cell-cell interaction, and then resuspended by pipetting or vortexing. The resulting resuspension can then be examined in a hemacytometer using a light microscope. Clusters consisting of CTL bound to target cells are called "conjugates" (Berke, et a.,l 1975). Conjugates of effector and target cells are immediately obvious (Figure 3.1A). The target cell population is selected such that target cells can be readily distinguished from lymphocytes, usually on the basis of size; most tumor populations have a diameter 2-3 times larger than even activated lymphocytes. Alternatively, the targets can be stained with fluorescent or other dyes and the conjugation assessed by fluorescence microscopy (Figure 3.1B). Suspension conjugate assays allow direct visualization and quantitation of effector CTL within an immune cell population (Berke et al 1975).

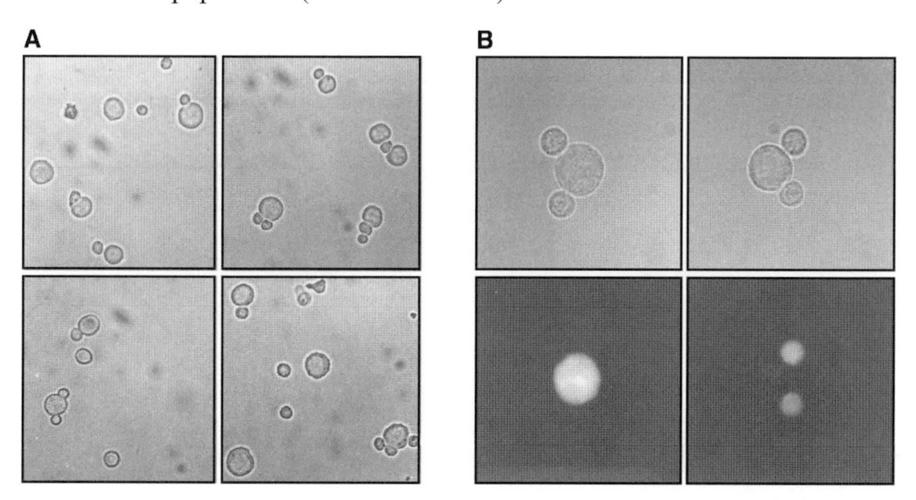

Figure 3-1. Conjugates of CTL with specific target cells. (A) light microscopy; CTL are the small cells. (B) fluorescence microscopy of fluorescein-labeled target cells conjugated with non-labeled CTL (left panels) and vice-versa (right panels).

Nylon-wool purified PEL are an excellent system for analyzing the conjugation process. After centrifugation and resuspension of PEL mixed with specific target cells, distinct antigen-specific cell clusters (conjugates) are readily observed by light microscopy (Berke et al., 1975; Martz et al., 1975). The conjugates consisted mainly (70%) of "doublets" – one PEL bound to one target cell (Figure 3.1, 3.2) There were fewer conjugates of one PEL and two or more target cells, or one target cell with two or more PEL, and on occasion even higher-order conjugates.

Microscopic monitoring of the fate of conjugates showed that one bound CTL is sufficient to kill a target cell, which has been used to estimate the proportion of CTL in lymphocytic populations (Zagury et al., 1975). There are some provisos, however. Not all CTL form strong conjugates, and some conjugating CTL may be disrupted by the resuspension procedure and thus fail to be counted or to lyse their targets within a given observation period. Although small, the proportion of conjugates involving more than one CTL or one target cell must also be taken into account. Nevertheless, conjugate-based analyses allowed the first direct visualization and quantitation of CTL, and provided some of the earliest information about their properties.

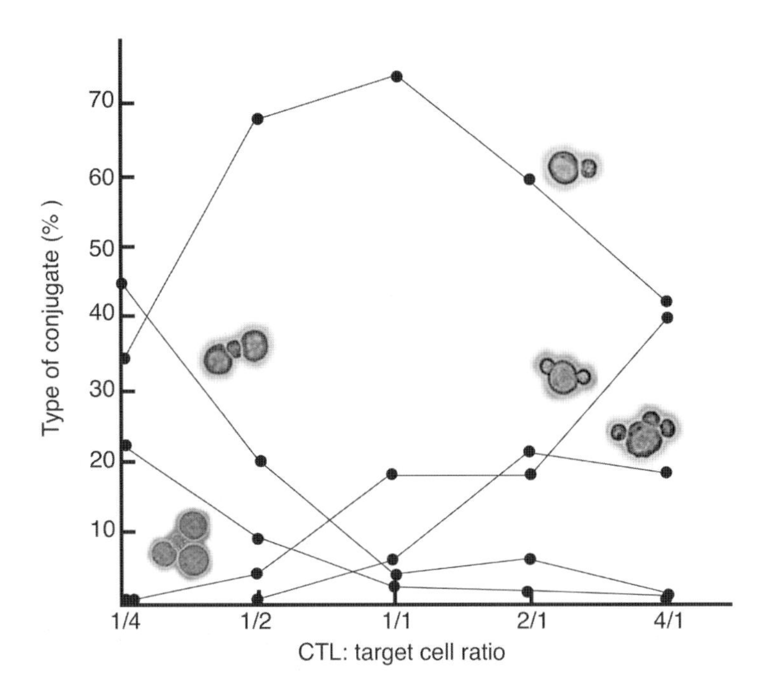

Figure 3-2. Statistics of conjugate formation. Proportion of CTL-to-target cell and type of conjugate formed. Mixtures containing CTL and target cells were co-centrifuged, resuspended and analyzed for the different types of conjugates present.

Martz refined his "post-dispersion assays" to further quantify and characterize CTL populations. In these assays, conjugates are first allowed to form under conditions that discourage target cell lysis (e.g., room temperature, Mg^{+2} but no Ca^{+2}; see below). The conjugates are then resuspended and dispersed into medium that allows lysis to proceed, but which prevents re-association of CTL recently dissociated from lysed target cells with new target cells. Slightly viscous dextran solutions have proven useful in this regard. The percentage of targets in conjugates at a given point in time is equated to the percentage of targets that lyse following dispersion to prevent new CTL-target interactions. This type of assay is capable of excellent quantitation and time resolution, showing half-maximal conjugation of PEL with their target cells in 1-2 minutes at room temperature (Martz, 1975).

Flow cytofluorometry has also been used to sort specific CTL-target cell conjugates and study single-cell kinetics of lysis (Berke, 1985; Perez et al., 1985; Denizot et al., 1986; Lebow et al., 1986; Liu et al 2002). This procedure is based on the simultaneous monitoring of cell scatter and either single-color fluorescence of pre-labeled CTL or target cells, or double-color fluorescence of conjugated effector and target cells. The method may, however, underestimate weakly bound conjugates, since flow cytometry subjects the conjugates to substantial shear force. However, it enables the processing of numerous cells and samples, thus increasing statistical precision.

Quantification of responding CTL to a variety of allogeneic, tumor antigen and intracellular parasites has traditionally relied on limiting dilution analysis (Moretta et al, 1983; Kabelitz et al, 1985), in which populations containing pre-CTL or memory CTL are seeded into microtiter plates in decreasing numbers, until the lowest number capable of giving rise to stable clones is determined. This technique is still in use today where gross estimates of CTL frequency are useful (Morishima et al, 2003), but this approach has a number of drawbacks that make it less than useful in many situations. Cloning efficiency is dependent on a number of factors having nothing to do with actual precursor frequency. Moreover, limiting dilution yields information about the total number of clones responding, not about individual clonal specificities. There is also the problem of variable affinities of T-cell receptors for individual antigenic determinants.

2.3 Soluble peptide-MHC tetramers in the analysis of T cell specificity

Monitoring the frequency of antigen-specific T cells in heterogenous lymphoid cell populations by limiting dilution, conjugation and controlled lytic assays all have distinct drawbacks, and they do not enable recovery of the actual effector cells. As has been discussed, CD8 and CD4 T cells recognize antigenic peptides of 8-21 amino acids that are bound to surface membrane MHC class I and class II molecules. In fact, the membrane-bound peptide-MHC complexes are the actual and only stimulatory antigens for the T-cell immune system, including CTL responses in pathological disorders such as transplantation, cancer, autoimmunity, and infectious diseases (Lee et al., 1999; Ogg et al., 1998a; Romero et al., 1998; 2004). Therefore, monitoring the interactions between known specific peptides and T cells procured from patients could reveal the state of immune responsiveness during pathological states such as malignancy, virus infection, autoimmunity and transplant rejection, in which T cell responses play an important role. The simplest approach to enumerating antigen-specific T cells directly would be to monitor the binding of soluble, monomeric peptide-MHC complexes to T cells, employing fluorescently tagged peptide or MHC molecules. However, the low intrinsic binding affinity of monomeric peptide-MHC molecules to T cells precluded development of a reliable binding assay to assess peptide-specific T cells.

A major step forward in circumventing these difficulties, and increasing the precision and information yield in quantitative assessments of antigen-specific T cells, came with the development of the so-called peptide-MHC class I tetramer technology (Altman et al., 1996; Ogg et al., 1998b; Meidenbauer et al., 2003). In its most commonly used form (Figure 3.3), the tetramer assay for T cells employs four class I recombinant MHC/β-2m monomers that have been refolded in the presence of the peptides of choice, biotinylated, and allowed to bind to fluorescein-labeled avidin. The fully loaded complex is then reacted with a population of cells presumed to contain precursor or memory CTL specific for the peptide/MHC complex defining the tetramer. Cell-surface-bound tetramers are then assessed or sorted by flow cytometry. Although initially presenting a number of technical difficulties, these have largely been worked out (Meidenbauer, et a.l, 2003), and the assay as now generally used offers unprecedented sensitivity and accuracy in defining antigen-specific T cell and CTL populations. Class I MHC tetramers can also be used to directly activate T cells, and may be sufficient in the absence of costimulatory signals (Wang et al, 2000; Cohen et al 2003). More recently, peptide-class II MHC tetramers have been developed to assess CD4 T cells. Finally, peptide-MHC dimers, tetramers and even pentamers are now commercially available for analyzing peptide-specific TCR on both CD4 and CD8 T cells in health and disease.

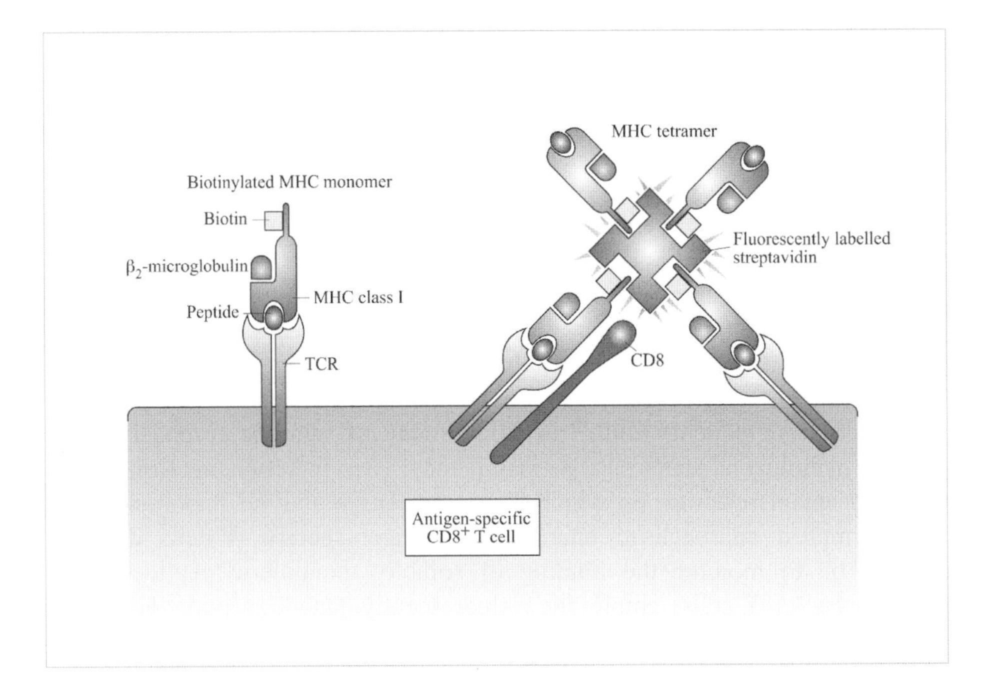

Figure 3-3. Peptide MHC tetramers: molecular tools for monitoring antigen-specific T cell receptors.

3 PHYSIOLOGICAL AND METABOLIC ASPECTS OF CONJUGATE FORMATION

Formation of a stable interaction between CTL and target cells is strictly Mg^{2+}-dependent (Stulting and Berke, 1973; Martz, 1975; Golstein and Smith, 1976). This requirement likely resides in the LFA-1/ICAM-1 interaction (Shaw et al. 1986; Dustin and Springer, 1989), presumably in the consensus metal-binding domains of the LFA-1α chain. In the presence of millimolar amounts of Mg^{2+} (physiological range), Ca^{2+} affects neither the rate nor the extent of adhesion, although Ca^{2+} can increase target binding in low Mg^{2+} concentrations (Martz, 1980).

No conjugation occurs at 0°, but significant and rapid conjugation can occur at room temperature (25°), a condition under which virtually no subsequent lysis occurs (Berke and Gabison, 1975). As the temperature is raised, the degree of binding increases (Figure. 3.4 A). The sensitivity of conjugation to changes in temperature suggests that the binding of sensitized lymphocytes to target cells is initiated via preformed receptors and ligands already present on the surface of the CTL and its target. However, destruction of target cells occurs only at temperatures conducive to active cellular metabolism. Interestingly, although no conjugation occurs at 0°, or in the presence of metabolic inhibitors, pre-formed conjugates remain stable under these conditions (Figure 3.4 B).

A

(a) Temperture (°C)					
	2	4	10	22	37
Number of conjugates x10^{-4}	5.4	12.4	22.4	34.0	27.0

(b) inhibitor (15mM)			
	N^-_3	CN^-	PBS
Number of conjugates x10^{-4}	2.4	3.2	30.0

B

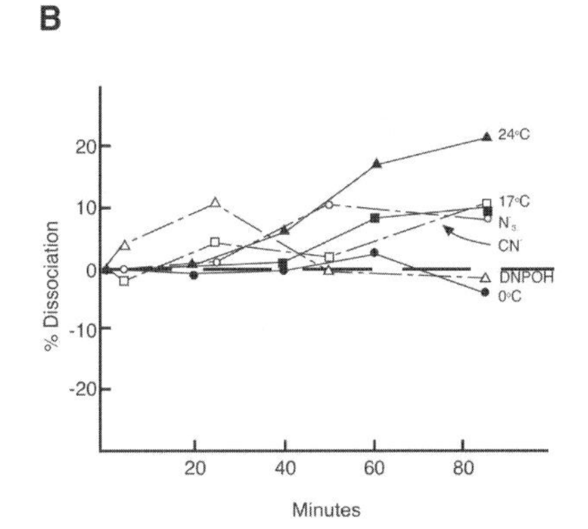

Figure 3-4. The energy of conjugation. Effects of temperature and metabolic inhibitors on conjugate formation (A) and dissociation (B).

During the early 1970s virtually every metabolic inhibitor was tested for its effect on CTL action. Obviously, any drug that interferes with CTL-target cell binding will *a priori* inhibit target cell killing. Studies with such inhibitors have been exhaustively reviewed by Martz (1977). Basically, anything that inhibits cellular metabolism (e.g., azide, cyanide, and DNP) or cell motility (cytochalasin) or the microtubule system (colchicine), blocks conjugation. The effects of these inhibitors on lysis, independent of their effect on conjugation, will be discussed in Chapter 5.

4 ULTRASTRUCTURE OF CTL-TARGET CELL CONJUGATES

Early electron microscopy of CTL-target cell conjugates revealed extensive and mutual plasma membrane interdigitation at the contact region (Kalina and Berke, 1976; Sanderson and Glauret 1977), but not cell fusion or cytoplasmic junctions. Although up to one-third of the surfaces of conjugating cells may be in close physical proximity, in fact only a relatively small proportion of the cell membranes at the contact area are involved in intimate contact at a given time. Initial engagement between CTL and target cells is likely followed by a "zipping-up" process between molecular components on either side of the contact point. Short points of contact, averaging 1500 Å in length, are the main form of junction at a given point. Periodic substructures are observed in some of the contact areas (Figure. 3.5; Kalina and Berke, 1976). The exact nature of these short, periodic substructures has not been ascertained, although there is an intriguing similarity between them and gap or septate junctions observed in other systems of interacting cells. A requirement for multifocal interactions over a large surface area has also been examined by following the interaction of CTL with artificially generated antigenic membranes. The binding was found to be dependent on both the size of the artificial membrane and the density of antigenic determinants (Mescher, 1995).

Grimm et al. (1979) described stretching and rupture of the target-cell membrane near areas of broad contact and suggested that these might represent the primary lytic event. An EM study of material fixed at 37° within 10 minutes of contact, under conditions giving very high rates of killing, provided a detailed description of finger-shaped projections of the CTL (Sanderson and Glauert, 1979). It was noted that projections pushed into the target cell and, in some cases, appeared to distort organelles, including the target cell nucleus, in their path. In all cases, the target cell membrane remained intact during CTL-target contact, ruling out physical damage as a cause of target cell death.

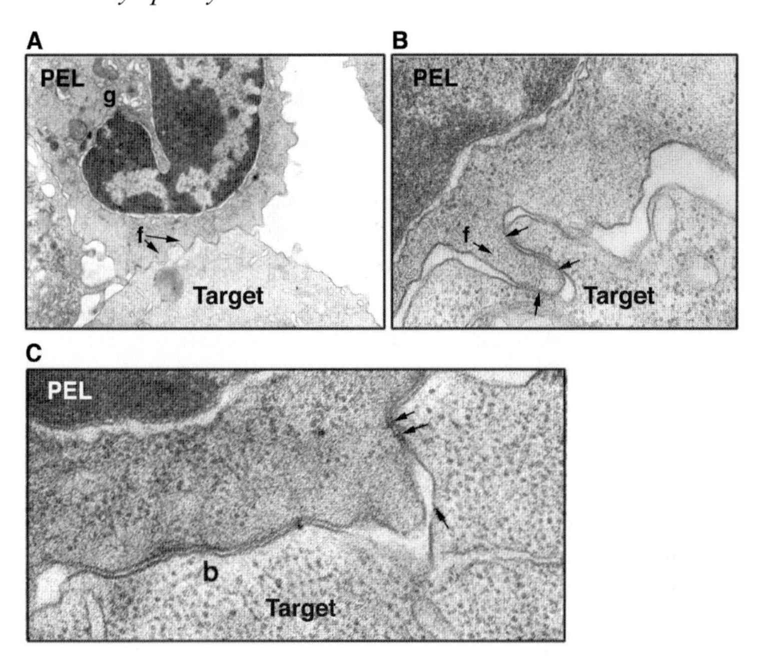

Figure 3-5. The CTL-target contact region. (A) Cytoplasmic projections of the PEL CTL contain microfilaments (f). Note location of the Golgi apparatus (g) at a distance from the contact region. x11,250. (B) At least three point contacts (arrows) exist between the interdigitating projections; only microfilaments (f) can be seen at the contact region in the CTL. x65,000. (C) Septa-like structures (arrows) in the intercellular space between the apposing membranes. b – broad contact area devoid of any intercellular structures. x 140,000.

The projections contain microfilaments and other organelles, and areas of close contact between the two plasma membranes were usually seen at the tip of the projection (Figure 3.6). The fact that projections are relatively difficult to find in sections suggests that they form rapidly and then disappear. This is reinforced by time-lapse films of fluorescently stained cells showing rapid movement of CTL microvilli and rapid movement of target cell organelles at the contact zone.

Although physical damage to the target cell membrane as a result of conjugation is not seen as a factor in target cell death, there have been several reports of intercellular transfer of target cell membrane components, including peptide-MHC, onto the CTL, perhaps as a result of localized target cell membrane destabilization. If these components are internalized, then reach the endoplasmic reticulum and get re-expressed by the CTL, they could conceivably be involved in attracting other CTL to attack and kill them ('fratricide'), which could be a way of controlling ongoing CTL reactions (Huang et al., 1999; Hanon et al., 2000; Hudrisier et al., 2001; Stinchcombe et al., 2001).

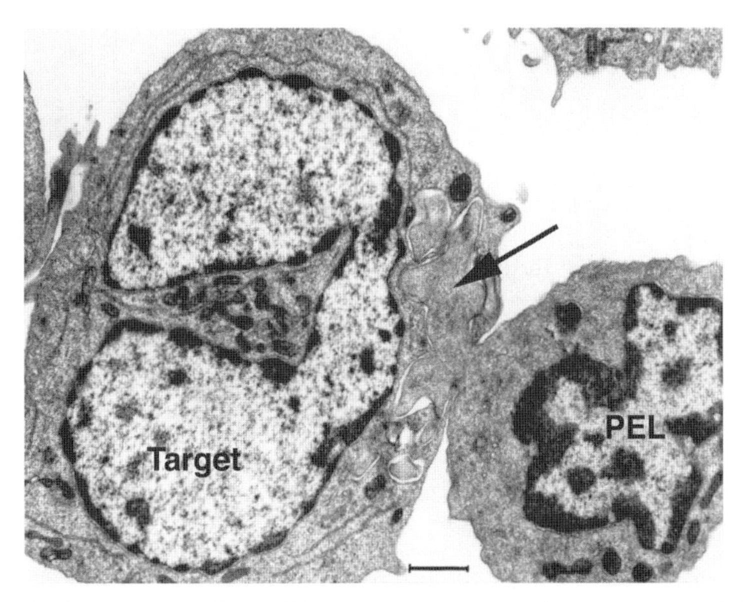

Figure 3-6. CTL protrusion. Note CTL protrusion penetrating the larger target cell (arrow). Bar represents 1000 nm. By D. Rosen

The extensive interdigitation of CTL and target cell membranes suggests that, in addition to the multitude of stereochemical interactions involved in binding, membrane-folding forces that form the interdigitating finger-like structures operate at the contact zone. This would be consistent with the metabolic energy and intact CTL cytoskeletal system (sensitive to the cytoskeleton disruptive drugs cytochalasin A and colchicine) requirements for CTL-target cell binding. The CTL projections involved in conjugation contain a network of fine fibrillar material and are devoid of ribosomes, granules, and other cellular organelles (Figures 3.5, 3.6).

A cardinal ultrastructural feature of CTL-target cell conjugates is orientation of the microtubule organizing center (MTOC) and the Golgi apparatus of the CTL toward the point of contact with the target cell (Geiger et al., 1982; Zagury et al., 1975) (Figure 3.7). Although the Golgi apparatus and MTOC need not be near the contact point for initial interaction between effector and target cells (see Figure 3.5 A), an active reorientation of the MTOC toward the target cell occurs shortly after contact (Geiger et al., 1982; Kupfer and Singer, 1989; Bykovskaja et al., 1978). The centriole, which is associated with the MTOC, has been observed localized in front of the nucleus, toward the leading edge of the plasma membrane of cells in motion (Albrecht-Buehler and Bushnell, 1979; Malech et al., 1977). The membrane in this area exhibits an increased protrusive and deformational potential that may render it more suitable for the formation of stable intercellular contacts with the target cell (Ryser et al., 1982).

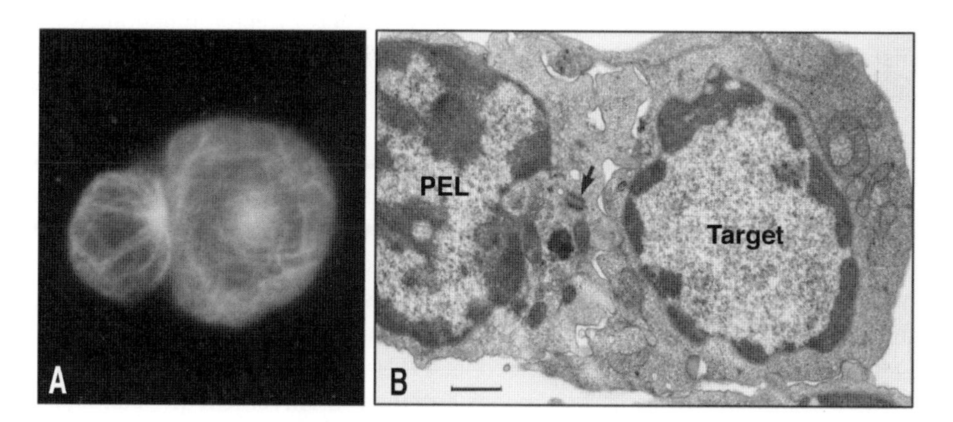

Figure 3-7. The CTL MTOC. (A) Indirect immunofluorescence labeling for tubulin of CTL-target cell conjugates fixed immediately after conjugate formation. The CTL MTOC is proximal to the contact area, whereas that of the larger target cell is randomly oriented. (B) TEM of CTL-target cell conjugate. The longitudinally sectioned centriole of the CTL (arrow) is localized at the contact region. (From Geiger et al 1982). Bar represent (A) 1200 nm, (B) 776 nm. By D. Rosen.

Localization of other CTL organelles such as lysosome-like granules in the vicinity of contact regions is consistent with the involvement of a secretory process in either conjugate formation or target cell lysis. The hypothesis that the location of the CTL Golgi apparatus directs the lethal hit in a vectorial fashion is also consistent with the observed sequential, rather than simultaneous, lysis of multiple target cells bound to a single CTL (Zagury et al., 1979). The region beneath the CTL membrane at the point of contact with the target cell shows an unusually high concentration of cytoskeletal elements such as talin and actin (Ryser et al., 1982; Kupfer et al., 1986), which are likely to play a role in the delivery of lytic signals. An important point that will be discussed in Chapter 4 is the relationship of CTL-target cell contacts, as seen in transmission electron microscopy, and the structure of the immunological synapse defined by confocal microscopy (Chapter 1).

5 MOLECULAR DYNAMICS OF CTL-TARGET BINDING

Although the specificity of CTL interaction with target cells resulting in conjugate formation is clearly dictated by the α and β chains of the T-cell receptor, this interaction alone is insufficient for stable target cell binding and CTL activation. A separate but complementary role for adhesion molecules in conjugate formation, and thus in CTL effector function, was first suggested by Dustin and Springer (1989) and co-workers, who described the involvement of lymphocyte function-associated antigen 1 (LFA-1) and Lyt-2,3 (later renamed CD8). LFA-1 is a heterodimeric (CD11a/CD18) β-2 integrin present on the surface of all T cells, which interacts with ICAM-1,-2, and -3 on partner cells. As discussed in Chapter 1, CD8 interacts with the conserved α3 domain of class I molecules. In addition to LFA-1 and CD8, the adhesion molecules VLA-3, -4, and -5 also help to stabilize T-cell–partner cell interactions through binding with their ligands: fibronectin and VCAM-1, fibronectin, and laminin, respectively. Other adhesion molecules implicated in CTL function include CD44 and MEL-14, which are primarily known for their involvement in leukocyte extravasation, but which play a role in T cell interactions with other cell types as well (Haynes et al, 1989)

It is presently unclear whether the conjugation of CTL and target cells begins with the T-cell receptor or the adhesion molecules. The latter could well be responsible for initial transient interactions observed between CTL and even non-specific target cells. O'Rourke et al. (1990, 1993) have assumed that surface expression of the adhesion molecules on resting CTL may be too low to initiate stable cell contact. They propose that initial contact is made via the T-cell receptor. What is clear is that surface expression of all adhesion molecules, including CD8, increases several fold after CTL encounter specifically recognized target cells. For example, surface CD8 expression in resting CTL is insufficient to initiate contact with class I-bearing cells not recognized through the T-cell receptor. However, after activation with TCR antibody, CTL will bind – albeit weakly – to any class I-positive cell, and this binding can be blocked by the CD8 antibody (O'Rourke et al., 1990). It is thus possible that the adhesion molecules simply stabilize cell contacts initiated by the TCR. However, CD8 antibodies do not block the lytic activity of some potent CTL such as PEL.

There is increasing evidence that adhesion molecules may also be involved in generating, complementing or sustaining portions of the activating signals developed through the immunological synapse. For example, CTL can still be generated in CD28 knockout mice (Shahinian et al., 1993). CD28 is an important part of the signal transduction system operating in all T cells, through its interaction with the B-7 ligand on partner cells. However, if LFA-1 is also lost, CD28-deficient mice cannot develop a CTL response (Shier et al., 1999). Ni et al. (1999) showed that LFA-1 is able to support many of the signaling functions of CD28. When CD8 is prevented from binding to class I by genetically disrupting the binding site in the class I α3 domain, CTL cannot be activated under any circumstance due to failure to phosphorylate the ζ-chain of CD3 (Purbhoo et al., 2001). It had previously been shown that CD8 regulates the function of the p56 kinase in CTL (Anel et al., 1996).

The relative strength of CTL-target cell adhesion has been estimated by varying the amount of shear applied before counting the conjugates (Bongrand et al., 1983). One method for doing this involves the use of laminar shear between parallel plates to dislodge CTL from a fibroblast target monolayer. Using this approach, Hubbard et al. (1990) showed that the specificity of CTL-target cell adhesion might be overlooked if too low a shear is applied. Both groups noted a wide range (100-fold) in the strengths of individual conjugates. The avidity of CTL-target cell conjugates has also been evaluated by determining the mechanical force required to separate conjugated CTL from specific and non-specific target cells, before the delivery of the lethal hit. The force (1.5×10^4 dynes/cm^2) required to break bonds between specifically conjugated cells is about 10 times that required to disassociate a non-specifically bound lymphocyte-target cell pair (Sung and Sung, 1986; Tozeren et al., 1989). Hence, most of the binding force, and probably the energy required to keep specifically bound CTL and target cells together, must come from interactions of the TCR/CD3 complex of the CTL with the MHC ligand on the target cell. Importantly, force and intercellular binding affinity (avidity) are not interchangeable, and force measurements do not simply determine energy (affinity), which is a thermodynamic parameter.

After lysing their target cells, most CTL can recycle, bind to and lyse additional target cells at the same rate (Zagury et al., 1975) for several more cycles. When the number of target cells exceeds that of the effector cells, some CTL, such as in vivo primed PEL, can lyse at an undiminished rate for up to 5 hr, demonstrating their recycling ability (Berke et al., 1972). Detachment of effector CTL from affected target cells is obviously a prerequisite for CTL to bind to and lyse additional target cells. Yet CTL-target cell binding appears to be an equilibrium process, as indicated by the reversal of specific cell adhesion between allogeneic CTL and target cells. The rate of this reversal depends on the affinity of CTL for bound and unbound target cells, and on the particular signals delivered to both the CTL and the target.

The specific point in the lytic cycle at which CTL detach from affected target cells, and the nature of the detachment signal(s), are not clear. An obvious source for a detachment signal is likely to come from the lytically affected target. We know that CTL are unable to form conjugates with cognate, but dead, target cells (Stulting et al., 1975; Schick and Berke, 1979). The exposure of target cells to low concentrations of formaldehyde still allows them to exclude vital dyes such as trypan blue and take up ^{51}Cr, and such cells are also able to conjugate specifically with CTL, and indeed to be lysed by CTL (Bubbers and Henney, 1975; Schick and Berke, 1978; Balk and Mescher, 1981). At higher concentrations of formaldehyde, they fail to conjugate with CTL. Hence CTL binding to its target requires preservation of at least some cellular activities of the target, suggesting that the target is not passive in conjugation.

Detachment from live target cells prior to their lysis would most likely require a signal different from that required for CTL to detach from dead or dying cells. Some sort of spontaneous detachment process, perhaps involving CTL cycling between strong and weak adhesion, has been suggested (Mescher et al., 1991). However, cinematography of individual CTL-target cell conjugates demonstrated that conjugated CTL do not detach spontaneously prior to target cell lysis (Zagury, unpublished). This has not been the case with in vitro-generated killer cells interacting with fibroblasts as targets, where a great deal of spontaneous detachment of previously bound CTL can be seen in the absence of overt killing (Ginsburg et al 1969). A transient rise of cAMP in effector CTL has been proposed as a signal for CTL detachment following the delivery of the "lethal hit" (Valituti et al., 1993), but this interesting proposal has not been followed up.

Mescher (1995) proposed a working model for the sequence of events occurring when a CTL contacts the surface of another cell, which assumes an initial weak and antigen non-specific adhesion between the cells mediated by LFA-1 and perhaps other adhesion molecules (Figure 3.8). If the TCR is not engaged, however, adhesion remains weak and the cells may dissociate after a short time. If the TCR is engaged, a signal is generated that results in up-regulation of additional adhesion molecules such as CD8, LFA-1 and VLA in the immediate vicinity, and adhesion between the surfaces increases.

This, in turn, will result in additional opportunities for the TCR to engage additional target cell peptide-MHC structures so that a spreading wave of adhesion and activation is propagated across the surface of the CTL, leading to the formation of tight CTL-target conjugates involving large regions of the cell surface (see Figures 3.5, 3.6). As the spreading wave of adhesion and signal generation moves outward and second messengers reach high levels in the region of initial contact, these may begin to deactivate adhesion in this region (Figure. 3.8). Thus, at an intermediate stage, the cell surface remains bound to the target via adhesions at the periphery of the spreading wave while contact in the central region decreases. Ultimately, Mescher suggested that as the leading edge of adhesion reaches the periphery of contact possible between the cells, it would be overtaken by the trailing wave of de-activation of adhesion. At this point, the CTL would release from the target and, with CD8 and other adhesion molecules retored to functionality in the non-activated state, repeat the cycle of adhesion and lysis upon interaction with another antigen-bearing cell. Target cells that have received a lethal hit will go on to die by apoptosis.

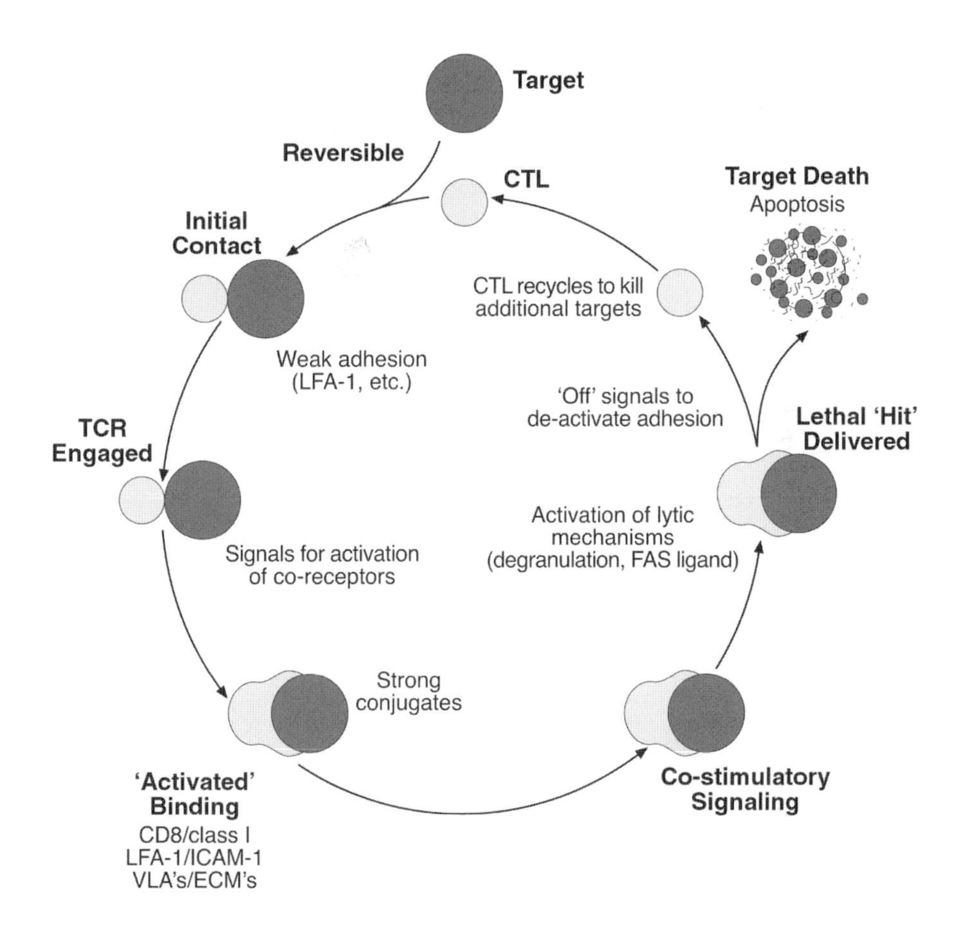

Figure 3-8. Proposed adhesion and signaling events in the cycle of CTL binding to, killing of, and dissociation from a target cell (after Mescher, 1995).

Chapter 4

CYTOTOXIC T LYMPHOCYTES

Target Cell Killing: Cellular Parameters

From the moment lymphocyte-mediated cytotoxicity was first defined, there were two major questions driving research in the field that its discovery engendered. The first was to understand how CTL (and, as would be appreciated later, the T-helper cells that promote the functional maturation of CTL) recognize the cells with which they interact, both during the sensitization phase of the CTL reaction, when the decision between self and non-self must be made, and later, when the CTL must interact with and kill non-self or compromised-self cells. In other words, what was the chemical nature of the T-cell receptor (and associated molecules) for antigen? The second problem, specific to the field of lymphocytotoxicity, concerned the nature of the lytic mechanism by which CTL, once sensitized, were able to destroy specifically recognized target cells. Interestingly, both of these problems began to be resolved at almost the same time, in the early 1980s, making this an extraordinarily exciting time in CTL immunology. But while the nature of the T-cell receptor was determined in considerable detail in just under two years, a full appreciation of the mechanism – mechanisms, as it turned out – used by CTL to destroy target cells would take a full decade and is still incomplete.

In the period between 1960 and 1980, while both of these questions underwent initial analyses in laboratories around the world, steady progress was made in defining the cellular parameters of CTL-mediated destruction of target cells. These studies were crucial in the search for a lytic mechanism, because they defined both the requirements for and the limitations of what a lytic mechanism must do, and thus what it represented. But before researchers could study the detailed cellular and molecular parameters of target cell killing by CTL, they needed a reliable way to measure the process they were studying.

1 QUANTITATION OF CMC: THE ^{51}CR-RELEASE ASSAY

When Govaerts first described lymphocytotoxicity in vitro, his findings were based on direct microscopic observations of the destruction of graft donor cells cultured in Petri dishes. This subjective method of estimating target cell death continued to be used for several years, but had obvious quantitative drawbacks. Uptake of the vital stain Trypan blue by killed (but not live) target cells was used in an attempt to increase quantitation, and worked reasonably well in many situations. However, this assay, which requires counting large numbers of stained vs. unstained cells in a light microscope, is tedious. Progress toward a complete definition of CTL-mediated lysis required a rapid, reproducible, and highly precise assay for recording target cell death. This was provided by the ^{51}Cr release assay.

The release of trapped intracellular radioactive chromium ($^{51}CrO_4^{-2}$), a weak gamma emitter with a half-life of 29 days, had been used to assess the viability of red blood cells in hemolytic anemias as early as 1950, and ascites tumor cell viability in 1958. Arnold Sanderson (1965) and Hans Wigzell (1965) first used ^{51}Cr release to quantitate the lysis of lymphocytes induced by antibody and complement. Over the next several years several authors adapted this assay for use in measuring lymphocyte-induced target cell death (Holm and Perlmann, 1967; Brunner et al., 1968; Berke et al., 1969). It proved to be rapid and highly accurate, and could be used for both adherent and non-adherent target cells. Moreover, it correlated well with other assays of target cell destruction such as Trypan blue uptake, loss of cell cloning potential, and the release of endogenous macromolecules.

An important advantage of the ^{51}Cr assay is that chromium released from lysed target cells is not reincorporated into CTL or into target cells that have not yet been killed, which would greatly complicate quantitation of the assay results. This failure to reincorporate released chromium is thought to be due to a reduction in the oxidation state of ^{51}Cr inside living cells. Alternatively, the released chromium may be bound to other released cellular components, which are not readily taken up by other cells. Martz (1976a) found that greater than 90 percent of the ^{51}Cr released during target lysis by CTL was less than 4,000 daltons, but larger than ^{51}Cr itself. Some of the chromium initially taken up by target cells also binds to cellular components that are not released upon the death of the cell. The proportion of non-releasable chromium must thus be determined in each assay for precise quantitative determinations, for example, by measuring the radioactivity released from freeze-thawed or detergent-treated targets. Nevertheless, the ^{51}Cr-release assay has proved superior to all other methods of assessing target cell death, and is used by virtually every laboratory studying lymphocytotoxicity.

To quantitate lysis of adherent cell monolayers, cultures are incubated (usually overnight) with medium containing 1-5 μCi/ml $Na_2{}^{51}CrO_4$. The monolayers are then carefully washed several times, and a suspension of putative effector cells is plated over the labeled targets. After incubation, the overlying medium is harvested, centrifuged to remove any detached but unlysed target cells, and radioactivity is assessed in a gamma counter. Lytic assays of adherent fibroblasts usually take 16-24 hours, and are referred to as "long term" assays. Whether the longer time required for lysis is related to the fact that the cells are in a monolayer, or to the distinct nature of the cells themselves, is unclear.

Non-adherent target cells are labeled with 100-200 μCi ^{51}Cr /ml for 1-2h at 37°C and then washed. CTL populations are mixed with labeled target cells in varying proportions. Typically, assays are carried out in 96-well V- or U bottom microtiter plates in a volume of 200 μl. The plates are centrifuged to initiate the binding of CTL and target cells, and incubated for several hours at 37°C. At the end of the assay, identical quantities of supernate (usually 100 μl) are collected, and the radioactivity counted. Separate sets of control samples (no effector cells, or non-immune lymphocytes from a source similar to the effector cells) are included for determination of the total amount of ^{51}Cr in the assay system, and the amount of radioactivity that is released spontaneously from target cells during the assay period. The data are usually presented as the percentage of target cell lysis at varying E:T ratios (Figure 4.1).

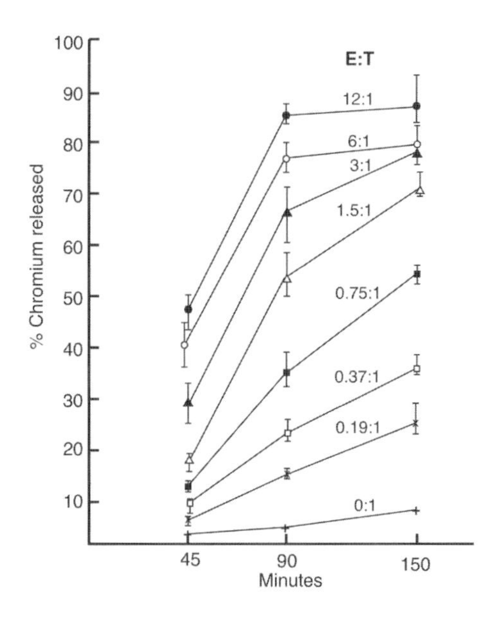

Figure 4-1. Kinetics of lysis by CTL. E:T, effector-to-target cell ratios (from Berke et al 1972a).

The most commonly used equation to calculate the percentage of lysis in either assay is:

$$\text{Percent lysis} = \frac{\text{E-S}}{\text{T-S}} \text{ x100}$$

where E = cpm released from an experimental sample during a fixed assay period; S = cpm released spontaneously from a control sample not incubated with killer cells during the same period; T = total cpm available for release. "Spontaneous release" may reflect ^{51}Cr that spontaneously leaks out of living cells, or it may reflect target cells that die naturally during the assay period. Either way, it is not radioactivity whose release is driven by CTL.

Lytic assays of co-centrifuged cell suspensions usually show substantial target cell destruction in 2-4 hours, and are referred to as "acute" lytic assays. Again, whether this is due to the nature of the target cells used, or the geometry of the assay, is unknown.

Caution must be exercised in the interpretation of ^{51}Cr-release data, especially when comparing degrees of target cell lysis by different effector cell populations. Inspection of the above equation suggests that high spontaneous release values (e.g. greater than 25%) could compromise the statistical value of the derived percent lysis. Moreover, the value for total cpm in the system (T) is subject to variability, depending on whether or not one corrects for radioactivity that is not truly available for release. Failure to include raw data for E, S, and T, in addition to the derived percent lysis values, prevents accurate interpretation and comparison of lysis among experimental groups (Stulting and Berke, 1973).

Another important aspect of CTL action pertains to variations in the cytolytic activity of lymphoid populations following immunization. This could reflect alteration in the frequency of effector CTL and/or their activity. A close correlation between the number of conjugate–forming cells and their cytocidal activity has been found after both primary and secondary alloimmunization in vivo (Figure 4.2), suggesting that a target-binding CTL is a fully effective killer. Hence, differences between cytocidal activity of CTL populations are primarily due to differences in the number of effectors capable of binding target cells, and less to the lytic activity of bound CTL. Comparing lytic activity of poorly cytocidal populations, when maximal lysis caused is low (e.g. under 30%), is always problematic and prone to serious statistical error. One of the better ways to compare the lytic activities of lymphoid populations that vary widely in lytic potency against a given target is to derive a quantity called lytic units (LU), which are inversely related to the number of effector cells required to induce lysis (Cerottini and Brunner, 1974). One LU is defined as the number of effector cells required to lyse a defined fraction (usually 30-50%) of the target cells, and is deduced from lytic data obtained with serial dilutions of effectors tested against a fixed number of target cells in a fixed assay period. The number of LU per 10^6 cells is then calculated for each population, and is directly comparable between populations.

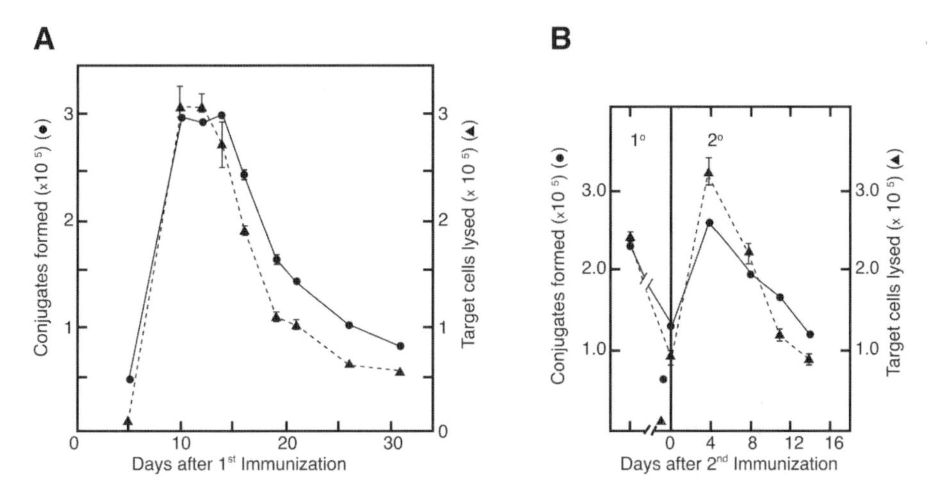

Figure 4-2. Time course of CTL induction in vivo. Conjugation and target cell lysis by CTL after primary (A) and secondary (B) intraperitoneal immunization (Berke et al., 1975).

An additional procedure to measure lymphocytotoxicity that applies where extended incubation (6–18h) is required is the end-labeling method, often with ^{35}S-methionine to quantify the number of residual viable cells incorporating the label into proteins. With the end-labeling procedure problems emerging from high spontaneous ^{51}Cr-release as a result of extended incubation are circumvented. Fluorogenic methods that measure cell viability are often used as well, as is FACS-based analysis to measure lymphocytotoxicity in conjugates (Lebow et al., 1986). Finally there is the JAM test of cell death, which measures DNA fragmentation of cells whose DNA was pre-labeled by ^{3}H-thymidine (Matzinger, 1991); it is noteworthy that DNA fragmentation, a hallmark of apoptosis, precedes ^{51}Cr-release in CTL-mediated lysis. Hence the JAM test, in fact, is a measure of CTL-induced target cell apoptosis.

It should be pointed out that ^{51}Cr release measures only one aspect of target cell damage by killer lymphocytes. Killing by cytotoxic effector cells induces apoptotic cell damage and death in target cells, including DNA fragmentation, and in fact this damage precedes that caused by plasma membrane disruption resulting in ^{51}Cr release. Nevertheless, with a convenient, reproducible lytic assay in hand, researchers were able to make rapid progress in defining the cellular events in both the CTL and target cell as the lytic sequence unfolded. We review the major findings from their efforts in the remaining sections of this chapter.

2 THE CTL IN CELL-MEDIATED CYTOLYSIS

2.1 Architecture and polarity

Detailed observations of individual CTL-target cell conjugates before, during and after lysis suggest that CTL lytic action is polarized, both from a functional and cytological point of view. Cytotoxicity itself is strictly unidirectional; it affects only the bound target, without adversely affecting the effector CTL itself. Furthermore, although a CTL can bind a number of targets simultaneously (see Figure 4.2), lysis of multiply-bound target cells occurs sequentially rather than simultaneously (Zagury et al., 1979). These data suggest that cytotoxicity is deployed by the CTL in a highly focused and polarized fashion. However, the very formation of conjugates containing two or more target cells bound to one CTL (Berke et al., 1975) demands that all of the CTL receptors cannot be focused at the site of ongoing lysis, and that it is the lethal hit itself, rather than the distribution of receptors, that is polarized.

This functional polarization of lysis is reflected at the cytological level as well. As discussed in Chapter 3, a well-developed Golgi apparatus, numerous granular and lysosome-like structures, and a microtubule organizing center (MTOC) are found in the CTL in the region of contact with a target cell to which it is bound (Zagury 1975; Geiger et al 1982). No such reorganization of organelles takes place in the target cell, even when the target is itself a specifically recognized CTL. In the case of multiply bound targets, the MTOC and associated structures move from one target cell to the next as sequential lysis proceeds (Kupfer et al., 1985). It has been proposed that reorientation of the Golgi/MTOC complex toward the target-binding site focuses the cytolytic secretory process to that site, and that dispersion of the Golgi elements would de-focus it (Kupfer et al., 1985). The inhibitory effect of microtubule disruption of killing induced by various drugs (e.g. colchicine and nocodazole) could thus be due to a ‘de-focusing’ of Golgi apparatus-derived secretion at the CTL-target cell interface.

2.2 CTL killing as a Ca^{2+}-dependent event

Although Ca^{2+} is neither required nor sufficient for CTL-target cell binding (Stulting and Berke 1972), it is absolutely required for one of the mechanisms of CTL-mediated lysis, granule exocytosis, although it is not required for a second major cytolytic pathway mediated through target cell Fas molecules (Chapter 5). For the first twenty years or so of research into lymphocytotoxicity, lysis by CTL was categorically declared to be a highly Ca^{2+}-dependent event (e.g., Plaut et al., 1976; Golstein and Smith, 1976), presumably (we would now conclude) because the lytic activity of primary CTL is heavily biased toward the exocytic pathway. Occasional reports of Ca^{2+}-independent killing were generally ignored. Engagement of the TCR complex with cognate antigen – whatever the final mode of lysis by the CTL - triggers an immediate increase in CTL intracellular Ca^{2+} to several hundred nanomolar (Poenie et al., 1987). These increases are partly the result of release from internal Ca^{2+} stores, but they also depend on the influx of external Ca^{2+} for prolonged elevation and full CTL activation (Gray et al., 1987; Engelhard et al., 1988; Haverstick et al., 1991). Ca^{2+} influx is also seen in the target cell in the course of CTL-mediated lysis (Tirosh and Berke, 1985a, b; Poenie et al., 1987; see below).

2.3 Programming for lysis

A major step forward in describing lysis of targets by CTL came with work showing that this process occurs in distinct stages (Berke et al., 1972b; Wagner and Röllinghof 1974; Martz, 1975). Martz found that after co-centrifuging CTL and ^{51}Cr-labeled target cells at 4°C, and then raising the temperature to 37°C, stable conjugates would not form in the presence of EDTA, because of the requirement for Mg^{2+} (Stulting and Berke,1973) and no lysis ensued because of the absence of Ca^{+2}. When conjugates were disrupted with EDTA at various times, and extra Ca^{2+} added to promote lysis, Martz found that no ^{51}Cr was released if the conjugates were disrupted during the first two minutes of coincubation. Between two and ten minutes, the amount of ^{51}Cr released gradually increased to a maximal level. Dissociation and dispersal of conjugates after ten minutes had no impact on the final amount of ^{51}Cr released. This critical experiment defined a stage, between two and ten minutes after conjugate formation, and before the onset of lysis as measured by ^{51}Cr release, which became known as the "programming for lysis" stage. It was assumed that this was the stage during which the "lethal hit" – whatever it is that CTL do to target cells that causes them to die – was ultimately and irreversibly delivered.

A further refinement of this experiment with other markers for plasma membrane damage was also informative. When the same basic approach was used with ^{86}Rb and ^{14}C-nicotinamide as trapped markers, Martz found that ^{86}Rb (a K^{+} analog) was released from target cells essentially as soon as active programming for lysis began (Martz, 1976b), similar to earlier observations by Henney (1973). Nicotinamide began to be released about halfway through the programming stage, whereas ^{51}Cr was not released until nearly thirty minutes after conjugate formation – well beyond the programming stage. Membrane depolarization – as measured by the efflux of ^{86}Rb – occurs very early in the lytic sequence, and in Martz's estimation is coincident with the delivery of the lethal hit. The relevance of this to the overall lytic sequence is difficult to judge, since target cell lysis, as we will see in the next chapter, is molecularly very complex.

It would seem from the foregoing that ^{86}Rb might be the preferred marker for following the onset of action. However, its very high rate of spontaneous release makes its usefulness limited. Moreover, as a potassium analog, ^{86}Rb is most likely shuttled in and out of cells via specific pumps and channels. Thus, enhanced egress of ^{86}Rb during cytolysis could reflect an altered operation of these shuttling mechanisms in response to metabolic stress, rather than damage to the plasma membrane.

Unraveling what happens during the programming for lysis stage became *the* holy grail of the field of cell-mediated cytotoxicity. Yet this phase of cytotoxicity research was reminiscent of the classical "blind men and elephants" fable: lacking any coherent vision of what the lytic mechanism might be, researchers were limited to compiling lists of peripheral descriptions of correlations. At first, none of these gave any insight at all into the underlying mechanisms. But ultimately, all had to be accounted for.

In terms of temperature, cytolysis as measured by ^{51}Cr release is undetectable at 4°, and barely detectable at 15° when conjugation is essentially complete. It rises slowly to about 24°, and then increases steeply through 37° (Figure 4.3). The temperature dependence of programming for lysis and of lysis proper is about four times greater than the temperature dependence of conjugate formation (Berke et al., 1972b; 1975; Martz, 1975). Lowering the temperature below 15°C, even at advanced stages of CTL-target interaction, halts ^{51}Cr release completely. Lysis returns to its original rate once the temperature is restored to 37°C (Berke et al., 1972b; Martz and Benacerraf, 1975). Thus, there appears to be an intermediate stage, where targets are irreversibly programmed to lyse, and may even have proceeded some distance down the pathway to lysis, yet still exhibit characteristics of intact and functioning cells, e.g. retention of ^{51}Cr-labeled constituents and the exclusion of vital dyes like Trypan blue and eosin.

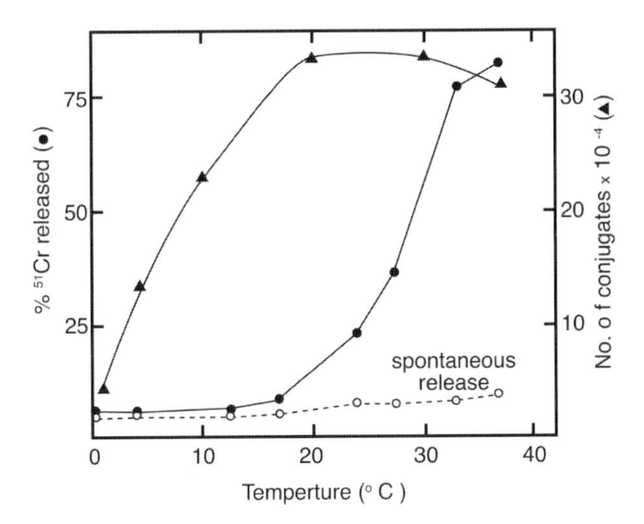

Figure 4-3. The influence of temperature on CTL-target conjugation and lysis (After Berke and Gabison 1975).

Except for the observation that the CTL must be viable, and the effects of temperature, little is known of specific metabolic requirements for cytolysis. Classical studies with metabolic inhibitors are difficult to interpret, because a viable and motile cell is required for cytolysis; obviously any drug that compromises cellular metabolism or motility will suppress its lytic function. Drugs that inhibit the initial step – conjugate formation – thus inhibit lysis only indirectly. On the other hand, *failure* to inhibit lysis with drugs has enabled the delineation of a number of metabolic pathways that must be unrelated to lytic expression. For example, CTL do not need to proliferate in order to lyse target cells, nor do they need to synthesize DNA, RNA or protein (Brunner et al., 1968; Thorn and Henney, 1976a), at least for an initial round of lysis.

Lysis of target cells by primary CTL utilizing the exocytosis pathway requires Ca^{2+} during the programming for lysis stage. Completion of the lytic cycle beyond the programming stage is, however, Ca^{2+}-independent (Martz et al., 1974). As noted earlier, engagement of the CTL via its antigen receptor invariably triggers an increase in intracellular Ca^{2+} concentration. Part of the Ca^{2+} is derived from internal stores, but the majority comes from the influx of extracellular Ca^{2+}. For target cell lysis dependent entirely on the exocytic pathway in CTL, lysis will not proceed in the absence of exogenous Ca^{2+}. The reason for this is clear: not only is reorientation of the MTOC and associated structures to the CTL-target cell interface dependent on Ca^{2+} (Kupfer et al., 1985), but also granule exocytosis itself and the action of perforin are Ca^{2+}-dependent.

2.4 CTL recycling

In early studies on CTL killing, it was assumed that individual killer cells would detach from killed target cells and recycle to fresh targets, but this was based on indirect evidence. Using highly lytic PEL incubated with a number of target cells far exceeding the number of CTL, vigorous and continual lysis of targets far beyond the number of possible CTL was observed, and the rate of lysis did not diminish during 5 h of incubation (Berke et al 1972c). These observations strongly suggested that CTL are neither inactivated nor lysed during the killing process, and ultimately showed that each CTL could kill on average more than six consecutive target cells without losing potency. The first visual demonstration that a single CTL could kill more than one target cell sequentially involved micromanipulation. Using a unique liquid microchamber enabling direct handling and simultaneous visualization of individual CTL-target conjugates, Zagury et al. (1975) showed that after the target cell in a 1:1 conjugate was lysed, a second target brought into contact with the same CTL was also lysed. A third target was then brought into contact with the same CTL, and was again lysed. Recycling of killer CTL has also been deduced from microcinematography of the unperturbed lytic action of CTL (Ginsburg et al., 1969; Sanderson, 1981).

2.5 Sparing of "innocent bystander" target cells

One of the properties of primary CTL with major implications for the underlying lytic mechanism is a general failure to lyse third party bystander cells, i.e. non-recognized targets mixed among cognate targets in a lytic assay (Berke et al., 1972a; Cerottini and Brunner, 1974; Martz, 1975b; Wei and Lindquist, 1981) Bystander lysis is only rarely seen with primary CTL, whether generated in vivo or in vitro, even when assays are carried out (as is the norm) in densely packed pellets in which CTL and target cells are co-centrifuged. Even multiple target cells bound to a single CTL are not killed simultaneously, but rather sequentially (Zagury et al., 1979). Sparing of "bystander" targets in this case is presumably due to the reorientation of the Golgi/MTOC toward only one target cell at a time, which is directed by signals generated in the immunological synapse.

The apparent lack of bystander lysis by CTL was for years a barrier to consideration of a secretory mechanism for CTL killing. However, as evidence for such a mechanism mounted in the early 1980s, various proposals to account for the lack of bystander killing were put forth. The rather tight "pockets" formed by extensive CTL-target cell membrane interdigitations (Figure 3.6) were seen as possibly limiting diffusion of any soluble mediator from the immediate vicinity of the CTL. It was also postulated that soluble mediators might have extremely short half-lives. As we will see in the next chapter, both hypotheses (particularly the latter) are probably true.

Occasional exceptions to the strict antigen specificity of lysis were noted early on (Streilen, 1972; Prehn, 1973; Brocker et al., 1977), but they were rare enough to be generally ignored. However, as investigators began working with CTL clones in the 1980s, increasing exceptions were noted. Fleischer, for example, observed non-specific lysis by a human CD8 CTL clone (Fleischer, 1986). Interestingly, not all bystander targets were killed, leading him to propose that the target must be critical in determining whether bystander lysis would proceed, and that the mechanism could be different from that normally operating in target-specific lysis. These turned out to be prescient conclusions. Bystander lysis seemed to be particularly prevalent in CD4 CTL clones (Tite and Janeway, 1984; Nakamura et al., 1986), and in the immune response to tumor cells (Paul and Lopez, 1987; Barker et al., 1989; Woods et al., 1989).

Subsequent research would show that the lytic mechanism operating in innocent bystander lysis is in nearly all cases the Fas/Fas-ligand pathway present in CD8 and CD4 CTL, as well as NK cells. We will discuss this pathway and its involvement in bystander lysis in Chapter 5.

2.6 Are CTL resistant to their own killing mechanism?

The ability of CTL to survive a lytic interaction with undiminished lytic activity, and recycle to kill additional targets, raises the possibility that CTL are resistant to their own cytolytic mediators. This issue was first addressed by Pierre Golstein in 1974 in his "polarity of lysis" experiments, later confirmed and extended by others (Golstein, 1974; Kuppers and Henney, 1977; Fishelson and Berke, 1978). The basic scheme of these experiments was to mix A anti-B CTL with B anti-C CTL, where A, B, and C are three mutually allogeneic strains of mice. The B anti-C CTL in this case serve as a cognate target cell for the A anti-B CTL. Because the in vivo-derived cell populations used in these experiments were heterogeneous, it was not possible to assess directly (e.g., by ^{51}Cr release) whether the B anti-C CTL were actually killed. Rather, after co-incubation for several hours, the residual ability of B anti-C CTL to kill C-type target cells (of non-CTL origin) was measured. It was found that the lytic capacity of B anti-C CTL was greatly reduced under these conditions. Although Golstein was careful not to draw conclusions regarding the viability of the B anti-C CTL in these experiments, in practice his results led to the generally accepted view that CTL are sensitive to their own lytic mechanism.

However, when these experiments were examined some years later using cloned CTL as targets, a different conclusion was reached. Cloned CTL populations consist of essentially pure CTL, and they thus can be labeled with ^{51}Cr and used as targets in cytotoxicity assays. Blakely et al., (1987) found that cloned CTL used as target cells were absolutely refractory to lysis, in all effector-target combinations, even in the presence of lectins known to mediate lymphocytotoxicity in the absence of cognate target cell recognition. These investigators showed that cloned CTL were also resistant to direct lysis by isolated CTL granules containing perforin, one of the mediators of CTL lysis (Chapter 5). The same conclusions were reached by Eisen and colleagues (Kranz and Eisen, 1987; Eisen et al., 1987; Verret et al., 1987), Skinner and Marbrook, 1987; Golstein's group (Luciani et al., 1986), Shinkai et al., 1988; Liu et al., 1989; Lanzavecchia, 1986; and Müller and Tschopp, 1994. However, both Luciani et al., and Blakely et al,. presented data suggesting that primary CTL, generated in MLC, *are* able to lyse cognate CTL clones as targets. Eisen's laboratory examined enriched CD8 cells from a 6-day MLC culture; he found these cells also to be highly resistant to CTL-mediated lysis and to granule contents (Verret et al., 1987), and suggested that primary CTL are probably genuinely immune to lysis.

In follow-up experiments with cloned CTL, Gorman et al., (1988) showed that while completely resistant to lysis as determined by ^{51}Cr release, cloned CTL used as target cells *are* functionally inactivated, just as the earlier investigators had claimed. Target CTL were also inactivated functionally by purified perforin.

But the story was not yet finished. Schick and Berke (1990) examined the same question using in vivo-generated cytolytic PEL as target cells, as well as PEL-blasts – PEL that had been maintained in long-term culture in the presence of TCGF as an IL-2 source. They found that both CTL populations were readily lysed by antigen-specific PEL. Although these CTL populations, particularly the in vivo-generated primary PEL, were not 100 percent CTL, the fact that between eighty and ninety percent of the total target CTL populations were lysed in these experiments strongly suggests that PEL were killed by effector CTL that recognized them as targets.

If CTL are in fact resistant to their own lytic mechanism, what might the basis for such resistance be? Various proposals have been put forth, but by and large they remain speculative. Kupfer et al., (1986) suggested but provided no evidence that the unusually high concentration of talin beneath the CTL membrane at the site of CTL-TC contact might confer resistance of the CTL to its cytolytic agent. Several reports from Müller-Eberhard suggested that homologous restriction factor (HRF), which protects cells within a given species from attack by that species' own terminal (cytolytic) complement components (Zalman et al., 1988), might be responsible for self-protection of perforin-secreting cells (Müller-Eberhard, 1986; Martin et al., 1988; Zalman et al., 1988). However, two papers from Young's laboratory challenged this notion (Jiang et al., 1988, 1989) and it is not now generally accepted.

Thus, the question of the sensitivity of CTL to their own lytic mechanism remains unsettled. Golstein later suggested (Luciani et al., 1986) that long-term cloned CTL lines might become selected over time to become insensitive to their own killing mechanism. Yet his data suggest they *are* still sensitive to killing by primary CTL, and as far as we know, the killing mechanisms in primary and cloned CTL are identical. If CTL release perforin into enclosed intercellular spaces, as they seem to do during target cell lysis, and yet recycle to kill additional targets, then by definition they must be resistant to at least perforin. But the basis for this resistance is simply not clear at present. A recent report from Henkart suggests that the protease cathepsin B, co-secreted with perforin and granzymes from lytic granules, binds to the CTL surface proximal to the contact region with the target and thus protects the CTL from potentially adverse effects of secreted perforin (Balaji et al., 2002). If confirmed, this could explain CTL refractoriness to its own secreted lytic mechanism, and possibly the apparent refractoriness of some CTL lines discussed above.

Overexpression of serpins in CTL and NK cells may be an alternative mechanisn for protection against self-annihilation during the killing process. For example, serine proteinase inhibitor (PI)-9, with a reactive center P1 (Glu)-P1, is a natural antagonist of granzyme B and is expressed in high levels in CTL. PI-9 can protect CTL from its own fatal arsenal and potentially enhance their vitality. Serpins have also been implicated in the sparing of antigen-presenting dendritic cells, as well as in tumor escape from CTL or NK cell attack.

3 THE TARGET CELL DURING CELL-MEDIATED CYTOLYSIS

3.1 Morphological parameters

From the very beginning of the study of cell-mediated cytotoxicity, there was a sense that target cell death induced by CTL was distinct from necrotic cell death, such as is seen after lysis by complement, or after starvation of cells or from oxygen deprivation. In a remarkable series of early studies on the rejection of transplantable tumors, Kidd and Toolan (1950) suggested that lymphocytes could be directly cytotoxic to other cells. They also concluded that killing by lymphocytes and death by antibody plus complement were likely due to distinct mechanisms, based on a comparison of the morphology of tumor cells dying *in situ* in proximity to lymphocytes, with similar tumor cells dying in culture after treatment with antibody and complement. Later studies confirmed this impression. Ultrastructural changes in target cells occur in several stages after contact with a killer cell. Early events detected after ten or fifteen minutes in adherent target cells, such as withdrawal of cell processes and loss of surface microvilli with cell rounding and, after a variable interval of time, plasma membrane blebbing, DNA fragmentation and mitochondrial swelling (Wilson et al., 1963; Ginsburg et al., 1969; Sanderson, 1977; Liepins et al., 1977; Matter, 1979), are indicative of cytoskeletal lesions. Interestingly, CTL-induced rapid retraction of normally adherent target cells can occur under conditions that do not lead to cell death (Abrams and Russell, 1989, 1991)

Zagury and co-workers also studied the lytic history of individual CTL-target conjugates (Zagury et al., 1975). Using single-cell manipulation enabling examination of individual CTL-target conjugates under phase contrast and electron microscopy, they were able to discern the fine structure of proven effector CTL (those previously observed to have killed a target). The following sequence of events was observed during the killing process, under phase contrast microscopy, verifying target viability by staining with Trypan blue (Figure 4.4). Before the onset of killing, conjugated target cells were not stained and appeared refractive in phase contrast microscopy; the cell contour was well defined. At an early stage of lysis, the target cell became stained by Trypan blue, and its contour ill-defined. Within a few minutes the size of the target cell increased, and its nucleus moved toward the killer-target interface. A short time later, the nucleus migrated to the center of the cell, the target began to deteriorate, and the killer cell detached and moved away. The duration of the lytic process was examined with 180 isolated killer-target conjugates, consisting of a single CTL bound to a single target cell. Some conjugated target cells were lysed within a few minutes, while others took up to two hours. Most conjugated targets were eventually lysed, indicating that at least 80 percent of conjugated CTL were competent killer cells.

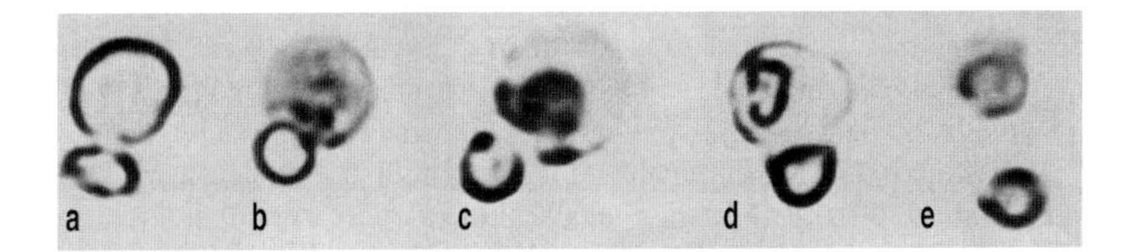

Figure 4-4. Killing of conjugated target cells. Lysis was determined by phase contrast microscopy and by Trypan blue (0.2%) uptake. Sequence of events prior to (a), during (b-d) and after (e) the lytic process. (From Zagury et al, 1975).

Colin Sanderson (1977) described the "boiling" of the plasma membrane and the blebbing and shedding of cytoplasmic particles of target cells dying after a CTL attack (Figure 4.5). He termed this phenomenon "zeosis" and drew analogies to "natural" death occurring in tumor populations. Zeiosis is an interesting but poorly understood phenomenon, apparently overlooked (or defined differently) for a long time in both time-lapse and electron microscopy of lymphocyte killing. Zeiosis is likely caused by massive changes in the cytoskeletal system, since agents that disrupt the cytoskeleton induce similar membrane blebbing.

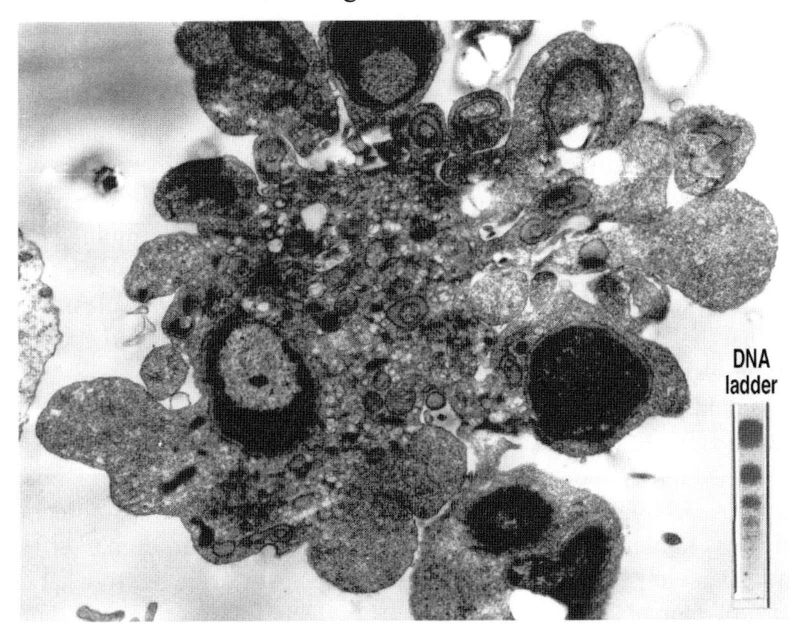

Figure 4-5. Apoptosis of target cell killed by CTL. By D. Rosen. Insert: Agarose gel electrophoresis of DNA from dying target cells.

Independently, Kerr and colleagues concluded that the morphology of cells dying after a CTL attack was similar to examples of "natural" cell death in populations that they had previously termed apoptosis (Don et al., 1977). Apoptosis was characterized by cytoplasmic blebbing, but more importantly, by the condensation of chromatin material within the nucleus and DNA fragmentation (Figure 4.5, insert). Wyllie extended these studies and proposed a unified "apoptotic" mechanism for a "natural" or "programmed" death resulting from a variety of causes (Wyllie et al., 1984). Almost certainly, Ginsburg's description of target cell death (Ginsburg et al., 1969) and Sanderson's zeiosis reflected various aspects of apoptotic cell death. The notion that CTL-induced target cell death was clearly different from complement-mediated cell death, and quite possibly apoptosis, shaped the search for the underlying mechanism of cell-mediated lysis.

3.2 Plasma membrane integrity

As a result of cognitive interaction with a CTL, the target cell undergoes a progressive series of membrane permeability changes ending in major leakage of cell contents. These changes can be followed by monitoring the release of markers of varying molecular size from the damaged target cell (Henney, 1973b; Martz et al., 1974). As noted earlier, changes in the permeability of the target cell membrane (measured by e.g. enhanced ^{86}Rb efflux) are specifically induced almost immediately after exposure to CTL. Larger markers escape from target cells only later, after lag periods increasing with the molecular size of the indicator. The release of ^{51}Cr, which may well be bound to various macromolecular cell components, some releasable, others not, is usually not evident before 10-15 minutes (Martz, 1976b). These findings suggested that the initial lesion allows a rapid exchange of ions and small molecules, but not of macromolecules. It seems likely that the latter escape only after the secondary effects on the target cell occur, perhaps resulting from disordered osmotic regulation.

It should be noted here that while the focus of early studies was on plasma membrane leakage, we now know that target cell death as a result of killer cell action is apoptotic in nature, with the target undergoing extensive membrane blebbing and the formation of apoptotic bodies. The relation of this process to membrane permeability changes inferred from marker release has not been well studied. It is difficult to know, for example, to what extent release of trapped markers such as ^{51}Cr reflects leakage through damaged cell membranes, and how much it represents ^{51}Cr trapped in apoptotic bodies that do not spin down at the end of the assay. The latter will appear to be "released", but may still be within membrane-bounded bodies.

The eventual demise of the target cell was originally thought to be caused by "colloid osmotic forces" resulting from a net water influx to compensate for the initial loss of ions and small molecules and retention of proteins trapped intracellularly. This conclusion was based on observations that both macromolecular effluxes from the damaged target cell and plasma membrane destruction could be prevented by exogenous high-molecular weight dextrans (Henney, 1974a; Ferluga and Allison, 1974). The minimum size of dextran molecules that afforded such protection was approximately 40,000 Dalton, suggesting that the initial CTL-induced lesion was approximately 90 Å in diameter (Henney, 1974a).

The progressive nature of the lytic lesion induced by CTL, the suppression of its development by solutions of high osmotic pressure, and the apparent initial size of the lesion are all remarkably parallel features of complement-induced permeability changes in tumor cells (Green et al., 1959ab). This led to the early proposal that CTL insert terminal (lytic) complement components, or similar molecules, into the target cell membrane, ultimately causing osmotic lysis (Henney, 1974). We will examine this possibility more closely in Chapter 5.

Damage to the membrane of any cell is also accompanied by an influx of Ca^{2+} down a 10,000-fold concentration gradient, and this could be a potentially damaging event as well. Pre-lytic and persistent elevation of target cell intracellular Ca^{2+} was first shown using PEL (Tirosh and Berke, 1985a, b). This phenomenon was later observed with other CTL systems (Poenie et al., 1987; Allbritton et al., 1988; Hassin et al., 1987), natural killer (NK) cells (McConkey et al., 1990a), as well as in other systems of programmed cell death (McConkey et al., 1990b; Campbell 1987; Orrenius et al., 1989). Buffering of $[Ca^{2+}]_i$ inside the target cell by the Ca^{2+}-chelator Quinn-2AM retarded both their cytolysis and DNA fragmentation (McConkey et al., 1990a). Target cell Ca^{2+} elevation occurs in some targets upon CTL attack even in the absence of free medium Ca^{2+} in the medium (Tirosh and Berke, 1985), presumably due to Ca^{2+} released from internal stores (e.g. mitochondria, sarcoplasmic reticulum) upon intracellular acidification (Russell, 1983). A persistent increase in Ca^{2+} could induce harmful internal processes, including ATP wasteage (due to activation of actomyosin, as well as damage to mitochondria), DNA fragmentation (activation of endonucleases), and bleb formation (through activation of proteases that cleave the cytoskeleton) (Orrenius et al., 1989).

Tirosh and Berke (1985) suggested that increased intracellular Ca^{2+} in the target was responsible for activating an endogenous pathway resulting in the nuclear lesion. Certainly Ca^{2+} has been implicated in the suicide pathways of a variety of systems (Jewell et al., 1982; McConkey et al., 1990), and Poenie and colleagues (1987) have measured such increases in targets under attack by CTL. The fact that intracellular Ca^{2+} chelators can inhibit DNA damage (Hameed et al., 1989) also supports the Ca^{2+} hypothesis. Although it is likely that creation of the nuclear lesion does in fact require Ca^{2+}, it is not possible to correlate absolute increases in intracellular Ca^{2+} with a nuclear lesion or even lysis (Mercep et al., 1989). On the other hand, Ca^{2+} influx into the effector CTL is physiological and stimulatory; it involves micromolar concentrations that are not harmful to cells. The roles of Ca^{2+} in lymphocytotoxocity will be discussed further in Chapter 5.

3.3 Nuclear lesions and DNA fragmentation

Russell and Dobos (1980) confirmed earlier results on marker release from damaged target cells, including the release of nuclear material in CTL-mediated lysis, but not in complement-mediated lysis. Upon closer examination, they observed onset of DNA fragmentation well before the onset of ^{51}Cr release. Analysis of the DNA shortly after CTL attack revealed initially high-molecular-weight (but cleaved) forms that were progressively cleaved over the next 30-60 min into lower-molecular-weight species (Russell et al., 1982). Eventually, DNA fragments of repeating units of 180-200bp are observed (Russell, 1983; Duke et al., 1983), suggesting an endonuclease attack at exposed sites between nucleosomal particles. In contrast, in cells killed by complement or other agents inducing necrotic cell death, DNA is at best partially fragmented and requires treatment with proteases or nucleases to be released from cellular debris.

As we will discuss below and in more detail in the next chapter, all lytic pathways used by CTL result in damage to the target cell that is essentially apoptotic in nature, and involves degradation of DNA. Most early ultrastructural studies of CTL-mediated lysis clearly showed features that we would now recognize as apoptosis before the term and the concept were generally recognized. The pre-lytic nuclear damage observed by Russell prompted him to propose that CTL-mediated lysis is the result of induced "internal disintegration" of target cells, rather than of a membranolytic process (Russell, 1983). Classically defined apoptosis requires the active participation, including protein and RNA synthesis, of the dying cell. CTL-mediated lysis deviates from "classical" programmed cell death in that protein and macromolecule synthesis of the target are not usually required to enable its demise.

There is no question that a cell whose DNA has been reduced to multiples of 200-bp fragments will eventually die. However, even before the discovery of the CTL-induced nuclear lesion, it had already been shown that CTL could "kill" at least some enucleated cells (Fishelson and Berke, 1978; Siliciano and Henney, 1978). Thus, CTL-induced damage to the plasma membrane, resulting in leakage of cytoplasmic constituents, can occur independently of nuclear damage. This has recently been confirmed in experiments with redirected lysis showing that anuclear mammalian red blood cells can also be lysed by CTL (Lanzavecchia and Staerz, 1987). Ratner and Clark (1991) showed that this form of lysis requires no active participation of the red cell target. And finally, studies have shown that apoptosis induced by means other than CTL can also be demonstrated in anuclear cells (Jacobson et al., 1994; Schultze-Osthof et al., 1994).

It has been proposed that fragmentation of DNA, which always occurs in CTL-mediated lysis, may be the primary biological function of CTL. Given that the primary role of CTL is protection from intracellular parasites such as viruses, the nuclear lesion may reflect a defense against the release of infectious material from compromised cells. Interestingly, CTL are able to destroy replicative viral DNA within target cells, through limited DNA fragmentation, under conditions in which the cells themselves are not destroyed (Martz and Howell, 1989). According to this view, physical destruction of the target cell may be secondary to the destruction of the DNA.

4 ANTIGEN-NON-SPECIFIC LYSIS BY CTL

Although lysis of target cells by CTL is, under normal circumstances, absolutely dependent on TCR-mediated recognition by the CTL of MHC molecules on the target cell surface, a variety of means have been devised to by-pass this requirement. In many cases, these studies have provided insights into the nature of both recognition and cytolytic mechanisms utilized by CTL and other killer lymphocytes in the destruction of target cells. In the sections that follow, we briefly describe several of these systems.

4.1 Lectin-dependent cell-mediated cytotoxicity (LDCC)

As discussed in Chapter 2, in the early 1970s it was discovered that certain mitogenic lectins could actually drive T cells beyond the proliferative stage to become cytotoxic. However, as originally described, this lysis could only be seen if the same lectin used for activation was included in the cytotoxicity assay. This cytotoxicity, unlike classical CTL-mediated killing, resulted in the death of any target cell, even those bearing MHC antigens identical to the CTL itself. Interestingly, only mitogenic lectins capable of causing T cells to proliferate (e.g., ConA, PHA,and LCA), were able to mediate this type of non-specific cytotoxicity, called lectin-dependent cellular cytotoxicity (LDCC) It was subsequently found that MHC-specific

CTL generated by ordinary means in vivo or in vitro would also kill any target cell – including self-target cells – if one of these mitogenic lectins was included in the assay. It was soon shown that although all T cells proliferate in response to these lectins, only those cells that were precursors to "classical" (CD8) CTL supported LDCC after proliferation (Bevan and Cohen, 1975; Clark, 1975; Bonavida and Bradley, 1976). There have been occasional reports of CD4 CTL mediating LDCC, but in general these have not been well studied (Chu et al., 1990; Benvenuto et al., 1991; Gallo et al., 1991).

Initial thinking about the mode of action of lectins in overriding the strict specificity of CTL was that they simply acted as agglutinins to bring the CTL and target cell close enough together so that cytolysis could proceed (Forman and Möller, 1973; Bevan and Cohn, 1975; Bonavida and Bradley, 1976). But that seemed unlikely, since in that case non-mitogenic (e.g., SBA, WGA, and PNA) but equally agglutinating lectins should also mediate cytolysis, which they do not. Thus, it was proposed that in addition to "gluing" the CTL and target together, mitogenic lectins must also be delivering an activating signal into the CTL as well, which made sense since they must be doing the same thing in triggering proliferation (Green et al., 1978; Parker and Martz, 1980). In the case of specifically recognized targets, activation occurs via the T-cell receptor, but lacking such recognition, non-specific CTL-target binding via lectin suffices. As discussed in Chapter 2, activation of the pre-CTL by lectin most likely occurs through cross-linking of the T-cell receptor. But what initiated killing in LDCC?

Insight into the mechanisms involved in LDCC was provided by experiments carried out in the early 1980s (Berke et al., 1981a,b; 1982; 1983). If target cells were pretreated with a mitogenic lectin, and exposed to untreated CTL, lysis (and conjugate formation) took place as usual in LDCC, regardless of the MHC of the target. But if the CTL was pretreated with the lectin, no lysis (or conjugation) was observed, unless the target itself was treated with lectin (Figure 4.6).

This further nullified the notion of lectin acting simply as a bridge between the effector and target. It was also found that target cell MHC molecules are likely to be involved in the lytic process, since removal or blocking of the MHC blocked lysis of the targets, even though the target cells still bound lectin very well. But if the CTL receptor does not recognize the target cell MHC, what role could the latter be playing in LDCC? And if lectins were simply acting to cross-link the CTL antigen receptor, as was thought to happen when lectin stimulated proliferation in resting T cells, why would removal of MHC have any effect?

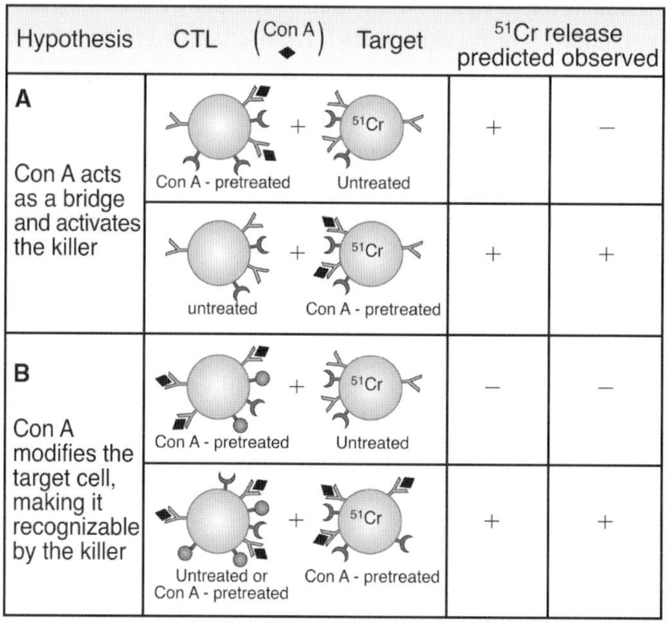

Hypothesis	CTL (Con A) Target	^{51}Cr release predicted	observed
A Con A acts as a bridge and activates the killer	Con A - pretreated + Untreated (^{51}Cr)	+	−
	untreated + Con A - pretreated (^{51}Cr)	+	+
B Con A modifies the target cell, making it recognizable by the killer	Con A - pretreated + Untreated (^{51}Cr)	−	−
	Untreated or Con A - pretreated + Con A - pretreated (^{51}Cr)	+	+

Figure 4-6. How Concanavalin A (Con A) mediates killing by CTL. See text for details.

A model for LDCC killing was proposed in which lectin interacted with target cell surface MHC and perhaps other glycoproteins, causing them to cluster and allowing any given CTL class I CTL receptor to interact with them (as well as with the lectin) in a non-specific fashion (Berke and Clark, 1982). It was proposed that in both direct CTL killing and in LDCC, the lethal hit involved distortion (through cross-linking) of target cell class I proteins, leading to membrane disruption and leakage of cellular contents. Although this model proved ultimately not to be correct, it was among the first to propose that CTL might induce lethal changes in the target cell merely through interaction with target cell membrane structures, not requiring a secreted lytic agent and pore formation, which in fact proved to be the case for one of the two recognized lytic mechanisms used by CTL (the Fas/FasL system (Chapter 5)).

4.2 Oxidation-dependent cell-mediated cytotoxicity (ODCC)

Another system in which specificity of target cell killing is over-ridden is based on oxidation of the target cell surface (Novogrodsky, 1975). Oxidation-dependent CTL-mediated cytotoxicity (ODCC) is induced by the generation of aldehyde groups with $NaIO_4$ or with neuraminidase and galactose oxidase treatment, which then form Schiff bases with free amino groups. As in LDCC, only activated CTL function as effector cells (Schmitt-Verhulst and Shearer, 1976). In both LDCC and ODCC, target structures include papain-sensitive glycosylated molecules (Gorman et al., 1987) as well as MHC class I molecules (Berke et al., 1981b).

Both LDCC and ODCC were studied in attempts to understand CTL-target cell interactions that lead to lysis. Once the T-cell receptor was defined, and "normal" lytic interactions were worked out in great detail for CTL and unmodified target cells, interest in both of these systems waned. However, they are still useful for detecting the presence of CTL in a population where the target specificity is unknown, because any CTL (but only CTL) will lyse targets modified with lectin or by oxidation. Unfortunately, some potentially interesting questions in these two systems, such as the exact nature of the structures involved in recognition, activation, Ca^{2+}-dependency, and the possible involvement of Fas and other lytic systems, have never been resolved.

4.3 Antibody-redirected lysis

In addition to stimulating unprimed T-cell populations to produce CTL, antibodies to the T-cell receptor complex can also be used to redirect CTL from the target recognized by the CTL receptor to other targets. Perhaps the simplest example of this is the ability of CTL to kill B cell hybridomas producing anti-CTL receptor antibodies, whether these are against the α/β chains of the TCR, or against the CD3 component of the TCR (Hoffman et al., 1985; Boylston and May, 1986). Presumably the interaction of the TCR complex with cross-linking TCR antibodies mimics the signal provided by specific antigen cross-linking of receptors. TCR antibodies can be chemically attached to otherwise unrecognized target cells (Kranz et al., 1984), or attached via interaction of the Fc region of the antibody with Fc receptors on putative targets (Leeuwenberg et al., 1985; Staerz and Bevan, 1985; Leo et al., 1986). The cytolysis exerted by CTL against these target cells is equivalent to that directed toward specifically recognized targets, and is equally effective with CD8 and CD4 CTL (Mentzer et al., 1985). The possible involvement of target cell class I molecules in antibody-redirected lysis has never been evaulated.

Triggering cytolysis by CTL, using antibodies against T-cell surface molecules other than the α/β receptor and CD3, are variable (Kranz et al., 1984; Mentzer et al., 1985; Oberdan et al., 1987; Seth et al., 1991) but probably can occur. Another strategy involves the use of so-called heteroconjugate antibodies, in which an antibody against the TCR is coupled to an antibody against a putative target cell antigen (Perez et al., 1985; Staerz et al., 1985). This approach has been intensely studied as a way to target essentially any activated T cell in the body to any tumor cell for which an antibody exists (Perez et al., 1986; van Ravenswaay et al., 1993). Another approach for redirecting cytocidal effector cells (both CTL and NK cells) is by transfecting them with chimeric receptor genes with single-chain antibody specificity (Eshhar and Schwartzbaum, 2002).

Chapter 5

CYTOTOXIC T LYMPHOCYTES

Target cell killing: Molecular mechanisms

1 EARLY STUDIES

The discovery by Govaerts of lymphocytes able to kill graft cells in vitro immediately triggered speculation about the molecular mechanism(s) underlying destruction of the targeted cells. Experimental approaches to this question fell into two broad camps: some looked for a fixed, physical property of the killer cell itself to explain cytotoxicity; others searched for the release of one or more soluble cytotoxic factors. Those following the first path initially adapted the tried and true biochemistry approach of breaking killer cells down into their component parts and testing each separately. The earliest of these studies analyzed various crude fractions of disrupted killer cell populations, but the results were generally negative. It appeared that viable effector cells were required to cause lysis, since heated, freeze-thawed or sonicated cells uniformly failed to exhibit lytic activity (Rosenau and Moon, 1961; Wilson, 1963; Ginsburg et al., 1969).

Later reports suggested that damage to target cells might be mediated directly by the killer cell plasma membrane. For example, one report claimed that purified plasma membranes derived from the lymph nodes of contact-sensitized mice were cytotoxic toward tumor cells in vitro (Ferluga and Allison, 1974). Unfortunately, these studies were neither controlled for normal lymphocytes or non-lymphoid cells, nor were the plasma membrane preparations well characterized. A somewhat related study suggested that CTL-mediated cytotoxicity might be due to the action of membrane-associated phospholipase (Frye and Friou, 1975). This enzyme, by removing one fatty acid chain from target cell phosphatidylcholine, could generate lysolecithin, a strong detergent potentially capable of destroying a target cell membrane. Although the reported results seemed encouraging, this study was not followed up by others, and no further claims regarding cytotoxic properties of isolated killer cell membranes were forthcoming.

A compelling argument against CTL membranes being *a priori* cytotoxic came from a version of the killer anti-killer experiments discussed in the last chapter. Kuppers and Henney (1977) presented data showing that when A anti-B CTL were incubated with B anti-C CTL, the latter lost their ability to lyse C target cells, but the former were undiminished in their ability to lyse additional B targets. Obviously whatever happened in this co-incubation of CTL occurred only in the direction of specific target recognition. Yet the membranes of both are in equally intimate contact. Henney concluded that CTL must be activated via their specific antigen receptor in order to exert an effector function on target cells, a view that turned out to be correct.

Others explored the possibility that inter-cytoplasmic connections between CTL and target cells, which could enable intercellular communication in isolation from the external milieu, might be the key to killer cell-induced lysis. For example, Sura et al., (1967) and Selin et al., (1971) reported the transfer of cytoplasmic contents or fluorescent probes, respectively. However, further ultrastructural and tracer studies failed to detect any obvious cytoplasmic junctions between CTL and their targets that would allow an exchange in either direction (Kalina and Berke, 1976; Sanderson et al., 1977). Freeze-fracture studies of fused human erythrocyte ghosts had revealed regions deficient in intra-membrane particles in the protoplasmic face of the fusion region (Zakai et al., 1977), which might represent a membrane-destabilizing event. Similar freeze-fracture analyses of CTL-target contact regions failed to reveal such alterations in the distribution of intra-membrane particles at the contact regions (Niedermeyer and Berke, unpublished). Interestingly however, a functional role for inter-cytoplasmic junctions, enabling transfer of materials presumed to be involved in inducing lysis, has recently been proposed (Trambas and Griffiths 2003). Confirmation of this claim will be required.

The second major approach to exploring cytotoxic mechanisms focussed on possible release of soluble cytolytic factors. One of the main arguments against soluble mediators was the failure of killer cells, separated by Millipore membranes from their target cells, to kill them in vivo or in vitro (Weaver et al., 1955; Wilson, 1963). However, this could equally be seen as blockage of a target-cell binding event needed to trigger the release of soluble lytic factors, rather than evidence against the involvement of such factors per se in lysis.

Since killer lymphocytes were clearly antigen specific, and antibody was the only known antigen-specific immune molecule at the time[2], it was natural to look for a role of soluble complement components, working together with killer cell surface antibody, in mediating cytotoxicity. However, complement could not be detected in killer cell populations, or in media in which they had killed target cells, and antibody to complement components did not interfere with the killing process (Henney and Mayer, 1971). Nevertheless, a possible involvement of complement in cell-mediated cytotoxicity, at some level and in some form, continued to intrigue investigators for many years.

Our understanding of how the complement (C) system works was advanced considerably with the description of the membrane attack complex (MAC), comprising the terminal C5b-9 components of the complement cascade. These components assemble to form a pore structure containing an outer hydrophobic surface and an inner polar channel that inserts into the membrane (Ramm et al., 1983; see Figure 5.1 A). Negative staining and EM freeze-fracture studies indicated an aqueous channel passing through the fully assembled complement MAC. Numerous functional studies showed that such structures inserted into membranes allowed the exchange of water molecules and ions, and most likely killed cells via colloid osmotic lysis.

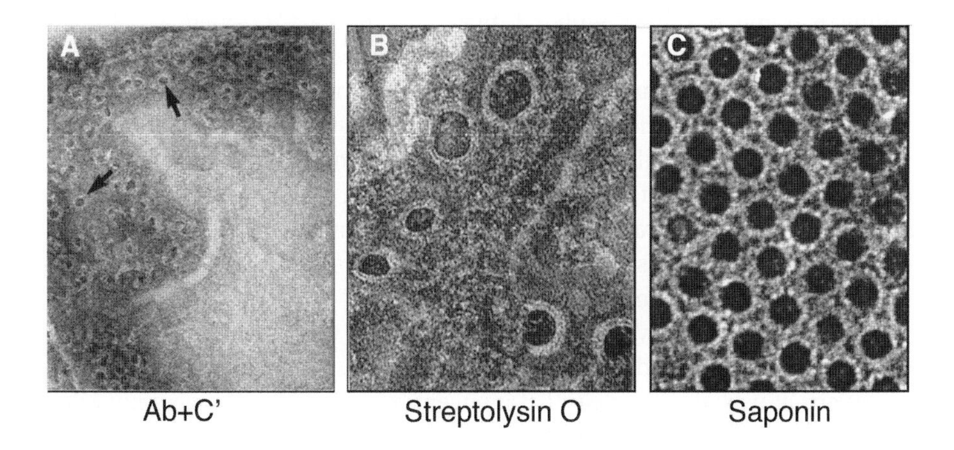

Figure 5-1. Pore structures in biological membranes. (A) Lysis of target cells by antibody (Ab) and complement; negative staining EM, by Dalia Rosen. (B) Lysis by streptolysin O. (C) Liver cell membranes treated with saponin (B and C are based on Dourmashkin et al., (1963) J. Roy. Micro. Soc 4:215

[2] The title of Govaert's original paper was in fact "Cellular antibodies in kidney homotransplantation."

However, despite repeated attempts by several laboratories, none of the components of the MAC could be detected in cytotoxic effector cells, or in the medium of a lytic assay. Despite this failure, many continued to be intrigued by "the complement connection." Mayer, among others, still pointed out similarities in cell-mediated and complement-mediated membrane damage as late as 1978 (Mayer et al., 1978). There would turn out to be fundamental differences between these two cytolytic mechanisms, but these would not be apparent for several more years.

Others continued to search in other directions for soluble mediators. Ruddle and Waksman (1967) and Granger and Kolb (1968) described soluble cytolytic factors released only by activated lymphocytes which they called "lymphotoxin". Lymphotoxin (LT) is produced by lymphocytes under a variety of circumstances, including activation by specific antigen or by mitogenic plant lectins such as PHA. Unfortunately, although most of the cell populations tested contained cytotoxic cells as defined in standard lytic assays, many did not, and the relevance of soluble factors released by such a wide range of cell types in relation to classical "Govaerts killing" was not clear.

Moreover, a number of facts about lymphocyte-mediated lysis quickly proved inconsistent with a role for lymphotoxin as the exclusive, or even major mediator of acute cell-mediated target-cell killing. The kinetics of killing was much slower with LT than with direct CTL-mediated killing. This could be overcome by inducing enhanced pinocytosis in target cells, but otherwise killing was much less vigorous than killing by CTL (Schmid et al., 1985). Moreover, kinetic evidence had suggested that target-cell destruction results from single collisions between killer lymphocytes and target cells (Wilson, 1963; Berke et al., 1969; Henney, 1971). If a soluble mediator were secreted into the culture medium, lysis should quickly become independent of such collisions. Close analysis had shown this not to be the case, even after extensive incubation periods.

Also troublesome was the fact that LT was non-specific in its action; once released, it killed any and all target cells. Yet by the time of its discovery, it was already clear that innocent bystander targets were not killed in the course of CTL assays. It was difficult to reconcile such specificity with a soluble, non-specific mediator released into the culture medium. It was also not clear why a soluble mediator would not kill the CTL itself. Finally, it was found that the cytolytic activity of killer cell populations could be dissociated experimentally from their ability to produce soluble mediators. Treatment of CTL with certain drugs (e.g., cholera toxin, enterotoxin, colchicine, and vinblastine) ablates lytic activity, but does not affect their ability to release soluble mediators such as LT.

Current thinking about LT, which is a member of the TNF family of cytokines, is that it is primarily a mediator of delayed hypersensitivity reactions mediated by both CD4 and CD8 cells. Although LT is widespread among cells of the immune system (Ware et al., 1992), and can be cytotoxic to certain cell types in long-term assays (>16 hours), it is not thought to contribute significantly to the acute cytolysis mediated by alloimmune CTL. (Cytotoxicity mediated by TNF-related molecules in general will be discussed at the end of this chapter.)

A number of researchers had been intrigued by the presence in most cytotoxic lymphocytes (both CTL and NK cells) of extensive cytoplasmic granules (Figure 5.2). Peripheral blood NK cells had initially been characterized as large granular lymphocytes (LGL) (Timonen et al., 1981). In NK cells, these granules are expressed constitutively, but in CTL they are seen only after stimulation and maturation of CTL precursors in response to cognate antigen or other activating agents; no granules are apparent in CTL precursors. It was initially not clear if and how these or similar but not identical lysosome-like granules observed by EM contributed to the cytolytic process (Bykovskaya et al., 1978; Zagury et al., 1975). Nevertheless, in the early 1980s several laboratories began intensively investigating the possibility that exocytosis of these granules and their contents could be related to the cytolytic process.

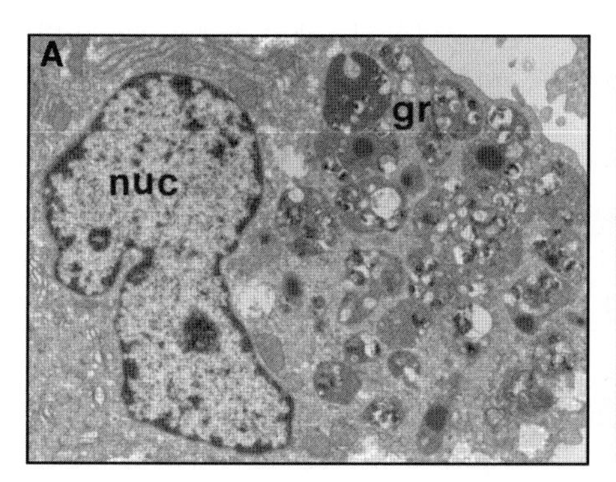

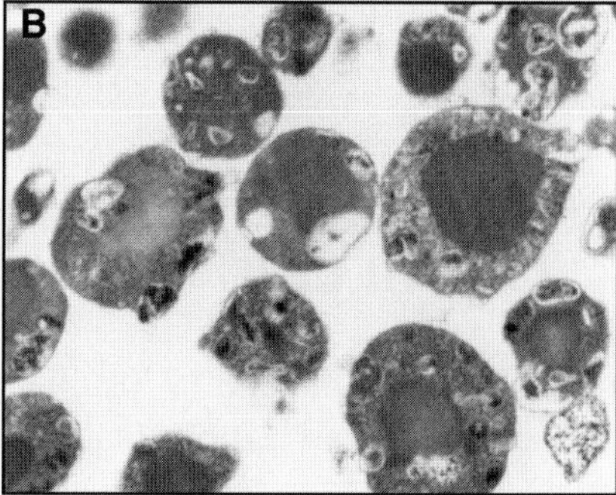

Figure 5-2. CTL granules. (A) CTL with numerous cytoplasmic granules (gr); (B) isolated granules, showing electron dense core and diffuse cortex (by D. Rosen).

The known Ca^{2+}-dependency of exocytosis was seen to be consistent with the general Ca^{2+}-dependency of lymphocyte-mediated killing (although see below, "Evidence inconsistent with exocytosis of perforin…"), and with observations made earlier that effector cell cytoplasmic structures potentially associated with exocytosis, including the Golgi apparatus, the microtubule organizing center (MTOC) and the granules themselves, rearrange at the interface between the effector and target cell after conjugation (Thiernesse et al., 1977; Bykovskaja et al., 1978a, 1978b; Geiger et al., 1982; Kupfer et al., 1983; Carpen et al., 1982). Bykovskaja et al. (1978a) and later Zagury (1982) also predicted a secretory event during cytolysis based on the deposition of apparent granule-associated material at the effector-target interface during cytolysis.

Secretion (exocytosis) occurs as a result of a membrane fusion process in which an intracellular granule ("regulated secretion") or vesicle ('constitutive secretion') fuses with the plasma membrane, releasing its contents into the extracellular space. CTL and NK cells use both secretory pathways. Studies on the effects on CTL-mediated lysis of drugs known to influence secretion support the involvement of secretory event(s) in CTL/NK-mediated lysis. For example, colchicine, which affects the microtubular and secretory systems, markedly inhibits cytolysis (Henney, 1977). Monensin, believed to influence secretion by its action on the Golgi complex, has also been shown to be an effective inhibitor of cytotoxicity (Carpen et al., 1981). Concanamycin A, which neutralizes intragranule pH and exocytosis of granular contents, is a powerful inhibitor of lysis. It was argued however, that the effects of at least monensin, and perhaps colchcine and nocodazole as well on CTL-mediated lysis could as easily be explained by inhibition of cellular mobility, which could inhibit the initial binding step of the killer and target rather than subsequent lytic events, with the same outcome in terms of ^{51}Cr release. Nevertheless, evidence that directed release of granule contents is involved in killer cell lytic function continued to build.

2 PERFORIN-GRANULE EXOCYTOSIS MODEL FOR CTL KILLING

Beginning in the late 1970s, Pierre Henkart and his co-workers embarked on a series of investigations that led them to propose the first coherent model for an exocytosis-based mechanism of lymphocyte-mediated cytotoxicity. Although most of these studies were carried out with various NK cell lines, we discuss them here, rather than in Chapter 6, because the exocytosis model deduced in NK cells was quickly shown to apply to CTL-mediated cytotoxicity as well.

Henkart had proposed as early as 1975, following studies of the interaction of killer cells with artificial membranes, that lymphocytotoxicity could involve the formation of channels or pores in the target cell membrane (Henkart and Blumenthal, 1975). In a paper published in 1980 (Dourmashkin et al., 1980), it was reported that the plasma membranes of red blood cells killed by NK cells, in a variant of cell-mediated cytotoxicity called antibody-dependent cell-mediated cytotoxicity (ADCC; see Chapter 6), contained ring structures similar to, although slightly larger than those seen in complement-mediated lysis of red blood cells (see Figure 5.1). The size range of pores seen after lysis was consistent with the target membrane permeability data Henkart had published earlier that year, which also had prompted him to suggest that ADCC most likely resulted in pore formation (Simone and Henkart, 1980). The ultrastructural data presented in the paper with Dourmashkin showed tubular structures apparently inserted into, and in some profiles seemingly jutting above, the target cell membrane.

At the first International Workshop on Mechanisms of Cell-Mediated Cytotoxicity, held in 1981, Henkart presented extensive electron microscopic evidence supporting the involvement of tubular structures, and proposed that they were stored in cytoplasmic granules in cytotoxic effector cells, being released only upon contact with appropriately recognized target cells (Henkart and Henkart, 1982). Subsequent detailed microscopy and micro-cinematography suggested that in CTL bound to target cells, the granules undergo a kinesin–driven reorientation, in the presence of Ca^{2+}, from a random distribution in the CTL to the CTL-target cell interface after CTL-target conjugation (Yannelli et al., 1986; Kupfer and Singer, 1989; Griffiths, 1995).

Henkart's group called the putative granule protein that forms these lytic structures cytolysin. Assembly of cytolysin into tubules on the target cell membrane was found to be Ca^{2+}-dependent, although prolonged exposure of tubules to Ca^{2+} rendered them non-lytic to red blood cells. To account for the lack of innocent bystander killing in lymphocytotoxicity, expected for secreted mediators of lysis, it was proposed that the secreted materials were not soluble, and functioned only by inserting immediately into membranes of the bound target; functional cytolysis thus could not diffuse away from the enclosed "pockets" formed by interdigitating effector-target membranes. To accommodate data showing that effector cells were not killed during target cell lysis, it was proposed that effector cells must have either a constitutive or secretion-induced resistance to their own secreted materials (Henkart et al., 1984; Millard et al., 1984), but as we have seen that is still an unresolved question.

Henkart's reports attracted the attention of Eckhard Podack, who had worked extensively on the terminal complement components, and their assembly into the membrane attack complex. Podack and Dennert (1983) and Dennert and Podack (1983) described tubular complexes virtually identical to those of Dourmashkin and Henkart in electron micrographs of target cells undergoing lysis by both CTL and cloned NK cells. Their results generally agreed with those of Henkart regarding the structure of the pores, the requirement for immediate assembly on or in target cell membranes, and likely storage of the pore-forming material in cytoplasmic granules. The requirement for assembly into a cell membrane in order to be functional is also typical of the MAC.

Podack named his pore-forming substance perforin, the name now generally used for the granule-associated, pore-forming protein released during cell-mediated cytolysis. Podack focused most of his attention on the structure and function of perforin protein subunits, and comparison of this system with the MAC. His group was also among the first to clone the gene for cytolysin/perforin (Shinkai et al., 1989; Lichtenheld et al., 1989; Kwon et al., 1989; Lowrey et al., 1989). As we will discuss shortly, the gene for perforin shows moderate homology with the genes for complement components C8 and C9.

Thus it seemed that the basic complement mechanism of pore formation and membrane destruction had been retained, although in a somewhat altered form, to carry out the membranolytic functions of cytotoxic effector cells. The existence of parallel and evolutionarily related sets of molecules to carry out essentially identical functions was not without precedence in the immune system. Antigen receptors for T lymphocytes are constructed from sets of gene segments virtually identical to those used for B cells, and are assembled by the same mechanisms. Yet these two gene systems are completely non-interacting, presumably because of distinct needs for regulation. It was not difficult to imagine a similar evolutionary rationale behind the gradual segregation of two separate systems for membrane destruction. But in fact, as we will see, the notion of a simple complement-like, pore-forming role for perforin in mediating killing did not, as initially proposed, account for all the known parameters of both CTL- and NK cell-mediated lysis.

2.1 Cytotoxic granules and their contents

The cytoplasmic granules noted in NK cells by earlier investigators, and proposed by Henkart and Podack as storage sites for the pore-forming protein cytolysin/perforin, were quickly shown to exist in a wide range of killer lymphocytes of mouse, rat, and human origin. Cellular sources include NK cell lines, large granular lymphocyte (LGL) tumors, and CTL lines. Typical lytic granules are also present in activated CTL in vivo, particularly at early stages of immune responses to viruses and allografts, as well as in naive lymphocytes cultured with IL-2 (probably NK cells) or stimulated by antigens or mitogens (most likely T cells) in vitro (Henkart 1985; Podack 1986). The cytological origin of these structures has long been a subject of debate. While retaining many of the features of lysosomes, there are also important differences, and they are often referred to as "modified" or "bi-functional" lysosomes (Henkart and Henkart 1982, Burkhardt et al., 1990;

Stinchcombe et al., 2000). The low intragranular pH of these structures is believed to be maintained by a membrane proton pump-vacuolar-type ATPase that can be blocked by the antibiotic concanamycin A, known to block CTL and NK lysis (Kataoka et al., 1994). Granules can be harvested by disrupting granulated cells using nitrogen cavitation, which ruptures cellular membranes without destroying the granules, followed by separation of the granules on Percoll density gradients. Isolated granules lyse a variety of target cells in a nonspecific manner. Red blood cells are the most sensitive targets. Nucleated cells are also killed, but with a hundredfold or so less efficiency (Shiver and Henkart, 1991, 1992; Duke et al., 1989; Browne et al., 1999).

When a CTL binds to its target via the TCR, several events occur: release of intracellular stores of Ca^{2+}, followed by influx of extracellular Ca^{2+}; exocytosis of granule contents; and *de novo* synthesis of replacement granule-associated proteins. Newly synthesized proteins can either be used to fill new storage granules, or be secreted directly via the constitutive secretory pathway (Griffiths, 1995). This could be one means by which the lytic potential of the granule pathway can be sustained as CTL recycle from one target to another.

The intracellular increase in Ca^{2+} that occurs when the CTL antigen receptor complex engages with a cognate target cell was originally thought to be polarized, with differential concentrations of Ca^{2+} at the point of target cell contact and at the opposite side of the cell. It was suggested this might provide a means for guiding granules to the appropriate region of the CTL membrane for exocytosis (Poenie et al., 1987). However, subsequent studies have challenged this notion (Lyubchenko et al., 2001), and the nature and regulation of the pathway leading to reorientation of the MTOC and granules for exocytosis are still unclear.

Two structurally distinct domains, a dense core and an outer, multivesicular cortex, are observed in electron micrographs of granules (Peters et al., 1989) (see Figure 5.2). As mentioned, the granules (0.5-1 micron in diameter) may be unique, dual-function organelles, in which secretory and pre-lysosomal compartments are combined (Burkhardt et al., 1990). Lysosomal proteins such as cathepsin D and acid phosphatase are abundant in the cortical regions of granules. Importantly, secreted lysosomal enzymes appear not to take part in the lytic process. The major constituents of cytotoxic granules, from a cytolytic perspective, are the pore-forming protein perforin, and a collection of serine proteases referred to collectively as ‘granzymes’ (granule-associated enzymes), found in the granule core. The granzymes have been extensively characterized at both the gene and protein levels (reviewed in Smyth et al., 1996; Trapani, 1998; Trapani et al., 1998; Kam et al., 2000). Granzymes are serine proteases. They are synthesized as inactive pro-enzymes that are processed proteolytically to generate active enzymes. They are rich in basic amino acids, exhibit highly conserved amino acid sequences, and display features of other serine protease such as trypsin (Waugh et al., 2000). They are also expressed in non-cytolytic CD4 T cells, and, interestingly, in mast cells (Brunet et al., 1987). The release of serine protease activity (a property of granzymes) from cytotoxic effector cells during target cell lysis is widely used as an indicator of granule exocytosis (Pasternack and Eisen, 1985; Ostergaard et al., 1987), and is readily quantified in a colorimetric reaction involving the cleavage of a chromogenic BLT-ester substrate.

Two granzymes of particular interest in CTL biology are Granzyme A (GrzA) and Granzyme B (GrzB). GrzA cleaves near arginine or lysine, a characteristic of a tryptase. This activity is often monitored as a BLT (benzoyl lysine thioesther) esterase activity monitored colorimetrically. Activated GrzA initiates a sequence of events culminating in single-strand DNA breaks, and furthermore interferes with DNA repair, which leads to apoptotic cell death. GrzB is a potent activator of caspases 3 and 7-10 (see below), which also leads to apoptotic death. GrzB can also trigger caspase-independent target cell death by inducing mitochondrial rupture through cleavage of Bid.

Perforin and granzymes are accompanied within the core by the chaperone and calcium storage protein calreticulin. Calreticulin is co-released with perforin and granzymes during exocytosis, and is thought to play a role in regulating perforin function after, and possibly before, release (Andrin et al., 1998; Fraser et al., 2000). Chondroitin sulfate (Masson et al., 1990) may play a similar role. The proteoglycan-protease complexes minimize serine esterase-induced autodegradation, and remain intact even after exocytosis (Stevens et al., 1989; Spaeny-Dekker et al., 2000). Numerous other proteins, many of which are lysosomal in nature (Smyth et al., 2001), and even lipids, are also found in cytotoxic granules. One granule-associated protein called granulysin has cytocidal activity in humans, and has been proposed to play a role in the lysis of certain tumor target cells by at least NK cells (Krensky, 2000; Sekiya et al., 2002). Due to its bactericidal activity, granulysin may kill intracellular bacteria after lysis of an infected cell by perforin/granzymes (Stenger et al., 1998; Chapter 8). A murine counterpart to granulysin has not been found.

A detailed analysis of the immunological synapse of CTL during target cell killing, using confocal video microscopy of individual conjugates at 37°C, provided insight into the relationship of organellar reorientation, synapse formation, and granule release during the killing process (Stinchcombe et al., 2001; Trambas and Griffith, 2003) (see Figure 1.8). Initial contact between an activated CTL and a specifically recognized target cell induces within one minute movement of the MTOC, followed by the Golgi apparatus, and then by granules, toward the CTL-target interface. Presumably these first events must be triggered by randomly distributed TCR on the CTL surface, but a full response requires formation of an immunological synapse. Signal transduction molecules such as Lck and protein kinase C-theta, presumably along with CD3, CD28 and CD8, are surrounded in the signaling core of each synapse by a ring of LFA-1, which is associated with the cytoskeletal connector talin (Freiberg et al., 2002).

The granules are essentially in place at the CTL-target interface by three minutes, by which time the first granules are beginning to be released. The central portion of the signaling core of the synapse has an asymmetric structure, with some of the molecules being displaced to allow passage of exocytosing granules. The synapses face into the intercellular "pockets" described earlier, surrounded on all sides by stretches of tightly sealed CTL-target cell membranes. However, before most of the granules are released, the target cell membrane begins to show the blebbing behavior associated with apoptosis.

Although conceptually very attractive, it is difficult at present to reconcile some aspects of the proposed immunological synapse as defined by relatively low-resolution confocal microscopy (Freiberg et al., 2002) with previous analyses of CTL-target cell interactions revealed by electron microscopy (Chapter 3). The latter show that large portions – up to 30 percent – of the membranes of the CTL and target cells are in contact with one another. This interface between the two cells is characterized by extensive interdigitation of membrane processes, with areas of contact running up to 100-150 nanometers. At a molecular scale, such regions of contact would almost certainly encompass numerous synapses. Yet most published photos of such allegedly single synapses show them as being on a scale that would span nearly the entire area of CTL-target contact. Given the mechanical strength of CTL-target conjugates, and the multiple points of contact between the interacting cells, it seems *a priori* unlikely they would be mediated by a single TcR-class I synapse, no matter how many accessory molecules were used to buttress it. Rather, it would seem more likely that small multiple synapses, spread across many regions of CTL-target interdigitation, would create the type of adhesion we see ultrastructurally, and be measured in conjugate dissociation studies. Clearly, a further reconciliation of electron microscopic and confocal microscopic analyses of conjugates is in order.

2.2 Perforin – the "Big MAC" of CTL?

It has been shown repeatedly that purified perforin alone causes rapid, Ca^{2+}-dependent lysis of red blood cells. Perforin also induces the rapid disintegration of artificial liposomes, as determined by the release of trapped carboxyfluorescein (Henkart, 1985). As noted above, however, its lytic action against nucleated target cells is relatively slight. Calcium is necessary for the assembly, membrane attachment, and insertion of perforin subunits, and for their subsequent lytic activity (Ishiura et al., 1990). Binding of perforin to membranes is complete within 1 min at 0°. Upon polymerization, perforin forms stable, hollow tubular structures originally estimated by electron microscopy to have an internal diameter of 100-200 Å; later studies suggested a possible radius of about 50 Å (Peters et al., 1990). In the electron microscope these structures appear as rings embedded in the target cell plasma membrane and are similar to those of the MAC of complement. Various receptors have been proposed for perforin on target cell membranes, including phosphorylcholine (Tschopp et al., 1989) and platelet activating factor (Berthou et al., 2000), but no consensus has been reached on the need for or identity of a perforin receptor.

When perforin subunits penetrate the cell membrane, their hydrophobic residues presumably interact with fatty acyl chains of the membrane, with hydrophilic residues lining the interior of the pore. Perforin channels (pores) are voltage insensitive, ion non-selective, and heterogeneous in size (Young et al., 1986; Persechini et al., 1990). The size of the functional membrane pores induced by complement and by perforin was determined by studying the transport of hydrophilic fluorescent tracers through the pores using confocal laser scanning microscopy and fluorescence microphotolysis scanning (Sauer et al., 1991). Single-cell flux measurements revealed pores with a 50-Å functional radius in cells exposed to antibody plus complement and in sheep red blood cells exposed to perforin. These complement- and perforin-induced pores agree well with single-channel electrophysiological recordings. Interestingly, in contrast to electrophysiological measurements of MAC of complement, which show flickering (closing and opening of channels), no closing events for perforin pores were detected by fluorescence microscopic single-channel recording at a time resolution of 0.25 msec, a wide range of temperatures, and intracellular Ca^{2+} concentrations $[Ca^{2+}]^{i}$ (Peters et al., 1990). This indicates a fundamental difference in the channel permeability of the two systems.

The genes for both mouse and human perforin were cloned in 1989 (Kwon et al., 1989; Lowrey et al., 1989; Lichtenfeld et al., 1989; Shinkai et al., 1989.) The perforin gene in both species is a single-copy gene. In each case the gene consists of three exons, only the 3'-two of which encode the perforin protein; the 5' exon contains only regulatory elements (Podack et al., 1991). The murine gene maps to chromosome 10 (Trapani et al., 1990); the human gene was originally proposed to reside on chromosome 17 at locus 17q11-21 (Shinkai et al., 1989), but was subsequently found to be on chromosome 10 (Fink et al., 1992), at locus 10q22. The corresponding regions of mouse and human chromosome 10 are generally considered to be syntenic (Nadeau, 1989). Perforin is synthesized as a 70-kDa precursor, which is subsequently cleaved to the mature form (Uellner et al., 1997) consisting of 555 amino acids, with an estimated unglycosylated molecular weight of 60,000 and a potential glycosylated molecular weight of about 69,000 (Figure 5.3). A modest but definite structural similarity between perforin and the terminal components of the complement was confirmed by sequence analysis of perforin cDNA clones. In mice, residues 160-390 (the central 1/3) of perforin and 270-520 (the carboxyl-terminal 1/3) of complement component C9 exhibit appreciable homology. This perforin-homologous portion is also found in components C6 through C9. It contains six potential membrane-spanning/pore-forming domains and a cysteine-rich domain (murine perforin residues 355-390), homologous with urokinase and the epidermal growth factor (EGF) receptor. The cysteine-rich domain may account for the immunological cross reactivity between reduced perforin and components C6 through C9 (Shinkai et al., 1998; Tschopp et al., 1986). The amino- and carboxyl-terminal regions of perforin and C9 are divergent.

What exactly is the role of perforin in the killing of nucleated target cells by CTL and NK cells? In the case of red blood cells and perhaps artificial membranes, perforin would indeed seem to be functioning like the "Big MAC" of complement. But nucleated target cells, whose plasma membranes are not markedly more resistant than red cells to complement-mediated damage, seem barely affected by purified perforin. The simple model in which perforin punches a hole in nucleated target cell membranes, leading to osmotic lysis, had to be reconsidered. But despite intensive research for over a decade, it is still not known how perforin binding to membranes results in pore formation. Typical membrane spanning motifs of 20 amino acids, which are characteristic of alpha helical transmembrane domains, have not been detected. Alternatively, polymerization of perforin monomers could assemble as a beta "barrel' similar to some bacterial toxins (e.g. staph alpha toxin).

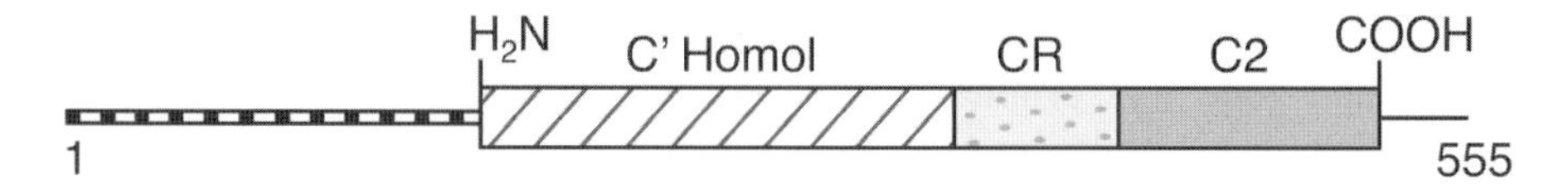

Figure 5-3. Schematic view of the perforin protein molecule. The central region of the molecule (C' Homology) shows significant but distant homology to complement components C6, C7, C8a, and C9 and contains a putative amphipathic helix at its left end. This is followed by a short cysteine-rich (CR) region of unknown significance, and then by the C2 homology domain implicated in calcium-dependent phospholipid interactions, and finally by the COOH peptide, which is removed by processing before granule storage (from Henkart, 1999).

2.3 The role of granzymes in perforin-mediated killing

In addition to being at best minimally lytic against nucleated target cells, perforin alone cannot account for all the known features of CTL-mediated target cell lysis. As indicated in the last chapter, John Russell had shown in 1983 that CTL-mediated lysis induces apoptosis-like damage to target cell nuclei, including DNA fragmentation, and that these effects are actually initiated before measurable ^{51}Cr release (Russell, 1983). Russell's findings were quickly confirmed by others (Duke et al., 1983; Cohen et al., 1985). When purified perforin alone, rather than intact granules, was tested in cytolytic assays, concentrations of perforin required to give measurable levels of nucleated target cell lysis failed to induce target cell DNA fragmentation. All the CTL tested caused DNA fragmentation in all the target cells used (Duke et al., 1989; Hayes et al., 1989). Damage to the nucleus and its contents is also not seen in complement-mediated damage to nucleated cells. Thus, purified perforin was less efficient than both CTL and complement in causing membrane damage to target cells, and like complement did not induce nuclear damage.

Interestingly, the more complex sequence of apoptotic events associated with CTL-mediated lysis is seen when intact cytotoxic granules are used as the lytic agent rather than purified perforin, suggesting that other granule components might play key roles in complete cytolysis. This possibility was pursued in experiments from Henkart's lab. Although purified perforin alone did not cause target cell DNA fragmentation, a combination of perforin and Granzyme A did (Hayes et al., 1989). The Henkart group subsequently transfected an expressable perforin gene into a non-cytotoxic rat basophilic leukemia line that could be degranulated by cross-linking its surface IgE receptors. The transfected cells lysed red blood cells coated with IgE, but did not lyse untreated innocent bystander RBC or nucleated target cells (Shiver and Henkart, 1991). However, if the basophilic cells were co-transfected with the genes for perforin and GrzA, nucleated targets were also lysed, and target cell DNA fragmentation (detected by the release of ^{125}I-labeled UdR) was observed (Shiver and Henkart, 1992). These experiments suggested that both perforin and granzymes were required for cell lysis and degradation of DNA, and implied that granzymes might enter target cells via a perforin pore.

Subsequent studies showed that both GrzA and GrzB could cooperate with perforin to induce apoptosis-like nuclear damage in target cells. In fact GrzB is the more potent of the two, inducing the early nuclear and DNA damage described by Russell. The relative roles of the two granzymes became clearer when the genes for each were disrupted in mice. When the gene for GrzA is knocked out, there is no impact on either ^{51}Cr release or the induction of DNA fragmentation by immune CTL (Ebnet et al., 1995). When the GrzB gene is disrupted, there is again no impact on ^{51}Cr release. DNA fragmentation is reduced in short term (four-hour) lytic assays, but goes to completion by sixteen hours or so, suggesting that loss of GrzB can be compensated by some other granule component (Heusel et al., 1994). That this other component is in fact GrzA is suggested by double knockout mice, negative for both GrzA and GrzB, where again ^{51}Cr release is normal but DNA fragmentation is essentially completely abolished (Simon et al., 1997). These data also reinforce the notion that nuclear damage is not a prerequisite for the plasma membrane leakage seen during exocytotic target cell killing, a point made clear by the fact that enucleated target cells can be killed (Berke and Fishelson, 1977; Siliciano and Henney, 1978; Nakajima et al., 1995).

These results initiated a rush by several laboratories to define exactly how granzymes A and B interact with perforin to induce nuclear damage. The initial thinking was that perforin simply provided a pore through which granzymes could enter the cell. But several studies have suggested that granzymes can enter cells independently of perforin, through what appears to be receptor-mediated endocytosis (e.g., Shi et al., 1997), although they remain trapped in the endosomal compartment in the absence of perforin or an appropriate endosomolytic agent (Froelich et al., 1996.) Recently, the mannose-6-phosphate receptor (MPR) has been proposed as a possible binding site for at least granzyme B (Motyka et al., 2000) although this mechanism has been questioned (Trapani et al., 2003). However, a granzyme entering a cell on its own still needs perforin to induce apoptosis (Pinkoski et al., 1998). Trapani has suggested that perforin may facilitate the escape of granzymes from the endosomal compartment, or help granzymes gain access to the nucleus (Blink et al., 1999; Jans et al., 1999; Trapani, 2000) (Figure 5.4), although this, too, has been challenged (Trambas and Griffiths, 2003).

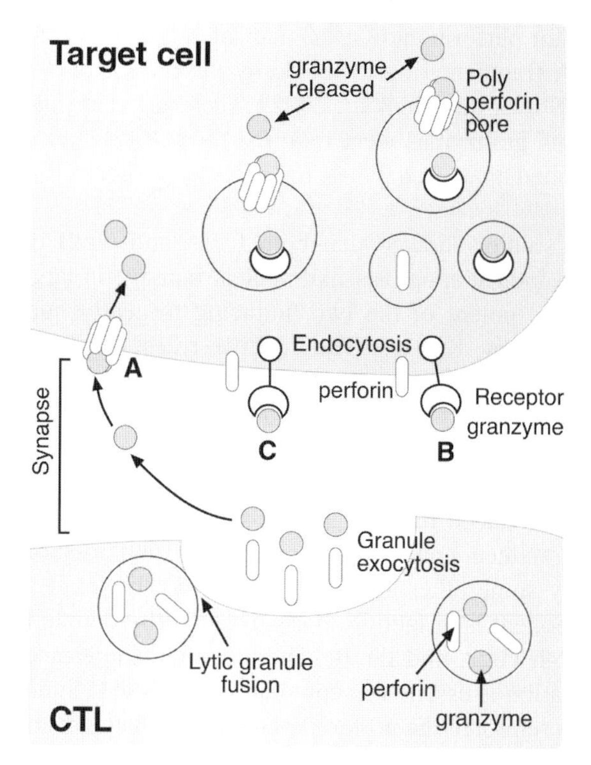

Figure 5-4. How secreted granzyme B enters the target cell. On CTL-target interaction, there is a directed exocytosis of CTL granules into the intercellular space. (A) The original view has been that perforin polymerized to form a pore in the target-cell membrane through which granzymes could pass. (B) More recently, the discovery of a receptor for granzyme B suggests this molecule might be taken up by receptor-mediated endocytosis, and that perforin acts to release endosome-entrapped granzymes into the cytosol of the target cell. (C) Alternatively, granzyme might bind to the cell surface such that granzyme uptake is stimulated by perforin-mediated damage to the membrane (after Barry & Bleackley 2002)

But what exactly do granzymes do once they gain access to the cytoplasm and/or nucleus? One attractive target for granzymes, especially for GrzB, would be the collection of proteases referred to collectively as 'caspases' – cysteine-dependent, aspartate-directed proteases (Earnshaw et al., 1999; Lieberman, 2003). Caspases have been known since the early 1990s to play a key role in the induction of apoptosis in cells under a variety of conditions. There are a dozen or so caspases in most eukaryotic species. They seem to act as a cascade, with some of the caspases having other caspases as their principal substrate ("activating caspases"), whereas others have targets influencing various portions of the apoptotic process ("effector caspases"). The full range of substrates acted on by effector caspases is still being worked out, but the general trend appears to be the destruction of proteins favoring survival of the cell, and activation of proteins involved in cellular disassembly. Caspases may also be involved in cell-cycle regulation (Los et al., 2001).

Caspases were of interest in the killing of target cells via the exocytosis pathway because GrzB is the only known enzyme, aside from the caspases themselves, with the highly unusual and characteristic caspase specificity involving cleavage on the carboxyl side of aspartate residues. Both intact granules and purified GrzB have been shown to activate caspases (Darmon et al., 1995, 1996; Talanian et al., 1997), and GrzB caspase activation can induce apoptotic nuclear damage in cell-free systems (Martin et al., 1996). Indeed, inhibitors of caspase activity are also potent inhibitors of nuclear damage inflicted by the exocytotic pathway of killing, but have no effect on target cell killing as measured by ^{51}Cr release (Sarin et al., 1997; Trapani et al., 1998), consistent with the mouse GrzB knockout experiments discussed earlier. The interactions of perforin and granzymes in the induction of apoptotic cell death in nucleated target cells are still being worked out, but the broad outlines of how it will work are beginning to emerge (Krammer, 2000; Russell and Ley, 2002; Barry and Bleackley, 2002).

But while caspases are clearly involved in the induction of apoptotic damage induced via the exocytotic pathway, they appear not to be involved in the pathway leading to plasma membrane damage, as revealed by ^{51}Cr release. As of this writing, the connection between perforin and plasma membrane damage has not been completely unraveled. Henkart's laboratory has shown that caspases *are* involved in the induction of *non-nuclear* apoptotic damage, including externalization of phosphatidylserine (PS) in the target cell plasma membrane, a loss of target cell inner-membrane mitochondrial potential, and the initiation of target plasma membrane blebbing (Sarin et al., 1998). All three of these features are also induced very early in CTL-mediated target cell death. PS externalization prepares target cells for phagocytosis by macrophages – the so-called 'eat me' signal (Savill et al., 1993; Martin et al., 1995) - and is unlikely to be involved in lysis per se, but the other two parameters very well could be. It is also possible that other granule components, perhaps some as yet unidentified granzyme, are involved in the pathways leading to ^{51}Cr release (Henkart, personal communication). On the other hand, the fact that perforin alone can cause at least some degree of ^{51}Cr release from nucleated target cells may indicate that perforin is in fact sufficient for the membrane damage revealed by ^{51}Cr release, and possibly other markers from target cells.

How important is induction of DNA fragmentation to the roles played by CTL in vivo? GrzA/GrzB-deficient mice are highly susceptible to ectromelia virus (Mullbacher et al., 1999a), but are at least partially resistant to cytomegalovirus (Reira et al., 2001; see also Chapter 9). The ability to reject allografts in vivo is unaffected by the lack of both granzymes (Davis et al.,

2001). These results suggest that in instances where cytolytic function of CTL (and NK cells) is presumed to be important in vivo, granzymes do not play a crucial role. Thus, it is possible that for in vivo cytotoxicity per se, the crucial role of perforin may be either in direct target cell membrane damage, or in facilitating the entrance of potentially lytic molecules such as granulysin or possibly cathepsins (Griffiths et al., 1995; Kaspar et al., 2001) into the cytoplasm of the target cell. We will assess the possible contributions of both perforin and granzymes to presumed lytic functions in vivo in Chapters 7-10.

3 EVIDENCE INCONSISTENT WITH GRANULE EXOCYTOSIS AS THE EXCLUSIVE LYTIC MECHANISM

3.1 Cytolysis in the absence of Ca^{2+}

All of the foregoing observations, taken together, put the stimulus-secretion or granule exocytosis model of lymphocyte-mediated cytolysis firmly on the map. Yet despite the attractiveness of perforin as a lytic mediator, and convincing evidence of its involvement in lymphocytotoxicity (Acha-Orbea et al., 1990; Shiver and Henkart, 1991), it soon became clear that this mode of lysis must be only part of the story. Several laboratories provided evidence for CTL-mediated killing under conditions where perforin seemed an unlikely candidate as a lytic mechanism. A major argument against perforin as the sole mediator of cytotoxicity was based on the ability of CTL to kill some target cells in the absence of Ca^{2+} (Tirosh and Berke, 1985a, b; Ostergaard et al., 1987). As discussed in Chapter 4, Ca^{2+} had been suggested earlier to be an absolute requirement for most killing reactions. As we have seen, Ca^{2+} is required both for the degranulation process, and for the membrane assembly and function of perforin complexes. Thus killing in the absence of calcium could not possibly involve perforin.

When cloned CTL killed target cells in the absence of extracellular Ca^{2+}, the effector cell MTOC was reoriented toward the target cell and the target cell underwent apoptosis, but there was no release of granule serine esterase activity, confirming that exocytosis had not taken place (Ostergaard et al., 1987; Trenn et al., 1987). Release of intracellular Ca^{2+} stores may have been sufficient to induce MTOC reorientation, but not to drive exocytosis (Haverstick et al., 1991). Other systems were described in which degranulation and killing could be uncoupled by various means (Lancki et al., 1989; Zanovello et al., 1989). Interestingly, only certain target cells could be killed in the absence of Ca^{2+} (Tirosh and Berke, 1985). As will be discussed below, Ca^{2+}-independent lysis depends on the presence of the Fas (CD95) protein on the target cell, and was one of the major pieces of supporting evidence for the existence of this alternative pathway in CTL.

3.2 Killing by primary PEL and PEL hybridomas

Compelling evidence for the existence of a non-perforin lytic pathway also came from the study of in vivo primed PEL and PEL hybridomas. In early EM studies on the interaction of CTL with target cells, a few single-membrane-enclosed, lysosome-type granules and a well-developed Golgi apparatus were observed in primary PEL (Zagury et al., 1975; Kalina and Berke, 1976). However, no lytic storage granules of the type subsequently described by the Henkart or Podack labs were seen in PEL (Figure 5.5).

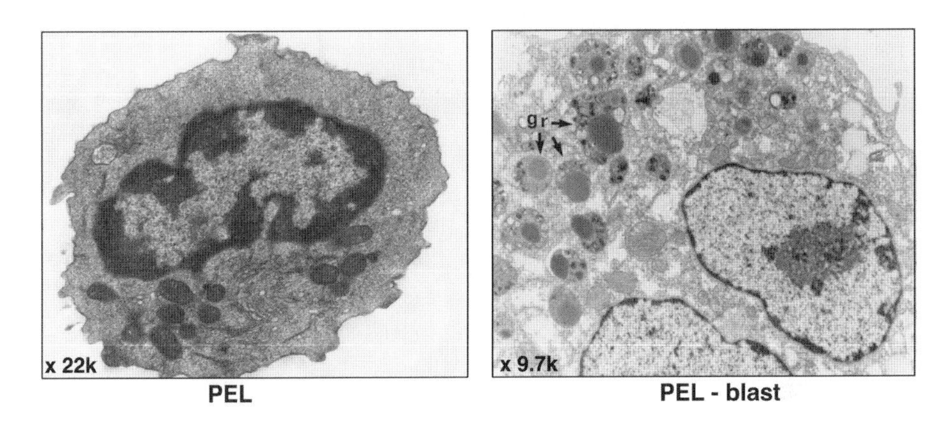

Figure 5-5. Induction of lytic granules in PEL. Electron micrographs of PEL and IL-2-induced PEL-blasts (PEB). Note abundance of mitochondria in PEL and granules (gr) in PEB.

At best only background levels of perforin or granzyme activities could be detected. Background levels of perforin and serine esterases in PEL was confirmed using monoclonal antibodies against perforin (Figure 5.6) and cDNA probes for the detection of of perforin and granzyme mRNA (Berke and Rosen, 1987, 1988; Berke et al., 1993a). Although no granules were seen by EM in PEL CTL, the MTOC and Golgi were reoriented towards the PEL-target cell interface, as in other CTL (Geiger et al., 1982). Furthermore, extensive electron microscopic studies, employing various negative staining techniques, could not confirm the presence of MAC-like ring structures – the hallmark of perforin action proposed by Henkart and by Podack – on membrane of cells lysed by in vivo-primed PEL or PEL hybridomas (Berke and Rosen, 1987, 1988) (Figure 5.7).

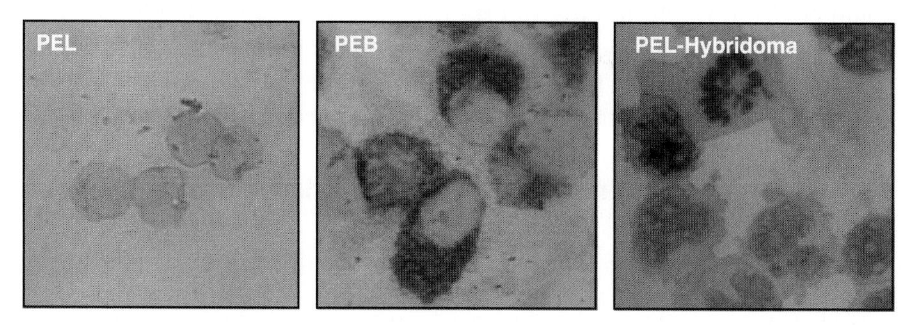

Figure 5-6. Potent CTLs devoid and rich in perforin. Immunoperoxidase staining with perforin antibody of (A) PEL; (B) PEB; and (C) CTL hybridoma. x100.

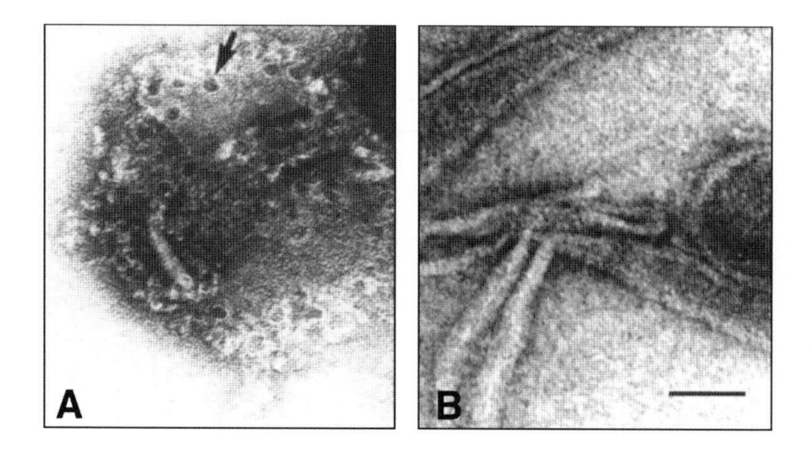

Figure 5-7. PEL kill target cells without forming perforin-like pores in their outer membrane. (A) Negatively stained target membranes after lysis by perforin-containing NK cells. Typical complement-like lesions are seen (arrows). (B) Negatively stained target membranes following PEL-induced lysis (85% lysis by ^{51}Cr-release assay). No rings are seen. Bar 44 nm. By D. Rosen.

Despite the deficiency of lytic granules and perforin and granzyme activity in highly lytic PEL, both these features can be induced in PEL by incubation with IL-2. IL-2 induces blast transformation and mitosis in PEL in the absence of antigen, and the transformed cells are referred to as PEL blasts, or "PEB" (Berke and Rosen, 1988; Figure 5.5). PEB, unlike PEL, exhibited large quantities of lytic granules, perforin and BLT-esterase activity (Figure 5.6). The expression of granzyme B mRNA paralleled the expression of perforin mRNA (Berke et al., 1993). Nevertheless, PEBs are not markedly different in lytic activity from primary PEL. Like PEL, but unlike PEBs, lytic PEL-hybridomas (Kaufmann and Berke 1983) showed no trace of lytic granules or BLT-esterase activity, either before or after activation with specific alloantigenic cells, or mitogens such as Concanavalin A or IL-2. Moreover, no expression of perforin or granzyme mRNA was detected by quantitative PCR analysis of the cytolytic PEL-hybridomas (Helgason et al., 1992; Berke et al 1993).

The vigorous cytotoxic activity of primary PEL led to the proposal that there must be an alternative, contact-mediated cytolytic process not involving the granule exocytosis pathway (Berke 1991). Why CD8 T cells expressing the perforin/granzyme lytic pathway should be selectively excluded from, or fail to develop in, the peritoneal cavity is an intriguing open question. It may be due to lack of provision of critical components of signal 2 required for CTL maturation in the course of the intraperitoneal response against MHC II-deficient tumors used to generate CTL PEL.

3.3 Cytotoxic function in perforin knockout mice

During the early 1990s, disagreement between those who felt perforin was a necessary and sufficient mechanism to explain all cell-mediated cytotoxicity, and those who were convinced there must be alternative lytic mechanisms, continued to grow. After several years of heated debates at meetings and workshops, it became apparent that the only way to resolve definitively the role of perforin in cell-mediated lysis would be to create a mouse lacking perforin. Several groups undertook this task at about the same time, using the technique of targeted gene disruption, and in 1994 five papers appeared reporting the first results of experiments with perforin-less mice (Kägi et al., 1994a, b; Kojima et al., 1994; Lowin et al., 1994; Walsh et al., 1994a).

The first paper, by Kägi et al. (1994a), showed that perforin-less mice infected intracerebrally with the lymphocytic choriomeningitis virus (LCMV) failed to produce CTL capable of destroying LCMV-infected fibroblasts in vitro. Such CTL are always detected in unaltered, LCMV-infected mice. The LCMV infection itself was not cleared in perforin-less mice, and the CTL-mediated cytopathology characteristic of this disease was absent. On the other hand, these authors also tested the ability of perforin-less mice to generate CTL capable of lysing uninfected allogeneic target cells. CTL generated in mixed leukocyte cultures, or in vivo as immune spleen cells or PEL were tested for their ability to lyse specific target cells in vitro. While the perforin-less mice, in comparison with unaltered mice, showed some reduction in cytotoxicity generated against allogeneic cells, there was in fact substantial cytotoxic function remaining in the perforin-less mice. The significance of this remaining cytotoxic function was debated vigorously in various meetings. Some thought the residual cytotoxic activity provided definitive proof for an alternate lytic mechanism, although those presenting the data initially played down this possibility.

Subsequent studies with perforin-less mice, by these and other authors, made it clear that an alternate mechanism does indeed exist, and that it is based on the Fas/Fas ligand system for inducing apoptosis that had recently been described by Rouvier et al., (1993; see below). Walsh et al., (1994a), while reproducing Kägi's findings with the LCMV response in vivo, and the cytolytic response against LCMV-infected fibroblast targets in vitro, went on to show that LCMV-infected mice lacking perforin exhibit vigorous cytolytic activity against non-adherent Fas-positive target cells in vitro. In fact, ^{51}Cr release seen in cytotoxicity assays using CTL from these mice toward Fas-positive cells was indistinguishable from that seen in CTL from mice with intact perforin genes (Clark et al., 1995; Glass et al., 1996). Thus, the failure of CTL from LCMV-infected perforin knockout mice to lyse virus-infected fibroblasts in vitro (Kägi et al., 1994a; Walsh et al., 1994a) is probably due to a lack of Fas expression by the fibroblasts used as target cells, rather than a failure of the mice to produce LCMV-specific CTL as some had initially concluded (Kägi et al., 1994a). And it turned out that primary PEL kill target cells primarily via the Fas pathway (Berke, 1995). Nevertheless, it was clearly demonstrated with the perforin knockout mice that in the exocytotic pathway for CTL killing, perforin is absolutely required for target cell lysis.

The availability of mice lacking perforin presents an interesting opportunity that unfortunately has not yet been exploited. The initial interest in perforin as a possible lytic agent was based in part on the appearance of complement-like ring structures on what were presumed to be target cell membranes after an attack by CTL, combined with the molecular homology between perforin and terminal complement components. It would be most interesting to see whether such complement-like structures are present when target cells are destroyed by CTL lacking perforin.

4 THE FAS/FASL SYSTEM IN CTL-MEDIATED LYSIS

In a key paper published in 1993, Rouvier et al., finally resolved the dilemma posed by Ca^{2+} - and perforin-independent CTL action. They showed that the ability of a CTL hybridoma previously demonstrated to kill target cells in a Ca^{2+}-independent manner correlated entirely with the presence or absence of the Fas protein on the surface of the target cell. The involvement of Fas (APO-1; CD95) in Ca^{2+}-free cytolysis by CTL was rapidly confirmed in a number of other laboratories, using a wide range of effector and target cell combinations including primary PEL, a PEL hybridoma, and CTL obtained from perforin knockout mice (Rouvier et al., 1993; Walsh et al., 1994a; Kägi et al., 1994b; Berke, 1995). Killing in acute cytolytic assays in the absence of perforin was shown to be absolutely dependent on the expression of Fas on the target cell, and of the Fas ligand (FasL) on the effector cell. This mode of killing could be blocked by a soluble form of the normally membrane-associated Fas protein (Fas-Fc), and by non-cytolytic Fas antibodies.

Fas is a member of the TNF family of death receptors and ligands that includes, among others, the receptors for TNFα, LT-α, and -β, CD-27L, CD-30L, CD–40L, and TRAIL (TNF-related apoptosis-inducing ligand). All members of the TNF receptor family described so far are type I membrane proteins, and share a fair degree of intra- and extracellular structural homology. The corresponding ligands, including FasL (Suda et al., 1993), are generally type II membrane proteins, belonging to the TNF family of membrane-bound and soluble protein molecules, and also share structural homology (ca. 25 percent). Many members of these two families are involved in the induction and regulation of apoptosis in a wide range of normal and transformed target cells, as well as other biological functions (Smith et al., 1994; Beutler and van Huffel, 1994; Gruss, 1996; Kitson et al., 1996; Nagata, 1997). When Fas is cross-linked by cell-bound FasL, or by a soluble FasL trimer, it binds via its cytoplasmic region to an adaptor molecule called FADD (Fas-associated death domain), which in turn binds to and activates caspase 8, initiating a cascade of events ending in the induction of apoptosis in the Fas-bearing cell (Nagata and Golstein, 1995; Nagata, 1998). The Fas lytic system employed by CTL and NK cells induces a fairly straightforward apoptotic death in target cells (Krammer, 2002; Wallach et al., 1999; Green , 1996).

All CTL appear capable of FasL expression, although in most cloned CTL its expression is normally not constitutive, requiring activation of the CTL through the TCR or by other means. Expression of FasL on primary CTL is difficult to assess because of heterogeneity of primary populations. The appearance of Fas-mediated killing in cloned CTL after stimulation with antigen or with PMA and ionomycin requires about 45 minutes, and is blocked by inhibitors of protein synthesis (Walsh et al., 1994a; Glass et al., 1996). There has been speculation that FasL may be delivered to the cell surface via the cytotoxic granules of CTL (Bossi and Griffiths, 1999). Once the CTL activating signal is removed, FasL is rapidly down-regulated (Glass et al., 1996). This is different from the perforin system, wherein once induced by primary activation, the cytotoxic granules are quite stable, and lysis by CTL clones is essentially immediate.

Unlike cloned CTL lines, FasL is expressed constitutively (independently of continuous antigen stimulation) in in vivo-primed PEL, which may be more representative of CTL in vivo (Li et al., 1998). Yet even PEL do not kill target cells not recognized through the T-cell receptor unless the target expresses unusually high levels of Fas (e.g. as a result of transfection with the *fas* gene). It may be that preformed FasL is stored intracellularly and translocates to the cell membrane only upon CTL stimulation by a cognate target (Figure 5.8). The recruitment of FasL requires neither new gene expression nor protein synthesis, and it can be inhibited by brefeldin A, which is known to disrupt the Golgi apparatus.

The Fas lytic system co-exists in all non-PEL CTL with the exocytosis/perforin system and their expression is not mutually exclusive. Moreover, non-cytotoxic cells, which express neither granules nor perforin, can be rendered cytotoxic (to Fas-positive target cells) by transfecting them with Fas ligand cDNA (Suda et al., 1993). NK cells (Arase et al., 1995; Oshimi et al., 1996; Zamai et al., 1998) and CD4 CTL (Vergelli et al., 1997) are also capable of using the Fas system to lyse target cells. In CD4 CTL, as in CD8 CTL, activation of cytolytic function with PMA and ionomycin, in particular, results in up-regulation of the Fas ligand, and non-specific lysis of Fas-positive targets (Kojima et al., 1997; Thilenius et al., 1999).

The Fas lytic pathway may account for most of the innocent bystander lysis seen with both the CD8 and CD4 CTL clones in acute lytic assays. Whether activated by specific antigen, lectin or PMA and ionomycin, cloned CTL lines express FasL within an hour, and are vigorously cytolytic toward Fas-positive target cells, regardless of strain or species (Burrows et al., 1993; Ohminami et al., 1999; Thilenius, et al., 1999; see also Chapter 4), in an LFA-1-dependent manner (Kojima et al., 1997). It is not entirely clear why innocent bystander killing is so readily apparent in CTL clones, yet is so rarely seen in primary CTL. For the first twenty years or so of study, killing by primary CTL seemed not only highly target cell-specific, but also highly Ca^{2+}-dependent (which Fas lysis is not). Fas-dependent lysis would also be of concern in vivo. Many tissues, such as the liver, are highly Fas-positive, and the presence of activated CTL in the liver – for example in response to a localized viral infection – could be expected to do considerable collateral damage to normal liver cells. The difference between cloned and primary CTL may lie in the fact that cloned CTL are repeatedly stimulated with antigen and IL-2, and may express unusually high levels of FasL. Moreover, experiments looking at Ca^{2+}-independent killing by CTL clones generally utilized target cells manipulated or selected for over-expression of Fas. The high degree of Ca^{2+}-dependent target specificity seen with primary CTL populations in vitro, and the highly selective action of CTL in vivo, may also reflect the phenomenon suggested by Li et al. (1998), that under normal conditions of Fas and FasL expression, Fas-mediated killing occurs only upon CTL-target interaction through the T-cell receptor.

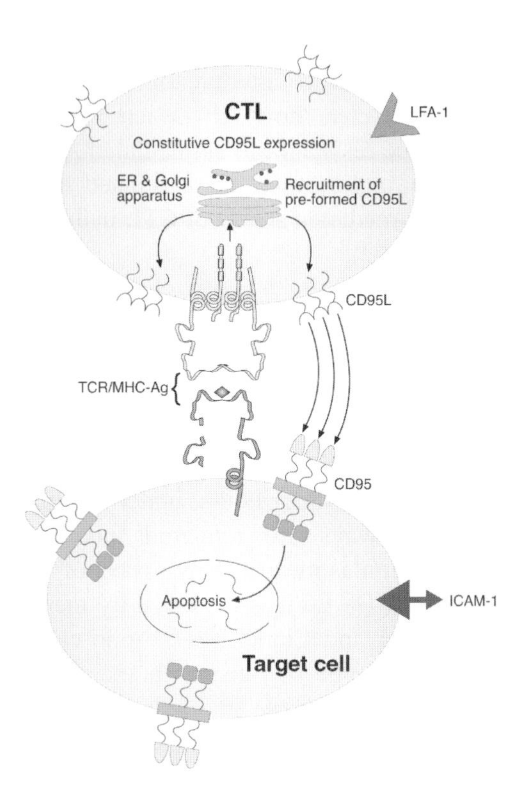

Figure 5-8. A model for CD95L expression and function in CTL conjugated to target cell. CD95L, either present in or recruited from the ER/Golgi complex to the CTL/target contact site under the influence of the TCR, triggers apoptotic death through CD95 receptors on the targets. No new CD95L gene expression is required.

The precise role of the Fas pathway in host defense is difficult to discern. It appears to play a small role in protection against viral infection, for example (Chapter 8), which is clearly a major function for the perforin lytic mechanism (Russell and Ley, 2002). On the other hand, there is abundant evidence that Fas plays a major role in lymphocyte homeostasis. The *lpr* mutation in MRL mice virtually abolishes Fas production (Watanabe-Fukunaga et al., 1992), and the *gld* mutation in the same strain produces a non-functional Fas ligand (Takahashi et al., 1994). Both mutant strains develop lymphadenopathy and splenomegaly early in life, which arises because they fail to clear (presumably through apoptosis) activated T cells at the end of activation cycles. Some T cells activated after exposure to antigen develop into memory cells, but the vast majority are destroyed. However, mice without a functioning Fas system fail to do so (Russell et al., 1993), and T cells accumulate in lymphoid tissue. This phenomenon of T-cell clearance after antigen activation, called activation-induced cell death

(AICD), has important implications for both immune homeostasis and immunological tolerance, and will be discussed in detail in Chapter 11. However, although AICD is clearly involved in down-regulating CD4 helper T-cell responses, it is not clear how AICD acts on CTL. In fact there is evidence that CTL may be resistant to Fas-mediated AICD (Ehl et al., 1996; Suzuki and Fink, 2000), and are possibly regulated by TNF (Zheng et al., 1995). Nevertheless, Fas-mediated AICD of CTL in a tumor setting has been demonstrated (Li et al., 2002).

5 THE ROLE OF TNFα IN CTL-MEDIATED LYSIS

Although it is generally accepted that the exocytosis and Fas lytic pathways account for all of the cytotoxicity seen in acute lytic assays, e.g., lysis of target cells seen in two-four hours, under conditions where both the exocytosis and Fas pathways are rendered inoperative, there is still significant target destruction as measured by assays of 16-24 hours or more. This slower cytolysis by CD8 CTL is mediated by membrane-bound TNFα (Perez et al., 1990; Ratner and Clark, 1993; Ando et al., 1997; Liu et al., 1999). The TNF family of type II transmembrane proteins induces a wide range of physiological responses in cells, including differentiation, proliferation and cell death via apoptosis (Beutler and van Huffel, 1994; Smyth and Johnstone, 2000). Interestingly, they can induce these responses while still embedded in the membrane of the cell that produced them, or as cleaved soluble factors, usually trimeric, which can act at some distance from the producing cell (Banner et al., 1993; Armitage, 1994). It is the 26 kD membrane-associated form of TNFα, rather than the 17 kD secreted form, which mediates slow cytolysis of target cells (Ratner and Clark, 1993). CD8 cells can also mediate slow lysis of bystander cells not recognized as cognate targets, although this is probably via soluble rather than membrane-bound TNF (Smyth and Sedgwick, 1998).

There is no evidence for the involvement of membrane TNFα in acute lytic assays (Jongneel, 1988; Ratner and Clark 1993). TNFα is also involved in slow target-cell lysis by CD4 T cells (Liu et al., 1992; Yasukawa et al., 1993) and macrophages (Monastra et al., 1996). The lytic pathway induced by TNFα is similar to that induced by lymphotoxin (TNFβ), and both share features in common with the Fas lytic pathway. The endpoint is once again apoptotic death of the target (Smyth and Johnstone, 2000).

The action of TNF ligands on their target cells – growth and differentiation vs. cell death – depends on the nature of the TNF receptor displayed on the target and on the associated cytoplasmic signal transduction molecules. TNFα binds to two distinct TNF receptors. TNFR1 is a 60-kD, type I transmembrane protein found mostly on epithelial cells; TNFR2 is a 120-kD transmembrane protein found predominantly on lymphoid cells. However, both receptors are found in both tissues, and in fact a single cell can display both types of receptor. TNFR1 can be activated by both soluble and membrane-bound TNFα, whereas TNFR2 receptors bind almost exclusively membrane-associated ligand (Grell et al., 1995).

The cytoplasmic tails of both TNF receptors have now been completely characterized with respect to their signaling motifs and interactions with other signal transduction molecules. One of the earliest of such motifs, and the one that concerns us most here is the so-called "death domain" (DD), the presence of which is absolutely required for the initiation of the apoptosis cascade in the targeted cell. DDs are found on the TNFR1 receptor itself, but not on the TNFR2 receptor (Armitage, 1994; Huang et al., 1996; Wallach et al., 1999). Like Fas-dependent acute CTL killing, cytolysis via either membrane or secreted TNFα is Ca^{2+}-independent. Expression of membrane TNFα and the release of secreted TNFα by CD8 T cells is enhanced several fold when the CD8 cell binds its cognate target (Ratner and Clark, 1993).

6 EXTRACELLULAR ATP AND CYTOTOXICITY

The plasma membrane-perturbing effects of extracellular ATP have been known for many years. Among other things, ATP can stimulate mitogenesis in thymocytes (Gregory and Kern, 1978) and cause permeability and pore formation in blood cells (Cockcroft and Gomperts, 1979; Steinberg et al., 1987; Buisman et al., 1988). Several facts about extracellular ATP attracted the attention of those studying the mechanisms by which CTL lyse target cells. When CTL are stimulated with TCR antibodies, they release extracellular ATP (Filippini et al., 1990a); extracellular ATP is lytic toward most commonly studied target cells (DiVirgilio et al., 1989, 1990). Moreover CTL are themselves resistant to the lytic effects of extracellular ATP, by virtue of a cell-surface ectoATPase activity (Filippini et al., 1990b). Interestingly, extracellular ATP induces DNA fragmentation in nucleated target cells, but not in CTL (Zanovello et al., 1990; Zheng et al., 1991). Concentrations in the range of 0.5-8.0 mM cause target cell lysis equal to that seen with CTL in ^{51}Cr release assays. One of the things originally driving interest in ATP as a lytic mechanism was the persistent report of Ca^{2+}-independent lysis by CTL, which seemed to obviate exocytosis as a mechanism of lysis. Lysis of most targets by ATP was found to be Ca^{2+}-independent (Filippini et al., 1990b; Zanovello et al., 1990). At the time of initial interest in ATP, the Fas lytic pathway had not yet been described.

While intriguing, the cytolytic effects of extracellular ATP are not always consistent with what might be expected from a CTL lytic mechanism. For example, target cells that are sensitive to CTL lysis are often resistant to ATP (Zanovello et al., 1990). In addition, when target cells were grown in the presence of increasing amounts of ATP to develop highly ATP-resistant sublines, such sublines were as sensitive as the parental EL4 line to lysis by CTL (Avery et al., 1992). Moreover, as noted earlier, CTL from mice lacking both perforin and Fas show no cytolytic activity in vitro, raising serious questions about the role of ATP in cytolysis. While some interest continued to be shown in a role for extracellular ATP in cytolysis (Blanchard et al., 1995; Dombrowski et al., 1995; Macino et al., 1996; Redegeld et al., 1997), attention was directed more toward possible roles in cell-cell interaction, or modulation of apoptotic pathways within target cells.

7 SUMMARY

In 1980 we had no idea how killer cells killed their targets; today we have two mechanisms to explain acute cytolysis (Figure 5.9), and another to explain so-called "slow lysis" (TNF-related molecules). Prior to 1980, everyone was looking for a gun, or a rope, or maybe a knife or poison. The perforin exocytosis mechanism looked initially like a smoking gun, but even by then it was clear that killer cells did not commit homicide against, but rather induced suicide in, the cells they did away with. All three mechanisms induce target cells to undergo apoptosis.

The suicide pathway(s) utilized in Fas-mediated and TNF killing are understood in considerable detail, thanks to the enormous energy invested in unraveling apoptosis generally in the past decade or so. The perforin exocytosis pathway must tap into this pathway at some level, but exactly where is still not understood, and presents a considerable challenge for future research.

The challenge now is to understand how killer cells operate in vivo – do they use the incredibly potent cytolytic mechanisms we have been studying for the past forty years or so, or are these simply in vitro artifacts, irrelevant holdovers from a previous epoch in our immunological history? We will address these questions in Chapters 7 through 11. But first, we will examine the second important category of killer cells – the (presumed) evolutionarily older natural killer (NK) cells. The path to understanding how they function in host defense has been no less tortuous – and no less exciting – than the unraveling of T killer cells.

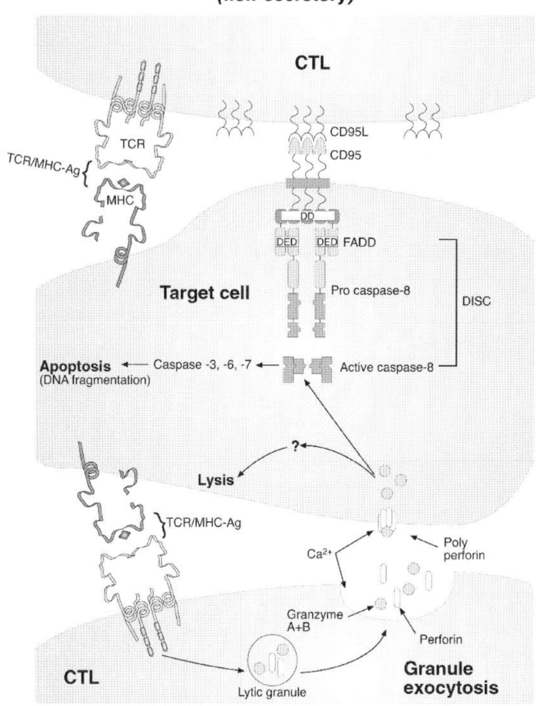

Figure 5-9. The two pathways of CTL-mediated lysis. A: Fas/FasL pathway of CTL action. Binding of the death-inducing Fas ligand (CD95L) of CTL to the target's death receptor Fas (CD95) leads to the formation of the death-inducing signaling complex (DISC). The initiator procaspase-8, which is recruited by FADD (Fas-associated death domain protein) to the DISC where it is activated by autocatalytic cleavage, ultimately leading to apoptotic cell death. B: perforin/granzyme pathway of CTL action. Upon CTL–target binding, there is a directed exocytosis of CTL granules into the intercellular space (see also legend to Fig 5.4).

Chapter 6

INNATE CELL-MEDIATED IMMUNITY

In trying to imagine a biological role for killer lymphocytes, aside from the obviously irrelevant one of graft rejection, many early researchers focused on a possible role in tumor surveillance and control. Tumors were usually found to be infiltrated with leukocytes, including lymphocytes, and the collective wisdom, supported in a few but admittedly not many cases by hard data, was that the degree of infiltration correlated inversely with progression of the tumor. This is a point of view still held (Eerola et al., 2000). When cell-mediated cytotoxicity in vitro was demonstrated in the 1960s, it seemed reasonable to ask whether animals mounted this type of immune response to their own tumors as well as to tissue grafts. Indeed, lymphocytes from tumor-bearing patients and experimental animals often displayed significant cell-mediated cytotoxicity toward their tumor cells *in vitro*. A great deal of effort was expended to show that cell-mediated tumor cytotoxicity, like cytotoxicity against allografts, was specific – in this case toward putative tumor antigens, rather than toward the MHC antigens controlling transplant rejection (Hellström et al., 1968; Bubenik et al., 1971; Baldwin et al., 1973; Hellström and Hellström, 1974).

However, it soon became apparent that cell-mediated cytotoxicity toward different tumors by killer lymphocytes was considerably less discriminating than allograft immunity. Moreover, as assays for cytotoxicity became more precise, an embarrassing fact emerged. As controls for their experiments, many researchers used lymphocytes from non-tumor-bearing animals. To their dismay, lymphocytes from non-tumor-bearing individuals often displayed as much or more cytotoxicity toward tumor targets as did lymphocytes from those bearing tumors. While at first dismissed as "background noise," or artifacts of the assays used, by the early seventies such findings could no longer be ignored (Hellström et al., 1970; Rosenberg et al., 1972). At a workshop held in 1972 to examine the various assays used to measure so-called tumor-specific cytotoxicity, the issue of anti-tumor cytotoxicity in control subjects came very much to the fore (see post-presentation discussions in National Cancer Institute Monograph 37, 1973.) A monumental study reported by Takasugi et al. in the same year (Takasugi et al., 1973), looking at anti-tumor cytotoxicity in 995 cancer patients and 1099 non-tumor-bearing controls, showed a wide range of anti-tumor cytotoxicity in both groups, with no obvious statistical difference between them. The implication was that many labs must have consciously or unconsciously selected lymphocyte sources with low anti-tumor cytotoxicity for controls in their experiments, assuming that, by definition, someone never exposed to a given tumor could not possibly be immune to it, and if they seemed to be, it must be an artifact.

Needless to say, these findings were not enthusiastically received by those searching for a role for "classical" – highly antigen-specific - killer cells in tumor immunity. But where some saw only disaster for their field, others saw an opportunity to discover something new, and it was quickly established that normal healthy individuals do indeed have a subpopulation of lymphocytes that, without any prior sensitization, will recognize, attack and destroy at least some tumor target cells in vitro (Herberman et al., 1975a,b; Kiessling et al., 1975a,b). Interference with these cells in vivo enhances growth of transplanted tumors in experimental animals (Habu et al., 1981). These effector cells became known as natural killer (NK) cells. They were assigned to that component of immunity referred to as innate immunity: the collection of infectious disease-resistant mechanisms that are genetically imprinted in each organism, and are fully functional independent of contact with environmental antigen.

It was established early on that NK cells also play a role in controlling viral infections. This can be inferred from the marked increase in viral pathogenesis in humans and experimental animals with impaired NK responses (Shellam et al., 1981; Bukowski et al., 1983; Biron et al., 1988). Some of the most effective target cells for assaying NK CMC were virally infected (Kiessling et al., 1975a). In healthy mice, the innate NK response to potential pathogens peaks several days earlier than the adaptive T-cell response (Welsh, 1978). Viral infection invariably results in an increase both in the absolute numbers and the activation state of NK cells, which although accumulating preferentially in infected organs also circulate throughout the body (Trinchieri, 1989).

It could well be argued that, as in the case of CTL, defense against viruses, rather than tumors, is the primary function of NK cells, but the exact role played by NK cells in viral infection is complex. In only a few cases can it be established that NK cells themselves play a key role in overcoming an infection (e.g., Bukowski et al., 1985). This could be through direct cytotoxic attack by NK cells against virally infected cells (Harfast et al., 1975; Santoli et al., 1978; Yasukawa and Zarling, 1983; Kurane et al., 1986; Tilden et al., 1986). Perhaps equally important, NK cells produce cytokines that interfere with viral replication, and help activate professional APC which in turn activate T cells (Zitvogel et al., 2002). NK cells may themselves provide important cell-cell costimulatory signals to CTL (Kos and Engelman, 1995, 1996), which develop several days after NK cells begin their attack. Beyond doubt, however, cytotoxic CD8 T cells are a much more potent and direct element in the ultimate clearance of viral infections than are NK cells (Chapter 9).

To make matters even more confusing, NK cells were also shown to kill normal human cells, such as fetal fibroblasts (Timonen and Saksela, 1977, 1978; Timonen et al., 1979a). They were also shown to play a role in rejection of bone marrow transplants (Miller, 1984; Bryson and Flanagan, 2000) as well as in the regulation of hematopoiesis (Trinchieri, 1989), in which cytotoxicity plays no role.

Minato et al (1981) recognized this complexity of NK cell function early on. They described at least four distinct types of NK activity; although their initial views have been modified somewhat, there are in fact at least four distinct NK activities that we will examine in this chapter: "classical" NK cells; natural cytotoxic (NC) cells; lymphokine-activated killer (LAK) cells; and natural killer cells with T-cell characteristics (NKT cells). It has now been proposed that NK cells even play a role in managing the fetal-maternal immune relationship (Moffett-King, 2002). NK cell biology is far more complex than its discoverers could possibly have imagined.

1 CELLULAR PROPERTIES OF NK CELLS

Like all blood cells, NK cells ultimately derive from CD34-positive hematopoeitic stem cells found in bone marrow, but their exact lineage and ontogeny remains to this day a source of controversy. Although their ultimate bone marrow origin has been clear for many years (Haller and Wigzell, 1977), mature NK cells are at best a small fraction of mature marrow cells. They are also a tiny fraction of thymus cells, and may be there simply as passengers in blood vessels. However, pluripotent cells found in at least the fetal thymus are able to give rise to NK cells in vitro (Carlyle and Zuñiga-Pflücker, 1998; Sato et al., 1999). The significance of this in vivo is unclear. Thymocytes treated with IL-2 develop wide-ranging cytotoxic activity toward a variety of NK targets (Ballas and Rasmussen, 1987; Ramsdell and Golub, 1987), but this is probably due to LAK activity (see below) rather than classical NK activity. In humans, NK cells appear in the fetal liver prior to development of a thymus (Sanchez et al., 1993). NK cell activity is completely normal in athymic nude mice (Herberman et al., 1975b; Kiesling et al., 1975b), which lack T cells with α/β receptors, and in mice with disrupted RAG-1 and –2 genes, the products of which are involved in TCR gene rearrangements (Reynolds et al., 1985; Shinkai et al., 1992; Mombaerts et al., 1992).

NK cells are found in spleen, where they may be several percent of total lymphocytes, and in the liver, but not in lymph nodes or thoracic duct (Perussia et al., 1983). In humans, they may be up to 15 percent of lymphocytes circulating in peripheral blood. NK cells can be found in most tissues of the body, including liver and intestinal epithelia, and are present in peritoneal exudate lymphocytes (Trinchieri, 1989). Herberman showed early on (Herberman et al., 1973) that mouse NK cells lacked the θ antigen, and were thus distinguishable from killer cells responsible for allograft rejection. Subsequent findings further supported a distinction between CTL and NK cells; the latter are not MHC-restricted in their expression of cytotoxicity, and do not express the TCR-associated cell surface CD3 complex, although they do express the CD3 ζ chain (see below). Although traditionally characterized as CD4- and CD8-negative, a small fraction (probably less than five percent) of true NK cells in humans display the CD4 marker, and can in fact be infected by HIV (Valentin et al., 2002).

That NK cells are not related to B cells was shown by their lack of surface Ig (Peter et al., 1975; Jondal and Pross, 1975), and by experiments in which spleen cell populations expressing NK activity were depleted of B cells – the NK activity of the remaining cells was actually increased (Herberman et al., 1975b; Kiessling et al 1975b). Like CTL, NK cells are nonadherent to plastic and nylon wool, and are relatively nonphagocytic, making a relationship to macrophages unlikely (Kiessling et al., 1975b; Herberman et al., 1975b; Zarling et al., 1975; Sendo et al., 1975). The cells remaining after extensive depletion of both T and B cells, as well as adherent cells, were found to be highly lytic in NK cell assays. They contained greater than 90 percent lymphocytes (Kiessling et al., 1975b), suggesting (but not yet proving) that NK cells were a previously undetected type of lymphocyte.

NK cells express a number of markers useful in their identification (Table 6.1) Few (if any) of these are absolutely NK cell-specific, but taken together and in various combinations they allow reasonable identification of NK cells and NK cell subsets. The nomenclature of these markers can be confusing, mostly because names kept changing as new insights were gained into the roles the corresponding proteins play in NK cell function. CD56 (formerly known as Leu-19 or NKH-1) and CD57 (Leu 7; HNK-1) are commonly used to define NK cells in humans (Lanier et al., 1986; Nitta et al., 1989). CD56 is highly similar to the neural adhesion molecule NCAM, and most likely plays a role in NK cell adhesion to target cells or to vascular endothelium (Lanier et al., 1989). The C-type lectin NKR-P1C (NK-1.1) is strongly expressed in some, but not all, strains of mice (Glimcher et al., 1977; Koo et al., 1984; Ryan et al., 1992). The human homolog of NKR-P1C is CD161, a C-type lectin involved in activation of NK function (see below). There is no counterpart of CD56 in mice. Ly49A is a murine inhibitory receptor. CD16 (Leu-11), associated with the low-affinity Fc receptor for IgG (FcγRIII) also expressed on granulocytes and monocytes (Perussia et al., 1984), is shared by both mice and humans, as is the neutral glycosphingolipid asialo GM_1 (Kasai et al., 1980; Young et al., 1980).

Table 6-1. Some common markers used to distinguish NK cells in mice and humans

Marker	Species	Characteristic	Function	Alternate name(s)
CD16	Both	Fc receptor	ADCC	Leu11; FCγRIII
Asialo Gm1	Both	Glycosphingolipid	-	-
MKR-P1C	Mouse	C-type lectin	Activation	NK1.1
LY49A	Mouse	C-type lectin	Inhibition	-
CD56	Human	Adhesion molecule	Conjugation ?	LEU 19; NKH-1
CD161	Human	C-type lectin	Activation ?	NKR-P1

In humans, resting NK cells can be conveniently divided into two subsets, defined by the relative densities of CD56 and CD16. Approximately 90 percent of resting NK cells are $CD56^{dim}$, $CD16^{bright}$ by flow cytometric analysis; the remainder are $CD56^{bright}$, $CD16^{dim\ or\ null}$. The $CD56^{dim}$ subset is more cytotoxic (Nagler et al., 1989; Cooper et al., 2001), containing more perforin and displaying a greater tendency to form conjugates with NK targets. These subsets have unique patterns of expression of monokine, lymphokine and chemokine receptors, as well as cytokine production (Table 6.2). In particular, the $CD56^{bright}$ subset constitutively expresses the high affinity IL-2 receptor (IL-2Rαβγ) and can be driven to proliferate by picomolar concentrations of IL-2, and by nanomolar concentrations of IL-15, which can also interact with the IL-2 receptor (Baume et al., 1992; Carson et al., 1994). Resting $CD56^{dim}$ NK cells increase their expression of CD56 after activation by various means, so caution must be exercised when assigning NK cells to $CD56^{dim/bright}$ subsets (Robertson et al., 1990; Caligiuri et al., 1993; Robertson et al., 1999). There is suggestive evidence that $CD56^{bright}$ NK cells may be developmental precursors to CD56dim cells, but this is still an open question (see Cooper et al., 2001 for discussion of this issue).

Timonen and coworkers were the first to make an association between NK activity and a low-density subset of lymphocytes called large granular lymphocytes (LGL) (Timonen et al., 1979a,b; 1981, 1982.) LGL represent between five and fifteen percent of circulating lymphocytes, depending on the species. Their large, non-adherent, low-density nature allows them to be enriched substantially by removing adherent cells on nylon wool, followed by discontinuous density centrifugation. NK activity against selected tumors correlates directly with the proportion of LGL in an effector population, and the cells that bind to NK-sensitive targets are by morphological criteria almost entirely LGL. Both spontaneous and interferon-enhanced lysis by NK cells can be attributed to LGL. Adsorption of purified LGL populations on plates coated with antigen-antibody complexes (which interact with Fc receptors) can deplete essentially all NK cytotoxicity.

Table 6-2. Phenotypes of Human CD56/CD16NK Cell (see also Cooper et al., 2000; Robertson, 2000).

Characteristic	CD56dim/CD16bright	CD56bright/CD16dim
Frequency (resting population)	90%	10%
Cytotoxicity (NKCC and ADCC)	High	Low
Monokine receptors	+	++
IL-1	+	++
IL-15	+	++
IL-18		
Lymphokine receptors		
IL-2	+ (low affinity)	++ (high affinity)
IL-4	+	+
IL-7	+	+
Chemokine receptors		
CCR7	+/-	++
CX_3CRI	+++	+/-
CXCR1	+++	+/-
CXCR2	++	-
CXCR3	+/-	+++
Other cytokine receptors		
IFN-α, β, γ	+ (?)	+
TGF-β	+ (?)	+
TNF-α	+	+
Cytokines produced		
IFN-γ	+/-	++++
TNF-α	+/-	++
TNF-β	+/-	+++
IL-10	+/-	+++
IL-13	+/-	+++
GM-CSF	+/-	+++

But not all NK cells are LGL. In humans there is a small population of agranular $CD3^-$, $CD56^{dim}$, $CD16^{bright}$ NK cells (Inverardi et al., 1991; Ortaldo et al., 1992a; J. Ortaldo, personal communication). They have the same target specificity range as LGL, release IFNγ upon stimulation, and are equally cytolytic in ^{51}Cr release assays. These cells contain perforin and granzymes, but spread diffusely throughout the cytoplasm rather than stored in granules. Perforin in these cells may possibly be membrane-associated (Ortaldo et al., 1995.) Upon incubation with IL-2, they enlarge, develop perforin-containing granules, and become indistinguishable from LGL. They may represent recently generated NK cells that have not yet encountered an activating signal.

It should also be pointed out that not all LGL exhibit NK activity. LGL morphology is a common characteristic of activated T lymphocytes as well, especially under the influence of IL-2 and other cytokines, rather than a distinct feature of NK cells.

Murine (Palyi et al., 1980; Dennert et al., 1981; Kedar et al., 1982) and human (Kornbluth et al., 1982; Hercend et al., 1982) NK cell clones were developed in the early 1980s, and have provided a vital source of cells for studying NK function. One important question such lines helped answer was the target cell range of individual NK cells. While the target range of whole NK populations was of course quite broad, this could well reflect a collection of NK subpopulations with more restricted or even unique target specificities. Dennert, however, found that individual NK clones displayed essentially the same broad target specificity as the populations from which they were derived (Dennert et al., 1981).

2 THE ROLE OF CYTOKINES IN NK CELL FUNCTION

In addition to being able to engage in cell-mediated cytotoxicity, NK cells interact with each other and with other cells of the immune system through a range of monokines, lymphokines, and chemokines (Cooper et al., 2001; Robertson, 2002). Dendritic cells, monocytes and macrophages, especially those that have interacted with pathogenic or stress-related stimuli, play an important role in NK activation, providing a range of critical stimulatory factors including IL-1, IL-10, IL-12, IL-15, and IL-18. IL-15 is particularly key for NK cell development and function (Dunne et al., 2001). It is also produced in the bone marrow, where it is critical for maturation of NK cells from CD34 precursors (Mrozek et al., 1996; Kennedy et al., 2000). IL-15 together with the lymphokine IL-2 stimulate both proliferation and cytotoxicity in NK cells, although by themselves do not alter cytokine production by NK cells.

NK cells are also strongly affected by chemokines (Campbell et al., 2001; Robertson, 2002), which regulate NK cell migration (Allarena et al., 1996; Campbell et al., 2001; Inngjerdingen et al., 2001) as well as cytotoxicity (Taub et al., 1995, 1996; Maghazachi et al., 1996; Yoneda et al., 2000). NK cells also produce a number of chemokines that interact with G-protein-linked serpentine receptors (Somersalo et al., 1994; Nieto et al., 1998; Oliva et al., 1998; Fehniger et al., 1998).

All three interferons (IFNα, -β and -γ), play a role in NK cell activation, although the role for IFNγ is probably relatively minor (Perussia et al., 1980; Lucero et al., 1981; Faltynek et al., 1986). IFNα and -β, for which NK cells have high-affinity receptors, are important proinflammatory cytokines, providing one of the earliest signals coming from virus-infected cells, for example. Interferons increase NK activity by several mechanisms, including increased frequency in a fixed population, accelerated kinetics of lysis, and increased recycling (Trinchieri, 1989), probably through up-regulation of the NK activating receptors discussed below. Interestingly, interferons acting on potential NK target cells often render them resistant to NK killing (Trinchieri et al., 1978, 1981; Welsh et al., 1981). This latter effect is on killing per se rather than recognition or binding by NK cells, and involves only direct killing by NK cells, and not ADCC (antibody-dependent cell-mediated cytotoxicity; see below). IFNγ treatment of target cells results in their inability to trigger Ca^{2+} influx into the NK cell, and to induce reorientation of the NK cell MTOC (Gronberg et al., 1988).

Transforming growth factor beta (TGF-β) has a marked inhibitory effect on NK cell functions. It reduces NK responsiveness to IFNα, resulting in reduced cytotoxicity and cytokine production (Rook et al., 1986; Bellone et al., 1995). TGF-β is produced by activated CTL, and may be used by them to down-regulate NK responses that occur early in viral infections, thus avoiding possible autoimmune damage from chronically activated NK cells (Kos and Engelman, 1996). LGL have a large number of high affinity receptors for TGF-β. TGF-β appears to exert its effect through alteration of IL-2 signaling pathways in NK cells (Ortaldo et al., 1991).

NK cells also release a number of cytokines, the most potent among them being gamma interferon (IFNγ), tumor necrosis factor-β, IL-10, IL-13, and granulocyte-macrophage colony-stimulating factor (GM-CSF) (Peters et al., 1986; Young and Ortaldo, 1987; Cuturi et al., 1987, 1989). These cytokines all play roles in the regulation of responsiveness by other elements of both the innate and adaptive immune systems. IFNγ stimulates both macrophages and T cells, and up-regulates class I and II MHC antigens on surrounding cells, making them more attractive targets for T cells (while simultaneously making them less attractive to NK cells – see below). Early cytokine studies were confused by the contamination of supposed NK populations with many other cell types, making the origin of some activation-associated cytokines uncertain. The production of TNFα by NK cells, for example, was long disputed but is now certain. Cytokine production is carried out predominantly by the CD56^{bright} subset, which as noted above is minimally cytotoxic. This is true whether NK cells are stimulated with cognate cells, LPS-stimulated macrophages (to supply NK-stimulatory monokines), or PMA and ionomycin (Cooper at al., 2001). Thus the role of the CD56^{bright} subset may be in regulation of overall NK biology and interaction with other cell types, rather than cytotoxicity.

3 NK CELL RECEPTORS AND REGULATION OF NK CELL FUNCTION

3.1 Inhibitory receptors

A key question concerning NK cells, from the very beginning, was how they recognize potential target cells. What possible structures could identify targets as diverse in nature as tumor cells, virally infected cells, and allogeneic bone marrow cells? T and B lymphocytes deal with an enormously diverse antigenic universe, and they meet this challenge by randomly assembling a large number of different receptors through recombination of moderately sized pools of gene fragments. The genetic rearrangements involved in assembling these receptors are easy to detect, but no such arrangements of these or any other genes have ever been detected in NK cells. So where would receptor diversity come from? What kind of receptor do NK cells use to recognize and bind to potential target cells, and what sorts of activating events, if any, follow from NK receptor engagement? The problem of target cell recognition plagued immunologists for nearly two decades after NK cells were first described, and began to be solved only when researchers studied the opposite problem: under what conditions do NK cells *fail* to kill target cells they would be expected to recognize?

Several labs had observed that many NK-sensitive tumor cells and virally infected cells displayed reduced levels of MHC molecules. In vivo, tumor cells with reduced expression of class I MHC were found to be less malignant than tumor cells with normal MHC expression (Karre et al., 1986). NK killing of target cells in vitro also seemed to be inversely correlated with target-cell surface expression of class I MHC molecules.

This suspicion was confirmed when it was shown that NK cells kill certain MHC class I-negative tumor cell lines, but kill poorly or fail to kill the same cell lines transfected with class I genes (Storkus et al 1987, 1989; Shimizu and DeMars, 1989.) It was later shown that many viruses cause down-regulation of class I molecules on infected cells, most likely as part of a strategy to by-pass presentation of viral peptides to CTL (Tortorella et al., 2000).

These observations raised the possibility that a key factor in the activation state of NK cells might be not so much the presence of antigens signaling that a particular cell was cancerous or otherwise compromised, but rather the lack of a normal self antigen – MHC class I – that suppresses NK activity. This gave rise to the concept of "missing self" to explain sensitivity to NK lysis (Ljunggren and Karre, 1990). If an NK cell encounters a potential activating signal on a cell, but that cell is expressing normal levels of class I protein on its surface, the NK cells will in most cases not become fully activated. But if class I is reduced or missing, the NK cell will rapidly gear up to produce cytokines and express its inherent lytic potential.

Implicit in the view that target cell class I MHC delivers an inhibitory signal to NK cells is the existence of MHC class I receptors on NK cells that moderate suppression of their own activation. Further exploration of this phenomenon led to the discovery of two distinct systems of NK cell inhibitory molecules, in both mice and humans: the C-type lectins, and the immunoglobulin superfamily inhibitory receptors (Figures 6.1, 6.2; Long, 1999; Ravetch and Lanier, 2000; Vilches and Parham, 2002). The first such receptor to be described was Ly49A in mice, a member of the small family of homodimeric C-type lectins in mice that interact with supertypic epitopes (shared by all alleles) of the $\alpha 1$ and $\alpha 2$ domains of mouse class I molecules (Karlhoffer et al., 1992; Brennan et al., 1994).

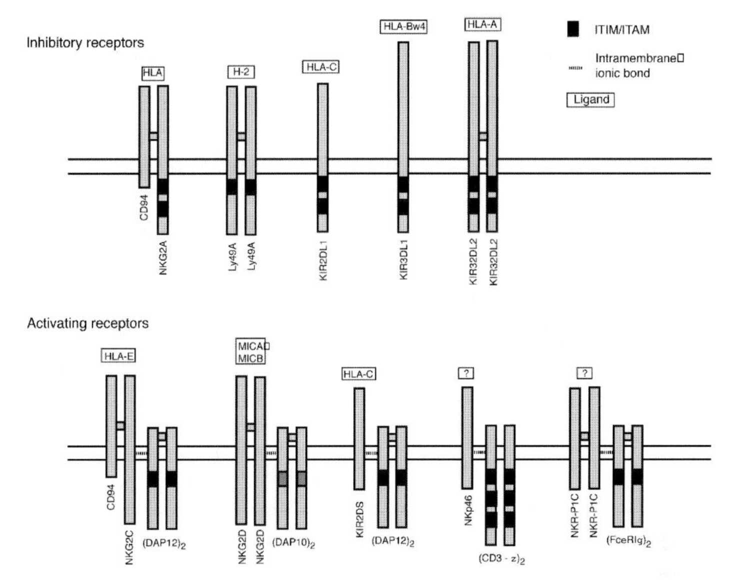

Figure 6-1. Some inhibitory and activating NK cell receptors. The inhibitory receptors all have one or more internal ITIMs. For most KIRs, it is not known whether they function as monomers or dimers. In the case of KIR3DL2, there is evidence that it interacts as a dimer with its ligand (Pende et al, 1996). Activating receptors have one or more ITAMs. In the case of DAP10, the ITAM sequence is slightly modified.

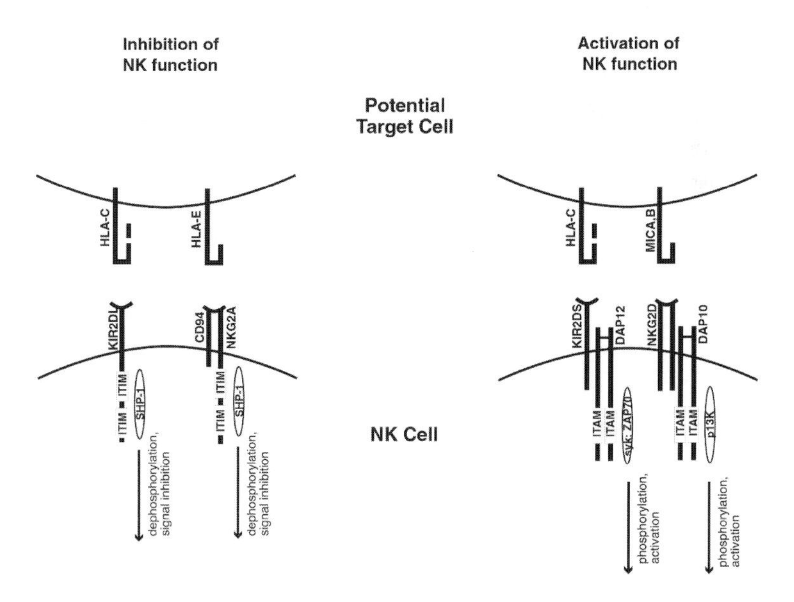

Figure 6-2. interaction of representative NK receptors with their corresponding ligands.

Despite their lectin-like nature, the Ly49 molecules bind to protein portions of class I rather than carbohydrates, and these protein regions are distinct from those recognized by T-cell receptors (Table 6.3a). Ly49 binding is thus not affected by antigenic peptide bound to class I (Natarajan, et al., 1999; Tormo et al., 1999). The repertoire of Ly49 proteins is somewhat larger than the underlying gene family due to alternative RNA splicing, and the various end-products are differentially distributed on overlapping sets of NK cells. Seven of the nine members of this family inhibit NK function when engaged with their respective class I ligands, but two of them, as we will see shortly, are involved in NK activation. Each of the genes within this family may also be multi-allelic within the species (Mehta et al., 2001).

Table 6-3a. Some common murine and human NK inhibiting receptors and their ligands.

C-type lectins	Exemplary ligands
Mouse	
Ly49A	H-2Dd
Ly49C	?
Ly49G	?
Human	
KIR2DL	HLA-C (with any peptide)
KIR3DL	HLA-A, B (with any peptide)
CD94/NKG2A dimers	
Mouse	Qa-1 (classical I α- and β-region peptides)
Human	HLA-E (classical I α- and β-region peptides)

The functionally corresponding but structurally distinct family in humans, KIR (killer-cell inhibitory receptors), is based on a dozen or so genes encoding receptors recognizing supertypic α1 and α2 epitopes of HLA classical class I molecules (HLA-A, -B and -C). KIR gene products are not C-type lectins; they are members of the immunoglobulin superfamily. But like Ly49 receptors, inhibitory KIRs prevent activation of human NK cells upon binding to class I molecules expressed on potential target cells, inhibiting both cytotoxicity and cytokine release. The KIR family also has members involved in activation rather than suppression.

Inhibiting members of the Ly49 and KIR families contain one or more copies of the six amino acid immunoreceptor tyrosine-based inhibitory motif (ITIM) sequence [(Ile/Val/Leu/Ser)-X-Tyr-X-X- (Leu/Val), where X is any amino acid] in their intracellular domains (Burshtyn et al., 1997). Clustering of both Ly49 and KIR receptors results in phosphorylation of the ITIM tyrosine by src kinases. This provides a site for docking of phosphatases containing a src homology domain, such as SHP-1 or SHIP. The phosphatases then engage in dephosphorylation of proteins phosphorylated as a result of NK cell activation, via pathways we will discuss shortly.

The second functional type of inhibitory receptor, present in both mice and humans, is the disulfide-linked heterodimeric CD94/NKG2A receptor (Lazetic et al., 1996; Brooks et al., 1997; Vance et al., 1998). This receptor is present on about half of NK cells; whether this represents a discrete stage in differentiation, or a stochastic distribution, is unclear. What distinguishes these receptors from those just discussed is that CD94/NKG2A interacts only with the $\alpha 1$ and $\alpha 2$ regions of the non-classical HLA-E and HLA-G molecules in humans, and Qa-1 in mice (Pende et al., 1997; Braud et al., 1998), rather than classical class I molecules. HLA-E and Qa-1 molecules are involved in the presentation only of peptides derived from the leader sequences of classical class I MHC proteins. This in effect provides a parallel system for monitoring the presence of classical class I molecules on cell surfaces. CD94 has a very short intracellular tail, but the longer NKG2A tail contains two ITIM sequences. When the tyrosines in these sequences are phosphorylated, NKG2A too can recruit SHP-1 or SHIP phosphatases, and interrupt NK cell activation. NKG2A is one of several NK receptors that appear to play an important role in regulating CD8 CTL function as well (Coles et al., 2000; Braud et al., 2003).

It has become apparent that multiple members of each class of inhibitory receptor can be expressed on the same NK cell, and subsets defined by individual receptors overlap considerably (Kubota et al., 1999; Hanke et al., 2001; Raulet, 2001). Multiple types of receptors are present on the same cell, as well as multiple members of the same class, and both alleles of a single gene locus may be expressed.

3.2 NK activating receptors

Release of inhibition alone seemed an unlikely explanation of NK-cell activation, and a search for activating receptors continued in parallel with defining the system of inhibitory receptors. As we have seen, virgin CTL that have not encountered the antigen recognized via their clonotypic receptors undergo a complex activation scheme after interaction with cognate antigen. They do not a priori express the lytic mechanisms, such as secretory granules and Fas ligand, evident in activated CTL; one to two days are required before CTL reach a mature, fully active lytic state. NK cells are thought to be lytically active as an end result of their cytological maturation, prior to encounter with a potential target cell. Secretory granules and Fas ligand (and other TNF-family ligands) are already in place. “Virgin” NK cells may thus be more akin to recently activated CTL, or memory CTL, which need a minimum of intracellular signaling to express cytotoxic function. Still, as with active but resting CTL (i.e., transiently not engaged with a target cell), NK cells need a target cell-associated signal to start the killing process, and to release cytokines. This is what we refer to when we speak of NK-cell "activation."

So how do NK cells recognize and engage with potential target cells, and direct their cytotoxic activity toward just these targets? Increasingly, the view is that NK cells are *potentially* capable of binding to, being activated by, and killing virtually any cell in the body, but the normal presence of class I MHC on healthy cells prevents this from happening through the inhibitory receptors just described. The nature of NK activating receptors became apparent only in the late 1990s. The number of such receptors on NK cells is undoubtedly large, and as of this writing is still growing (Lanier, 2001; Borrego et al., 2002). The ligands for these receptors include a range of different cell-surface structures, and are still not known in some cases.

In a few instances, such as Ly49-D and -H activating receptors in mice (Idris et al., 1999; Smith et al., 2000), and NKG2-C and -D (Smith et al., 1998; Bauer et al., 1999) and KIR2DS in humans (Moretta et al., 1995; Lanier, 1998), the receptors are members of the same families we have just seen for inhibitory receptors, and the activating ligand is a class I molecule, which ordinarily causes inhibition of NK functions (Table 6.3b). Why it would be advantageous to have class I MHC molecules act as inducers of both activating and inhibiting signals is still a matter of conjecture. Their affinity for class I products is rather low, and it is possible that their interactions with class I are secondary to some other, higher affinity molecular target. In humans, three immunoglobulin family receptors called NKp44, NKp46 and NKp30 are major activating receptors (Moretta et al, 2000). The ligands for these receptors are unknown.

Table 6-3b. Some common murine and human NK activating receptors, their ligands and transducers

Ligand	**Transducer**	
Mouse		
CD16	Fcγ	FcεRIγ
CD94/NKG2D	Rael, H60	DAP 10
NKR-PIC	?	FcεRIγ
Ly49D	H-2D^d	DAP 12
Ly49H	MCMV m157	DAP 12
Human		
CD16	Fcγ	CD3ζ, FCεRIg
CD94/NKG2C	HLA-E	DAP 12
CD94/NKG2D	MICA, MICB	DAP 10
NKp30	?	CD3ζ (?)
NKp44	?	DAP 12
NKp46	Influenza HA	CD3ζ, FcεRIg

	Ligand	Transducer
KIR2DS	HLS-c	DAP 12
2B4	CD48	SAP

The reasons why these molecules do not act as inhibitors are, however, clear: they lack ITIM sequences in their cytoplasmic tails. But they also do not contain the immunoreceptor tyrosine-based activation motif (ITAM) sequence [Tyr-X-X-(Leu/Ile)-$X_{6\text{-}8}$-Tyr-X-X-(Leu/Ile)] needed to communicate with submembranous signaling machinery. In fact all of the NK activating receptors identified so far do not themselves transduce an activation signal but require "adaptor" molecules bearing the ITAM motif. The most common adaptor molecules are the CD-3ζ chain associated with the T-cell receptor (Lanier et al., 1989b), and the FcεRIγ chain (Wirthmueller et al., 1992), both of which are found as disulfide-linked homodimers in NK cells, and the 12-kD disulfide-linked homodimeric molecule DAP 12. (Tomasello et al., 1998; Lanier and Bakker, 2000.) As in CTL activated by ITAM-bearing CD3 proteins, phosphorylation of ITAM tyrosines results in binding of ZAP-70 and/or syk protein kinases, although most DAP 12 adaptor proteins in NK cells use the latter. Interestingly, some mouse NK cells display CD28, the co-stimulatory receptor so crucial in T-cell function (Chapter 1). Although probably not a primary NK activation receptor, it plays an important auxiliary role in enhancing NK responses (Nandi et al., 1994; Chambers et al., 1996).

One of the more interesting NK activating receptors is NKG2D (Long, 2002; Vivier et al., 2002). In addition to being constitutively expressed on all NK cells in both mice and humans, NKG2D is also expressed on CD8 T-cells (Ho et al., 2002), γ/δ T cells (Wu et al, 2002) and NKT cells in both species. Unlike other members of the NKG2 family, NKG2D pairs with itself to form a homodimer to make an activating receptor. In T cells and NKT cells CD94/NKG2D acts as a costimulatory molecule, using DAP 10 as an adaptor molecule (Wu et al., 1999); in NK cells it can be either a primary activating receptor, with DAP 12 as an adaptor (Lanier et al., 1998), or a costimulator (Pende et al, 2002).

In mice, the ligands identified so far for NKG2D are the retinoic acid early-induced proteins (Rae1), and the minor histocompatibility antigen H60. Rae1 is expressed only on fetal tissues and tumors, and in normal cells only by retinoic acid treatment. Both ligands can stimulate tumor reactivity in NK cells (Diefenbach et al., 2001). The known activating ligands in humans are the MHC class I chain-related proteins A and B (MICA and MICB), which are class I-like molecules that do not associate with β-2m, and that do not require antigenic peptide fragments for surface expression. In general these ligands are thought to reflect a state of stress within the presenting cell, caused by any of a number of cellular derangements (e.g., parasitic infection, oncological transformation).

Almost all of the ligands (where known) for activating receptors on NK cells can be found on a wide range of normal cells, depending on their physiological state, as well as on so-called NK-sensitive targets; it is not the presence of molecular structures unique to oncologically or virally transformed cells that makes the latter NK-sensitive. Thus it is obvious why a system of negative regulatory receptors on NK cells is absolutely necessary if devastating autoimmunity via NK cells is to be avoided.

Interestingly, the genes for many of the NK-associated molecules we have been discussing map closely together in so-called NKC (NK gene complexes) in both mice and humans (Figure 6.3; see also Brown et al., 1997; Yokoyama and Plougastel, 2003). Most of these genes are expressed in cells other than NK cells; only the NKp30, p44, and p46 genes in humans are apparently NK-specific (Moretta et al., 2000). The KIR genes in humans are on a separate chromosome, in the midst of a complex containing other potential Ig-like inhibitory receptors not found on NK cells, with the possible exception of ILT2 (Colonna et al., 1997; Cosman et al., 1997). KIR genes have varying numbers of isotypes among individuals, as well as multiple alleles at each locus.

A fundamental difference between NK cells and CTL is that the former, after activation via various cytokine receptors or through engagement of surface receptors as just discussed, do not undergo significant proliferation and clonal expansion. Nor do they develop anything like a memory state as defined in T cells. There is no evidence that after exposure to a specific antigenic stimulus, the host response in terms of NK cells is any different, either quantitatively or qualitatively, than the first time the host responded to the same stimulus. But an unexpected similarity between these two cytotoxic effectors is that both use some of the same inhibitory receptors to regulate cytotoxic function (McMahon and Raulet, 2001; Braud et al., 2003). In fact, it has recently been shown that Qa-1 in mice can inhibit cytotoxicity in CD8 CTL, via CD94/NKG2A (Lohwasser et al., 2001). CD94/NKG2A is also found on CD4 cells, and can inhibit T-helper cell functions (Romero et al., 2001).

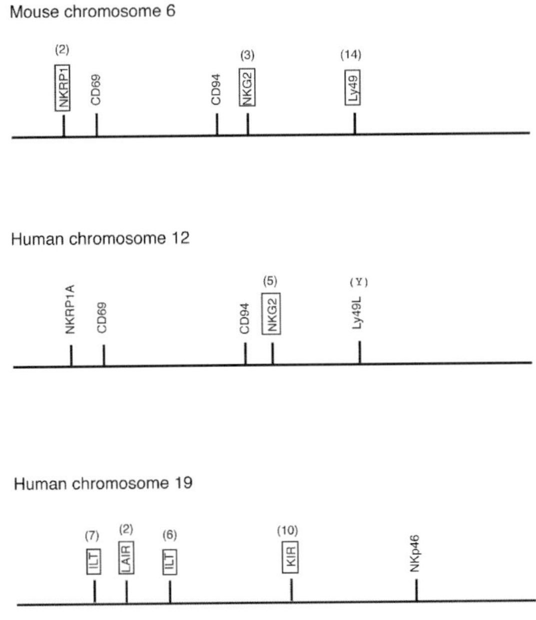

Figure 6-3. Mouse and human NKC gene regions. The genes for nearly all the activating and inhibitory NK receptors are encoded within a defined region on chromosome 6. The corresponding genes in humans are distributed between chromosomes 12 and 19. Mice do not have KIR genes, but the region syntenic with human KIR (mouse chromosome 7) contains the gene for NKp46 and KIR orthologs that encode activating receptors on dendritic cells, monocytes and B cells (Kubagawa et al., 1999; Ono et al., 1999).

4 DENDRITIC CELLS IN NK CELL ACTIVATION

We have seen so far that NK cells can be activated by cytokines – especially IL-2 and IL-15 – and by direct contact via membrane receptors with activating structures on potential target cells. With the exception of IFNα and β and IL-2, the amplifying effects of most cytokines require some degree of activation of the NK cell resulting in up-regulation of the corresponding receptors. Contact with cells bearing ligands for the activation receptors just discussed certainly provides one source of activation signals, but recent data suggest another cellular source for activation signals: dendritic cells. Dendritic cells have been studied for years as a source of activating signals for T cells. Dendritic cells lie at the interface between the innate and adaptive immune responses. They are equipped with so-called Toll-like receptors (reviewed in Janeway and Medzhitov, 2002) that recognize a wide range of pathogenic products, and are sensitive to stress signals. They are excellent APC for T cells, because they bear so many of the ligands required for cross-linking crucial T-cell activation receptors. It now seems they play an equally important role in activation of NK cells (Osada et al., 2001; Zitvogel, 2002; Moretta, 2002). NK cells can kill immature DC, which display very low levels of class I MHC (Wilson et al, 1999). But once activated during inflammation, DC can aid in the differentiation of NK cells through release of key cytokines. NK cells become more cytotoxic as a result of interacting with DC, and the DC also become further differentiated as a result of this interaction.

In mice bearing class I MHC-negative tumors, adoptive transfer of dendritic cells, especially in an activated state, led to enhanced NK anti-tumor immunity.

In vitro, exposure of NK cells to the same sources of dendritic cells resulted in enhanced cytotoxicity and cytokine release when the NK cells were subsequently presented with MHC-negative tumor target cells. This interaction required direct, physical contact, and was not mediated by dendritic-cell-secreted factors (Fernandez et al., 1999; Yu et al., 2001). The nature of the NK receptors involved, and their dendritic cell ligands, are not clear at present, but seem unlikely to be related to those already discussed. In more recent studies, it has become apparent that the NK-dendritic cell interaction is two-way, in that dendritic cells can themselves be driven to a higher state of activation by interaction with NK cells (Ferlazzo et al., 2002; Piccioli et al., 2002; Gerosa et al., 2002). This is one obvious point of interaction of the innate and adaptive immune systems mentioned earlier.

5 NK CELL BINDING TO TARGET CELLS

By the time of discovery of NK cells, the basic parameters of CTL-target cell interactions had been worked out, and those studies provided a framework that greatly accelerated gathering information about the interaction of NK cells and their targets. The first issue to address was the binding of NK cells to their targets. These studies were hampered by the fact that no one yet had any idea what the nature of an NK receptor for "antigen" was, or indeed, what "antigen" itself in this case might be. Nevertheless, the initial studies proceeded, aided greatly by the use of LGL, which could be isolated in large numbers in relatively pure form, as a source of NK cells. As with CTL, the binding of NK cells to tumor targets was found to be absolutely Mg^{2+}-dependent, with Ca^{2+} being neither necessary nor sufficient to support conjugation (Roder and Kiessling, 1978; Roder et al., 1978; Hiserodt et al., 1982).

Microscopically, the interaction of NK cells with their targets looks very similar to CTL-target cell interactions (Roder et al., 1978; Targan et al., 1980; Kang et al., 1987; Ortaldo et al., 1992b), with broad areas of surface interaction between the two cell types, and extensive interdigitation of effector and target cell membrane processes (Carpen et al., 1982). One aspect of target cell binding in the NK system turned out to be quite different from CTL. As noted in Chapter 3, CTL-target cell binding is slightly reduced at 22°, significantly reduced at 15°, and essentially non-existent at 0-4°. NK cells, on the other hand, bind target cells equally well at all three temperatures (Roder et al., 1978). The latter authors speculated that NK cells may require less membrane mobility of receptors involved in binding, although the basis for the difference in temperature effects on binding in the CTL and NK systems has never been worked out.

The same sorts of studies using metabolic inhibitors to dissect binding and lysis in the CTL system were also applied to NK cells, and again some interesting differences were found. In the CTL system, disruption of energy production inhibits both binding and lysis. With NK cells, however, inhibitors such as azide and DNP inhibit killing but have no effect on conjugation (summarized in Kiessling and Wigzell, 1979). Inhibitors of serine proteases, and the microtubule disrupter colchicine, also inhibit killing but have no effect on binding (Kiessling and Wigzell, 1979; Katz et al., 1982). Cytochalasin B, which disrupts microfilaments, blocked both binding and killing (Katz et al., 1982). γ-Interferon, which greatly enhances target cell killing (see below), has no effect on target cell binding (Roder et al., 1978). Thus the idea arose that in the NK system conjugation and killing are less strictly tied together metabolically than is the case with CTL; in particular, NK-target cell binding appears to be much less energy-dependent.

As with CTL, the involvement of adhesion molecules is crucial to NK cell interaction with targets. A role for LFA-1 was established early on (Hildreth et al., 1983; Chen et al., 1987), and as with CTL, the level of surface LFA-1 is increased after NK cell activation. In NK cell membranes, LFA-1 is physically associated with DNAM-1, which it phosphorylates, presumably as part of an NK activation pathway (Shibuya et al., 1999). The hyaluronidate-recognizing molecule CD44, involved in NK cell extravasation, also plays a role both in target cell adhesion, and as a costimulatory signal. Cross-linking of CD44 on NK cells increases their cytotoxic function (Galandrani et al., 1994). When either the LFA-1 or CD44 gene is knocked out in mice, NK cell adhesion to targets is greatly inhibited; when both are knocked out adhesion is completely abolished. However, the perforin and Fas pathways in even the double-knockout mice are intact and functional and release of IFNγ, TNFα and NKCF see below is unaffected in NK cells activated by noncellular means (Matsumoto et al., 1998, 2000).

Binding of NK cells to a susceptible target in the presence of Ca^{2+} induces reorientation of both the Golgi apparatus and the MTOC (Kupfer et al., 1985), consistent with bringing a secretory mechanism into play. Upon initial conjugation of an NK cell with a lysis-susceptible target, orientation of both organelles is essentially random. After one hour at 37°, both are oriented toward the target cell. There is also a movement of LGL cytoplasmic granules toward the region of target cell contact. The granules do not remain discrete and spherical, but rather become syncitial with one another, forming complex membrane-bound structures extending through the cytoplasm of the effector cell.

Reorientation of the MTOC is accompanied by formation of immunological synapses similar to those formed when CTL bind to target cells (Davis et al., 1999; Sancho et al., 2000; Vyas et al., 2001). When NK cells conjugate with lysis-susceptible targets (e.g., class I-negative tumor cells), the effector-cell side of the synapse contains adhesion molecules, activating ligands, reorganized lipid structures (Lou et al., 2000), plus all of the molecular machinery required for transmission of an activating signal. The downstream activating sequence leading to lysis involves MAPKs, especially ERK (Wei et al., 2000). In non-cytolytic conjugates, a synapse still forms, but it contains inhibitory rather than activating receptors (Erickson, et al., 1999), and the MTOC does not reorient. Key activating elements, such as SHP-1, PYK-2 and ZAP-70, are also missing. Like the CTL synapse, those formed by NK cells are relatively transient compared to CD4-APC synapses.

6 CELLULAR AND MOLECULAR PARAMETERS OF NK-MEDIATED CYTOLYSIS

One of the reasons NK-mediated cytotoxicity was not discovered for nearly ten years after in vitro lymphocytotoxicity had first been reported was the fact that the frequency of NK cells in the populations tested is substantially lower than the frequency of CTL in test populations. Especially with the unfractionated lymphocyte populations and freshly explanted tumor cells used as targets in the early experiments, significant lysis was difficult to see in assays of less than 16 hours. As had been the case with studies of CTL-mediated cytotoxicity, investigators quickly found target cells that were especially susceptible to the phenomenon they were studying, in this case nonspecific lysis of tumor target cells. The in vitro-grown murine Moloney-induced T-cell lymphoma YAC-1 quickly became a favorite of labs studying mouse NKCC (Kiessling et al., 1975a), while the human chronic myelogenous leukemia cell line K562 proved particularly useful for human studies (Jondal and Pross, 1975). But even with the most sensitive target cell lines, at the highest effector-target ratios, 6-12 hours was generally required to see substantial target lysis, compared with 2-4 hours for CTL lysis. However, once highly enriched LGL populations, and then NK cell clones, had been developed, the efficiency of killing by CTL and NK cells did not seem that different.

Almost from the beginning, the nature of NK cytotoxicity was clouded by the fact that there appeared to be two different ways NK cells could bind to and kill target cells. In one form of killing, NK cells seemed to recognize target cells directly, through an undefined cell-surface receptor recognizing an equally undefined structure on the target cell surface. In the second form of killing, NK cells use their cell-surface Fc receptor to attach to target cells coated with an antibody to any cell-surface structure. This form of killing was called antibody-dependent cell-mediated cytotoxicity (ADCC). The problem was that when primary in vivo sources of NK effector cells or tumor target cells were being studied, how could one be sure that some type of antibody was not bound to the surface of the target cell, resulting in ADCC rather than direct killing, or a mixture of ADCC and direct killing? And in that case, how much of the cytotoxicity seen was caused by NK cells, and how much by other Fc receptor-positive cell types? And finally, what was the relationship of direct NK killing and ADCC? Were they the same, or did each involve a different mechanism? The existence of an Fc receptor on NK cells was itself disputed initially (Herberman et al., 1975b; Kiessling et al., 1975b) but was eventually accepted (Oehler et al., 1978; West et al., 1977). It is now recognized that NK cells can engage in both direct killing and in ADCC, and in certain experimental situations this must be controlled for (Herberman 1981). $CD56^{dim}$ NK cells, with their higher levels of CD16, are most actively engaged in ADCC.

We now know that the receptor used to activate NK cells engaged in ADCC is in most cases the low-affinity FcγRIII receptor (CD16) (Perussia, 1998). This receptor also does not itself contain ITAM sequences, it associates with a γ-chain previously described as being associated with the FcεRIreceptor, but which is in fact shared by both Fc receptors (Table 6.3b). In humans, the CD3ζ chain can also function as a transducer molecule, although in mice CD16 is negatively regulated by association with the CD3ζ chain (Arase et al., 2001.) Once CD16 is cross-linked via the Fc tails protruding from an antibody-coated target cell, the events that follow – activation for lysis and release of cytokines – are indistinguishable from direct killing events mediated by cross-linking of other NK surface receptors. There has been a report that CD16 in humans may also be involved in direct (non-antibody-mediated) cytotoxicity as well as ADCC, although the ligand in such cases is unknown.

Following the lead of Martz in the CTL system, it was established that NK cells also pass through a stage of programming for lysis. After the Mg^{2+}-

dependent target binding stage, NK cells require a period of about 30 minutes in the presence of Ca^{2+} before lysis can begin. This is considerably longer than the programming stage required for CTL, and may be related to the slightly slower kinetics of killing in NK CMC. Contact with a susceptible target induces mobilization of internal Ca^{2+} stores, as well as an influx of extracellular Ca^{2+}, in both direct NK killing and in ADCC (Edwards et al., 1989; Castella et al., 1989). As with CTL killing, the role of Ca^{2+} influx into the target cell is intriguing, but its significance is largely unresolved. Nonetheless, both target cell killing and DNA fragmentation are inhibited if the intracellular increases of Ca^{2+} are selectively blocked in the target cell (McConkey et al., 1990; Oshimi et al., 1996).

After the programming stage is completed and the lethal hit is delivered, target cells will proceed to lyse in the absence of further contact with the NK cells (Hiserodt et al., 1982). This killer-cell-independent lysis (KCIL) stage is Ca^{2+}-independent but temperature-sensitive (Hiserodt et al., 1982; Sevilla et al., 1989). As with CTL-mediated killing, NK cells induce DNA fragmentation in target cells (Duke et al., 1986; Liu et al., 1989), and target cell death is apoptotic. And like CTL, NK cells recycle to kill multiple target cells in standard ^{51}Cr release assays (Ullberg and Jondal, 1981), although this may require a constant source of IL-2 (Abrams and Brahmi, 1986).

6.1 Natural killer cytotoxic factor(s)

One of the first proposals for a possible NK lytic mechanism came from Wright and Bonavida (1981, 1982). Reminiscent of earlier studies on lymphotoxin (TNFβ) in connection with CTL-mediated cytotoxicity, Wright and her coworkers found that the medium from an assay in which both mouse and human NK cells had just killed their targets contains a soluble factor capable of killing additional target cells in the absence of NK cells. Unlike TNFβ, this factor, called natural killer cell cytotoxic factor(s)[3] (NKCF), was induced only by NK-sensitive targets, and in turn lysed only NK-sensitive cells. An NKCF-insensitive variant of the standard NK target YAC, for example, was found also to be resistant to NK CMC (Wright and Bonavida, 1983b). IFNγ-treated target cells, which are resistant to lysis by NK cells, fail to induce release of NKCF (Wright and Bonavida, 1983c), although they are sensitive to NKCF (Gronberg et al., 1988).

Unfortunately, NKCF proved to be relatively unstable, its activity diminishing in a matter of hours at 37°, and after a few days at 4°. This lability, coupled with its relatively low activity, made biochemical characterization of NKCF difficult. It is at least in part protein in nature

[3] From the beginning, Wright and Bonavida pointed out that NKCF could well be a mixture of more than one activity. Others often referred to it as a single species.

(Bonavida and Wright, 1986; Uchida and Fukata, 1993). Attempts to define even its molecular size on Sephadex columns yielded values ranging from 5,000 to 40,000. TNFβ produced by NK cells has been eliminated as a candidate for NKCF. It does not lyse NKCF-sensitive targets, and NKCF does not lyse targets killed by TNF–β (Ortaldo et al., 1986a; Bialas et al., 1988). The situation with TNFα is less clear, and it is possible that at least some of the cytotoxicity exerted by NKCF is TNFα. For example, when NKCF kills the target cell U937, TNFα antibodies block a substantial portion of the lysis (Wright and Bonavida, 1987).

Several monoclonal antibodies were prepared using partially purified rat cytolysin (perforin), and screened for their ability to block NKCF killing of target cells. These mAb, which reacted with an approximately 12 kD protein from NKCF supernates, also physically blocked killing by NK cells (Ortaldo et al., 1987). Subsequently these same mAbs were shown to block lysis of red blood cells by cytotoxic granules isolated from NK cells (Winkler-Pickett et al., 1991). Since these were monoclonal antibodies, the suggestion was made that NKCF and perforin must contain cross-reactive epitopes. However, unlike perforin, NKCF kills target cells in the absence of Ca^{2+}.

The inability to define NKCF at the biochemical level, let alone isolate a corresponding gene(s), eventually led to abandonment of attempts to define a role for it in NK cell-mediated killing. It is probably a mixture of activities released by activated NK cells, possibly including TNFα and granulysin (Pena et al., 1997) and maybe even perforin. Perforin as we presently understand it is unlikely to be identical with NKCF; perforin, for example, would not be expected to exhibit the NK-sensitive target cell specificity that all investigators observed with NKCF, and perforin does not kill in the absence of Ca^{2+}. It may well be that in the future the molecular nature of NKCF will be unraveled, but for the present research into its identity has come to a halt.

6.2 Exocytosis, Fas, and other lytic mechanisms in NK CMC

A number of observations beginning in the late 1970s or so pointed to the possibility of a stimulus-secretion response for NK-mediated killing. The *beige* mutation in mice, which results in disruption of lysosome function and degranulation across all cell types, and is a model for Chediak-Higashi syndrome in humans (Spritz, 1998), leads to a substantial loss of NK CMC (Roder and Duwe, 1979). *Beige* mice lack NK cytotoxic function in all lymphoid compartments, against all NK targets tested, while other immune functions are largely unaffected (Roder et al., 1979b; but see Saxena et al., 1982). The rate of occurrence and metastasis of most tumors is increased in these mice (Talmadge et al., 1980; Gorelick et al., 1982; Haliotis et al., 1985). Although NK lytic function is absent, LGL cells from these mice show normal levels of NK target cell binding, suggesting the defect is in the lytic mechanism and not because NK cells per se are absent, or unable to recognize target cells. Chloroquine, a chemical inhibitor of lysosomal enzymes, induces a similar phenotype (Roder et al., 1980), which is accompanied by an apparent inability of lysosomes to fuse with the plasma membrane. Carpen et al. (1981, 1982) showed that NK CMC could also be inhibited by monensin, which disrupts vesicular traffic through the Golgi apparatus, and thus secretory processes.

As discussed in Chapter 5, some of the earliest observations leading to formulation of the granule exocytosis model of CMC were made in the NK system (Henkart and Henkart, 1982). Henkart felt that classical lysosomes per se were unlikely to contain all of the lytic agents involved in the killing process; the evidence he and others put forth in support of exocytosis of the granule structures visible in both LGL and mature CTL (with the exception of PEL) was reviewed in Chapter 5 and will not be repeated here. There appears to be nothing to distinguish the perforin and granzyme systems in NK cells and CTL (Henkart 1985; Trapani et al., 1994). Disruption of the perforin gene in mice drastically reduces NK CMC against some target cells without impairing other NK functions (Kägi et al., 1994a). Simultaneous disruption of granzymes A and B, while abolishing NK cell induction of DNA fragmentation in target cells, has no effect on the ability of NK cells to protect against either transplanted or carcinogen-induced tumor cells in vivo (Davis et al., 2001). Like some CTL, NK cells are themselves resistant to the lytic effects of perforin (Shinkai et al., 1988), and as with CTL, this resistance is unrelated to homologous restriction factor (Jiang et al., 1988).

The intense focus during the 1980s on granule exocytosis as an explanation of lymphocytotoxicity, and the role played by NK cells and clones in defining this mechanism, led to the assumption that exocytosis of perforin and granzymes must be a major factor in NK-induced killing. As discussed earlier, NK cell killing, like primary CTL killing, is highly Ca^{2+}-dependent, presumably reflecting a biased usage of the exocytotic pathway by most NK cells most of the time. Subsequent studies provided a more balanced view. It is now clear that an important contribution to NK cytotoxic activity is also provided by the TNF family of ligands and receptors, particularly Fas, TNFα, TRAIL (TNF-related apoptosis-inducing ligand), and possibly lymphotoxin.

Shortly after description in 1993 of the Fas system operating in CTL, several labs examined the possible involvement of Fas/FasL in NK-mediated CMC. Unlike CTL, where FasL is present only after activation by specifically recognized antigen or other means, FasL is constitutively present on naïve NK cells, although its level can be increased by various activation procedures (Montel et al., 1995; Arase et al., 1995; Oshimi et al., 1996; Zamai et al., 1998; Kashii et al., 1999). Although some investigators were unable to demonstrate Fas killing by NK cells (Kägi et al., 1995), others found that Fas ligand on NK cells induces apoptotic cell death in Fas-positive targets, and can do so in a Ca^{2+}-independent manner, obviating perforin as a lytic mechanism (Montel et al., 1995; Oshimi et al., 1996). The Fas pathway (along with exocytosis) is also utilized during ADCC by $CD16^{bright}$ NK cells (Oshimi et al., 1996). As with CTL, it is puzzling why Ca^{+2}-independent lysis by NK cells took so long to be discovered. The Fas status of most NK target cells has rarely been established, so it is not obvious in many cases that Fas-mediated lysis could have been expected. Alternatively, it may simply have been overlooked because of the strong and almost uniformly held belief that CMC was Ca^{2+}-dependent.

TRAIL is present constitutively on at least liver NK cells, and can be up-regulated on other NK cells by IFNα (Takeda et al., 2001). It may be constitutively expressed by immature circulating NK cells (Zamai et al, 1998). TRAIL-positive NK cells are active in lysing targets bearing an appropriate TRAIL ligand. Kashii et al. (1999) have suggested that multiple apoptosis-inducing ligands, particularly those present in low concentration on the NK cell surface, may be engaged simultaneously to generate lytic activity in NK cells. It is not clear at present what relation activation via TNF family ligands bears to activation via the receptors discussed earlier, such as CD94/NKG2D or Ly49D and H, or whether further activation beyond that which may be provided by Fas-ligand engagement, for example, is necessary for CMC. We also do not know whether Fas killing is negatively regulated by the inhibitory molecules discussed earlier, although seemingly it would be.

7 NATURAL CYTOTOXICITY (NC) CELLS

Osias Stutman first described in mice an apparent variant of NK cell activity, called natural cytotoxicity, in 1978. The imputed natural cytotoxic (NC) cells seemed possibly to be associated with a distinct lineage of killer lymphocytes (Stutman et al., 1978; Paige et al., 1978). NC activity was present at birth in mice, whereas NK activity was not apparent until about 5 weeks of age. A counterpart to NC cells in humans was described a few years later (Rola-Pleszczynski and Lieu, 1983). NC cells seemed to be more effective against solid tumor targets, whereas NK cells were usually directed against hematopoietic tumor cells. NC cells did not bear any of the markers associated with NK activity, such as NK1 or asialo-GM1 (Kumar et al.,

1979), and did not belong to the large granular lymphocyte class (Djeu et al., 1983). They were found to be somewhat more adherent than NK cells, suggesting a possible monocytic origin. While the cellular origin of NC killing, and its relation to NK, is still unresolved, Ortaldo et al. (1986a) showed that the lytic mechanism used by NC cells is TNFα, which had been shown not to be an effector mechanism for NK cells. Research into NC cells did not continue much past the mid-1980s, and the ontogenic origin and biological relevance of this form of cytotoxicity is difficult to assess.

8 LYMPHOKINE-ACTIVATED KILLER (LAK) CELLS

It was recognized early in the development of the NK story that in fact what were being described as tumor-specific NK cells had at best weak cytotoxic activity toward many freshly isolated syngeneic and allogeneic tumors (Vose et al., 1978; Vanky et al., 1981). It was frequently pointed out, as investigators struggled to understand the phenomenon they were studying, that cytotoxicity of these cells was being defined by reactivity in vitro toward a narrow selection of target cells whose relation to spontaneously arising tumors in vivo was often not clear. A new twist on this story came with the report in 1981 that human peripheral blood lymphocytes – the common source of human NK cells – cultured in the presence of high-dose IL-2 were able to lyse a wide range of both fresh and cultivated tumor cells, most of which were resistant to standard NK-mediated cytotoxicity (Lotze et al., 1981; Grimm et al., 1982). As with NK cells, there was no MHC restriction of the observed lysis. These results generated immediate interest, since IL-2 was a naturally occurring cytokine that might well be expected to be present at the site of any ongoing immune reaction involving T cells[4].

The cells resulting from IL-2 activation of PBL were referred to as lymphokine-activated killer cells, or LAK. Optimal activation of PBL was obtained after three-four days of culture in the presence of IL-2, and was dependent on cellular proliferation. The LAK cell precursors, present in both cancer patients and healthy controls, were non-adherent and non-phagocytic, and were thought initially to be distinct from both CTL precursors and NK cells (Grimm et al., 1983a). An identical set of IL-2-sensitive cells, giving rise to broadly cytolytic, antigen-non-specific killer cells, was identified among mouse spleen lymphocytes as well (Mazumder et al., 1983).

[4] The initial reports involved studies not with pure IL-2, but with supernates from reactions in which T cells had been activated by various means – so-called T-cell growth factor, or TCGF. It was quickly established that of the many possible cytokines in TCGF (IFNγ was certainly a possible candidate), IL-2 was the critical factor for generation of LAK cells.

The exact nature of the lymphocytic precursor to LAK cells was a matter of debate for several years, some investigators favoring NK cells and some T cells. Finally, various laboratories pooled their experimental results and came to the conclusion that the cells giving rise to LAK cells were almost entirely NK cells. While T cells could not be completely excluded, their contribution to overall LAK cytotoxicity in IL-2-activated human PBL or mouse spleen cell populations was judged to be minimal (Herberman et al., 1987; but see also Phillips and Lanier, 1986). One contributing factor was that while NK cells had been shown clearly to respond directly to IL-2 with increased lytic activity as well as cytokine production (Kuribayashi et al., 1981), T cells did not develop measurable cytotoxicity in response to IL-2 in the absence of some other triggering stimulus to up-regulate expression of the IL-2 receptor (Ortaldo et al., 1984). Moreover, the IL-2-responding population, both in terms of proliferation and development of cytotoxic function, was shown to be almost entirely from the Fc receptor-positive LGL fraction of lymphocytes (Ortaldo et al., 1984; Talmadge et al., 1986). Thymocytes exposed to IL-2 for several days develop strong LAK activity, but few of the resulting effectors are CD3-positive, and the cytotoxicity is directed toward NK-resistant as well as NK-sensitive targets (Ballas and Rasmussen, 1987; Ramsdell and Golub, 1987).

Detailed ultrastructural analysis of mature LAK cells showed them to be little different in appearance from mature CTL or NK cells. They have numerous cytoplasmic granules, which contain perforin and granzymes A and B. Their interaction with target cells, and the changes they induce in target cells, including DNA fragmentation, are also very similar to NK and CTL (Groscurth et al., 1990; Zychlinsky et al., 1990). Adhesion molecules play an important role in both conjugation and cytotoxicity (Gwin et al., 1996). One major difference between LAK cells and their NK progenitors, however, is the presence of a greatly enhanced Fas-based lytic mechanism in all LAK cells (Medvedev et al., 1997. LAK cells from perforin knockout mice lost the ability to lyse a substantial number of LAK-sensitive targets in short-term assays, the only LAK-sensitive targets being those that expressed the Fas antigen (Liu et al., 1995; Lee et al., 1996). In mice deficient in both Fas and perforin, LAK cells can still express cytotoxicity in longer-term assays, using TNFα (Lee et al., 1996) or lymphotoxin (Miyake et al., 1992) as effector mechanisms. Like CTL, LAK cells use the perforin and Fas-based lytic pathways simultaneously on Fas-positive targets. Granzymes A and B have been proposed to play a role in LAK-mediated cytotoxicity (as opposed simply to DNA fragmentation); unlike NK cells and CTL, LAK from Grz A/B knockout mice show significant reduction in lytic ability (Davis et al., 2001).

The potential use of IL-2 and LAK cells to inhibit tumor growth was explored in mice, with considerable success (summarized in Rosenberg, 1988). Visceral masses of most tumor types were greatly reduced, and combinations of IL-2 and LAK cells were particularly effective in controlling metastatic spread (Mulé et al., 1984; Mazumder and Rosenberg, 1984). The promising results in animal models formed a basis for clinical trials in humans, which began in the mid-1980s (Mazumder et al., 1984; Rosenberg, 1984; Lotze et al., 1985a,b; Rosenberg et al., 1985). In humans, a combination of LAK and intravenous IL-2 held initial promise, yielding significant reductions in tumor mass in patients with advanced cancers, particularly with hard to treat melanoma and renal cell carcinoma (Rosenberg et al., 1987; West et al., 1987). However, chronic exposure to elevated IL-2 resulted in serious alterations in vascular permeability (Rosenstein et al., 1986; Ettinghausen et al., 1988), which placed restrictions on its clinical use.

A variant of the LAK approach to tumor therapy involves the use of TIL (tumor-infiltrating lymphocytes). In at least some animal spontaneous tumor models it had become clear that progression or regression of the tumor correlated well with the cytotoxicity of lymphocytes harvested from the tumor itself. The ability to recover lymphocytes infiltrating a primary tumor had been developed earlier (Rosenberg and Terry, 1977), and eventually led to the idea of "adoptive immunotherapy", wherein activated killer cells isolated from one source and expanded in vitro are introduced into a second source to exert their effect. Initial studies in animals were highly encouraging (reviewed in Kedar and Weiss, 1987). In mice, for example, TIL proved to be 50-100 times more effective killing tumor cells in vitro and reversing tumor growth in vivo than standard LAK cells, and in combination with injected IL-2 and standard chemotherapy, led to complete tumor regression of the primary tumor and metastases in many cases (Rosenberg et al., 1986). In humans, IL-2-expanded TIL proved to be a mixture of cell types, including activated CD4, CD8 and NK cells. In most cases these cells killed target cells indiscriminately, much like standard NK cells. However, in the case of melanoma it proved possible to isolate CD8 T cells, and to derive clones thereof (Topalian et al., 1989; Fox et al., 1990), that were highly specific for melanoma targets. The current status and clinical outcomes of these and other studies with LAK and TIL will be discussed in Chapter 9.

9 NKT CELLS

Until recently, it would have been difficult to decide whether to place a section on NKT cells in a chapter on CTL, or a chapter on NK cells. While many questions doubtless remain about the developmental origin and functions of these cytotoxic lymphocytes, we place NKT cells in the present chapter because it is increasingly obvious they are part of the innate, rather than the adaptive, immune system.

It was apparent fairly early on that what everyone was calling NK cells was unlikely to be a homogeneous lymphocyte population (Lust et al., 1981; Minato et al., 1981; Ortaldo and Herberman, 1984). Examination of highly purified LGL preparations suggested that as many as half of the cells in these populations share a number of surface markers with both NK and T cells (Ritz et al., 1988), including (in humans) HNK1 or (in mice) NK1, as well as (in both species) CD4 and/or CD8 but not (at least initially) CD3 (Koo and Hatzfield, 1980; Ortaldo et al., 1981; Lanier et al., 1986).

However, when NK clones were generated from either LGL or whole circulating lymphocyte populations, and screened on the basis of NK markers and/or ability to kill a range of NK targets with differing MHC, some clones were found that not only displayed surface CD3, but rearranged and expressed TCR α/β genes as well (Hercend et al., 1983; Ritz et al., 1985; Lanier et al., 1986). The results with NK clones were difficult to interpret, since they were generated and maintained in high levels of IL-2, which as we have seen drives both NK cells and T cells into an LAK-like state, where both cell types express high levels of Fas ligand. CTL clones maintained in high levels of IL-2 also develop the ability to kill targets in an MHC-independent fashion, including multiple NK targets (Brooks et al., 1983). Expression of Fas ligand could be responsible in both cases for MHC-independent lysis of multiple target cells, depending on the Fas status of the targets. This was never determined at the time the NKT story was emerging, since Fas involvement in cytolysis would not be described until 1993. Nor was it determined whether IL-2 was driving the expression of NK markers on T cells, and vice-versa, or whether NKT clones represented a unique lineage.

The next stage in the NKT story came about when such cells were found in vivo, initially mostly in the thymus, but eventually in other tissues as well (Schmidt et al., 1986; Ballas and Rasmussen, 1990; Arase et al., 1992; Hayakawa et al., 1992). These cells were isolated and studied without cloning, avoiding questions raised by possible effects of excess levels of cytokines. There is some phenotypic heterogeneity among these cells depending on the tissues in which they reside (Godfrey et al., 2000), but all of them share some combination of the characteristics shown in Table 6.4. In mice, cytotoxic NKT cells are found in liver, bone marrow, thymus, and spleen, in more or less descending order. There are very few NKT cells in lymph nodes. In humans, the tissue distribution of NKT cells is less clear; they are found in liver, and comprise a percent or so of PBL (peripheral blood leukocytes). In both mice and humans, about half of NKT cells are double-negative (DN) with respect to CD4 and CD8 markers, with a similar proportion in each species being $CD4^+$, $CD8^-$ (Bendelac et al., 1997; Lee et al., 2002). There are probably no true $CD4^-$, $CD8^+$ NKT. While such cells can be found in the periphery, they are CD1-independent (Hammond et al., 1999; Stremmel et al., 2001), are probably generated extrathymically (Legendre et al., 1999), and may even represent normal T cells that have secondarily up-regulated the CD16 and NK1.1 (NKR-P1C) antigens (Slifka et al., 2000). NKT cells in mice also display NK regulatory molecules such as Ly49. Human NKT express CD56, CD16 and NKG2A.

Table 6-4. NKT cells in mice and humans

Phenotypic characteristic	Mouse	Human
NK surface markers	NK1; NKR-P1C; Ly49; CD16	CD56; CD161;NKR-P1a; NKG2A
T-cell surface markers	Thy-1; CD44; CD45B Ly6; CD4	CD4; CD28; CD45B,O
Cytokine secretion	IL-4; IFNγ ; TNFα ; GMCSF	IL-4; IFN; TNF; GMCSF
TCR expressed	Vα14Jα281/Vβ8	Vα24JαQ/Vβ11
Tissue distribution	Liver>BM>thymus>spleen> LN	PBL <1%

Both DN and $CD4^+$ NKT cells display T-cell receptors of the α/β type, but of greatly restricted heterogeneity. The vast majority of mouse NKT cells display a single α-chain variable region, $V\alpha_{14}J\alpha_{281}$, paired with a limited number of β-chains, mostly of the $V\beta_8$ family (Bendelac et al., 1997). The corresponding receptor in humans pairs $V\alpha_{24}J\alpha_Q$ with members of the $V\beta_{11}$ family (Dellabona et al., 1994; Exley et al., 1998). In both species these receptors are thought to interact with variants of the class I MHC-like molecule CD1. CD1 proteins are not encoded within the MHC gene complex, but CD1 genes are clearly related to MHC, and CD1 proteins pair with a β-2m molecule. In mice in which the β-2m gene has been disrupted, NKT cells do not develop (Stremmel et al., 2001).

The exact nature of what is presented by CD1 to NKT cells in vivo is unclear, but it is probably a glycolipid such as glycosylphosphatidylinositol (Joyce et al., 1998), or the marine-sponge glycolipid α-galoctosylceramide (α-GalCer) (Bendelac et al., 1997; Kawano et al., 1997; Burdin and Kronenberg, 1999) rather than a peptide. The antigen-binding portion of CD1 differs from classical class I binding sites in ways that allow it to accommodate large nonpolar structures (Gadola et al., 2002). As little as 1 ng of α-GalCer injected into mice causes a massive explosion in activated NKT cells (Singh et al., 1999). α-GalCer, although unlikely to be immunogically relevant in mammals, is of particular interest since it has been reported that this glycolipid can induce cell-mediated cytotoxicity against tumor cell lines (Kawano et al., 1999; Nicol et al., 2000). Indeed, there are reports that mice lacking NKT-restricted receptors have deficient anti-tumor immune responses (Cui et al., 1997).

The ontogenetic origins of NKT cells are still something of a mystery. Are they NK cells that acquire T-cell characteristics, T cells that acquire NK markers such as CD161 and Ly49 and CD16, or are they a totally unique lineage? When first described in the thymus, they looked very much like a special subset of thymic double-negative T cells that acquired certain NK markers, although not necessarily NK cell properties. But then it was questioned whether NKT cells even arose in the thymus, migrated there secondarily from bone marrow or liver (Sykes, 1990; Seki et al 1991), or were perhaps just contaminants present in blood vessels coursing through the thymus. On the other hand, NKT cells fail to develop in nude or neonatally thymectomized mice (Hammond et al., 1998; Coles and Raulet, 2000). More recent data have shown that NKT cells undergo a TCR-dependent selection in the thymus by cells expressing the CD1 molecule (MacDonald et al., 2001), but surprisingly, this molecule is expressed in the thymus on double-positive ($CD4^+CD8^+$) cortical thymocytes, rather than thymic epithelial cells (Bendelac et al., 1995a,b; Coles and Raulet, 2000). Whether this selection is biased by bound antigen is uncertain, but NKT cells fail to develop in CD1-knockout mice (Mendiratta et al., 1997; Chen et al., 1997) (Stremmel et al., 2001). A precursor to NKT cells has recently been identified in mouse thymus that is negative for NK markers, and has properties similar to T helper cells (Benlagha et al., 2002). If this cell were shown to originate in bone marrow, then NKT ontogeny would mimic that of T cells.

When appropriately stimulated, NKT cells rapidly secrete cytokines, principally IL-4 but also IL-10 as well as IFNγ and TNFα. In humans, there is differential production of these cytokines by $CD4^+$ and DN NKT cells. IL-4 is produced only by $CD4^+$ NKT; both subsets secrete IFNγ and TNFα (Gumperz et al., 2002). CD4 and DN NKT in humans also have different patterns of chemokine receptors (Lee et al., 2002). No such differences appear to exist in the corresponding mouse subsets (Bendelac et al., 1997). IFNγ secreted by NKT cells can have a potent antiproliferative effect on certain tumor cells in vivo (Kikuchi et al., 2001), and NKT cells are potent stimulators of NK cells, presumably through cytokine release (Carnaud et al., 1999; Eberl et al., 2000).

In addition to secreting cytokines that would influence the activity of other cells participating in both innate and adaptive immune responses, mouse NKT cells can engage in both direct and antibody-dependent cytotoxicity, especially after IL-2 activation (Ballas and Rasmussen, 1990; Koyasu, 1994). Humans also produce NKT cells cytotoxic for tumor target cells in a CD1d, α-GalCer-dependent manner (Metilitsa et al., 2001). Several reports suggest that exocytosis of perforin is a major lytic mechanism used by NKT both in vivo and in vitro (Kawano et al., 1999; Kodama et al., 1999; Brutkiewicz et al., 2002), but Fas ligand is constitutively expressed in NKT, particularly after IL-12 activation (Matsumoto et al., 2000), and NKT cells can engage in cytotoxicity against Fas-positive target cells (Arase et al., 1994). In general the cytotoxic mechanisms used by NKT cells do not differ significantly from those used by NK cells or CTL.

NKT cells have revealed an interesting aspect of interaction with dendritic cells (DC). As discussed earlier, DC are capable of providing activating signals to NK cells, although the precise nature of these signals is still being worked out. In the case of NKT cells, it has been found that while expression of class I on dendritic cells can suppress delivery of an activating signal to NKT cells, if the DC are stressed they can deliver an activating signal that overrides the class I inhibition, at least in terms of inducing cytokine production (Ikarashi et al., 2001). The basis of this "suppression of inhibition" appears to involve enhanced CD28-B7 interaction. Whether cytotoxicity is also enhanced is unclear, but normally signals that up-regulate cytokine production also enhance cytotoxic activity. These observations are also in harmony with the general view that NK cells respond most strongly to cells that express signs of having undergone recent stress reactions.

So are NKT cells NK cells, or T cells, both or neither? Although they look like LGL, they are probably thymic in origin, and thus ontogenically could be thought of as T cells. But their highly constricted TCR repertoire and restriction to CD1 obviates any role in adaptive immunity, which is a hallmark of both T and B cells. NKT cells are reminiscent of γ/δ T cells, which also have a restricted TCR repertoire and interact with CD1 (Chapter 2); both utilize the NKG2D activating receptor commonly found in NK cells. Both have been implicated as regulatory cells promoting various adaptive responses, as well as in microbial infections, stress, and a wide range of autoimmune diseases (Sharif et al., 2002), in addition to responses to tumors (Godfrey et al., 2000). In fact, γ/δ T cells could probably have been discussed in this chapter as well as in Chapter 2. Such highly specialized cellular subsets play an important and increasingly recognized role in innate immunity (Bendelac and Medzhitov, 2002), and from that point of view NKT cells are functionally closer to NK cells than to T cells.

10 CONCLUSIONS

Since the definition of the antigen receptors for B cells (1960s) and T cells (1980s), an enormous amount of investigative energy has been spent elucidating the mechanisms of adaptive immunity. Innate immunity was viewed as somewhat of an aside, important in its own right but perhaps less interesting mechanistically than adaptive immunity. But in recent years the focus of cellular immunology, if not reversed in the direction of innate immunity, has certainly seen a powerful surge of interest in that direction. It is now apparent that innate and adaptive immunity are not two evolutionarily and functionally distinct systems, but rather two highly interactive facets of cellular immune responses. While the cells involved generally have distinguishable histological origins, they share a great deal in terms of molecular mechanisms, including regulatory surface receptors, cytokines secreted and absorbed, and effector molecules such as perforin and FasL. The lytic potency of NK cells appears to be no less than that of CTL. Each branch can have both potentiating and inhibitory actions on the other, mediated through cytokines and dendritic cells (Moretta, 2002).

As with CTL, the original in vivo role perceived for NK cells – tumor control – is probably not the major raison d'être for these effectors; for both killer cell types, dealing with intracellular parasites appears to be a primary function. While some tumor cells do have reduced class I expression, in fact most do not, and class I expression on a putative target cell clearly inhibits NK-mediated cytotoxicity. On the other hand, a number of viruses do reduce class I on cells they infect as a means of escaping CTL-mediated destruction. However, for NK cells, the evidence that they use their lytic potential in ridding the body of compromised cells is underwhelming at best. As with CTL, the role of cytotoxicity in NK function in vivo assumes a more modest role than had been imagined a decade or two ago.

CONCLUSIONS

Chapter 7

THE ROLE OF CYTOTOXICITY IN ALLOGRAFT REJECTION IN VIVO

1 DELAYED HYPERSENSITIVITY, TISSUE TRANSPLANTATION, AND THE DISCOVERY OF CTL

The discovery of cell-mediated immunity, and ultimately of cytotoxic T lymphocytes, emerged from attempts to understand the immunological basis and mechanism of two phenomena discovered early in the history of immunology: delayed hypersensitivity reactions, and allograft rejection. Both of these phenomena were presumed by many investigators to be immunological in nature, showing the properties of specificity and memory, but for many years were misunderstood and open to other interpretations.

Hypersensitivity reactions were an unwelcome conundrum in the early days of immunology. Early studies of immunity, at the end of the nineteenth century, were understandably focused on its protective aspects, the ability of the immune system to quickly and efficiently clear infections by potentially disease-causing microorganisms. But shortly after the turn of the century, reports from the French scientists Paul Portier and Charles Richet suggested this might not always be the case. In a few cases, which were shown to be entirely reproducible, initial exposure to an antigen resulted in a state in which subsequent exposure to the same antigen resulted not in protection, but in morbidity or even death. For example, when a dog was given a sub-lethal but symptom-inducing injection of a particular bacterial toxin from which it fully recovered, and then several weeks later was given a second injection of the same toxin, the dog immediately went into shock and was dead within the hour (Portier and Richet, 1902). It was previous exposure to the toxin, in a regimen that in most other instances conferred immunity, that appeared to have created the problem. As can be imagined, the proposal that the immune system could also *cause* disease was not greeted with enthusiasm.

Nevertheless, numerous other examples of these negative reactions were uncovered, and came to be known as anaphylactic or hypersensitivity reactions. Moreover, a close study of such reactions in the ensuing years led to the realization that there are two different types of hypersensitivity, distinguished most notably by the kinetics of development of the response. Immediate hypersensitivity (IH) reactions develop within minutes of re-exposure to the provoking antigen, and reach a peak within a few hours at most. Portier and Richet's dog had clearly experienced an IH reaction. Delayed-type hypersensitivity (DTH) reactions, on the other hand, may not be apparent for 12-24 hours, and require 2-3 days to reach maximum intensity.

IH reactions, which came to include such readily recognizable maladies as allergy, and potentially lethal secondary reactions to bee stings or drugs such as penicillin, although not understood at first, seemed to fall within the developing paradigms of immunology generally. Initial exposure to a provoking antigen resulted in the production of serum substances (in humans, these turned out to be antibodies of the IgE subclass), which could be passively transferred from a hypersensitive to a normal individual, rendering the recipient selectively hypersensitive to the provoking antigen immediately after transfer. A similar phenomenon had been repeatedly observed in the passive transfer of positive (protective) immunity with serum, and was considered a hallmark of immune responses generally.

DTH reactions were found to underlie such well-known phenomena as contact sensitivity, certain fungal infections and the tuberculin reaction. The latter had been described already in 1891 (Koch, 1891) with the finding that tubercular guinea pigs developed inflammatory skin lesions when injected with substances obtained from cultures of *M. tuberculosis*. Animals made hypersensitive to a particular antigen were not hypersensitive in general but only to the provoking antigen. The problem posed by DTH reactions was that, although displaying the same specificity and memory properties as IH reactions, they could not be transferred with immune serum. Since serum antibodies were the only known immune mechanism in the first half of the twentieth century, many remained skeptical that DTH reactions were immunological in nature.

This puzzle was not solved until the 1940s, through experiments by Karl Landsteiner and Merrill Chase, who showed that both delayed cutaneous hypersensitivity to chemicals and the tuberculin reaction in guinea pigs could be transferred between animals using peritoneal lymphocytes, but not serum (Landsteiner and Chase, 1942; Chase, 1945). These were the first convincing reports that cells, as well as antibodies, could mediate a presumed immune phenomenon, but because of the ambiguous nature of DTH reactions it would be a number of years before the concept of cell-mediated immunity was rationalized and accepted by the immunology community at large.

Allograft rejection reactions posed a similar dilemma. From its inception, early in the twentieth century, the study of rejection of allogeneic tumors or normal tissue transplants in animals showed that this process displayed many of the characteristics of immune reactions, such as memory and specificity. For example, an inbred mouse that had previously rejected a tumor or skin from an allogeneic inbred mouse strain would reject a second graft from the same strain in a greatly accelerated manner. This accelerated "secondary response" phenomenon was well known to immunologists, and was typical of virtually every immune response that had been studied. In keeping with what was known about immune reactions generally, the secondary response was highly specific. The mouse just described, if secondarily transplanted with a tumor from an allogeneic strain unrelated to the first, rejected that graft with the slower kinetics typical of a primary reaction in an antigenically naive mouse. Gorer and Medawar would show some years later that an exchange of tumors between allogeneic animals was no different than the exchange of normal tissues between the same animals, an observation that set the field of tumor immunology back for several decades. We explore that story in Chapter 10.

So allograft rejection looked very much like an immune phenomenon. But as with DTH reactions, it was not possible to show that specific immunity to a tumor or skin allografts could be transferred from an immune animal to a non-immune animal with serum or any fraction thereof. The ability to carry out such "passive transfer" of protective immunity was one of the founding doctrines of immunology, and the failure to demonstrate it in tissue rejection led many to challenge the notion that rejection was immunological in nature. N. A. Mitchison unraveled this dilemma by showing that graft immunity, like DTH sensitivity, could be transferred between mice with immune lymphocytes rather than immune serum (Mitchison, 1953).

Mitchison's observation that graft reactivity could also be transferred with lymphocytes strengthened the notion of cellular mediators of immunity. Eventually it was shown that the cells responsible were under developmental control of the thymus, and allograft reactivity would be one of the hallmarks of this newly defined subset of "T" lymphocytes (Miller and Osoba, 1967; Manning et al., 1973). But at the time of Chase's and then Mitchison's seminal observations, a role for lymphocytes even as producers of antibodies had not yet been established, so the precise role lymphocytes might play in allograft rejection was not at all clear. And as we will see in this chapter, to this day a clear separation between the role of DTH and cytotoxicity in allograft reactions has been difficult to establish.

2 INITIATION OF ALLOGRAFT REJECTION IN VIVO

2.1 General requirements for primary CTL activation

In Chapter 2 we discussed some of the requirements for activation of CTL in virgin, resting CD8 T cells. The first step (signal 1) involves interaction of the CD8 T cell via its antigen receptor complex with cognate antigen. In the case of an allograft reaction, cognate antigen consists of a peptide-modified allogeneic class I molecule. Complete development of signal 1 requires the participation of an array of costimulatory interactions mediated through the immunological synapse. The result of signal 1 is up-regulation by the resting CD8 cell of receptors for lymphokines needed to drive it to the fully mature, cytolytic state. These lymphokines (signal 2) are supplied largely by CD4 cells activated at the same time by class II MHC antigens of the graft.

There is evidence that the requirements for activation of CD8 cells by cognate antigen are considerably less rigorous than for CD4 cells. For example, the number of CD8 TCR-MHC interactions required for activation is much less than for CD4 cells, approaching the level of 1-10 TCR contacts per CD8 cell (Brower et al., 1994; Kageyama et al., 1995; Sykulev et al., 1996). This may be because the CD8 molecule itself, a crucial costimulatory molecule, binds to class I MHC with a much greater affinity than that displayed in CD4-class II interactions (Garcia et al., 1996; Gao et al., 1997; Delon et al., 1998; Kern et al., 1998). CD8 cells also require a shorter contact time with antigen to achieve full activation than do CD4 cells (Mercado et al., 2000; Kaech and Ahmed, 2001; van Stipdonk et al., 2001), and yet in some situations they proliferate more vigorously (Foulds et al., 2002) and longer (Homann et al., 2001) than CD4 cells.

The CD28-B7 interaction so crucial in CD4 T-cell activation is apparently also less critical in activation of CD8 T cells, since CD28-knockout mice can still generate CTL, mount antiviral responses and reject skin grafts (Shahinian et al., 1993; Kawai et al., 1996), although the rejection of heart allografts may be impaired (Turka et al., 1992; Baliga et al., 1994). Even in cases where interference with CD28/B7 engagement shows an effect on development of allograft rejection or killer lymphocytes, the effect is rarely greater than fifty percent (Guerder et al., 1995; McAdam et al., 2000), and appears to be acting more on CD4 than CD8 cells (Newell et al., 1999). Although undoubtedly CD28 engagement normally enhances CD8-cell activation, it provides but one of several costimulatory pathways that can do so. CD40-CD40L interactions are also less crucial in CD8 activation (Whitmore et al., 1999). These properties of CD8 cells as they undergo functional maturation will be important as we attempt to understand how allograft reactions are initiated in vivo, and the role of cytotoxicity in allograft rejection.

2.2 The role of dendritic cells in triggering vascularized allograft rejection

Transplanted solid tumors rely on rapid establishment of a vascular connection with the host for survival. This connection is made differently depending on the transplant. Transplanted organs such as kidney or liver come with their own intact vascular network, and are surgically connected to the host circulatory system at the time of transplant. Skin transplants also come with a resident vasculature, but for practical reasons cannot be surgically connected to the host circulation. In that case, as we will see, the resulting graft vasculature consists of host vessels that rapidly infiltrate the transplant (most likely in response to signals such as VEGF emanating from ischemic transplant cells), most of which quickly anastamose with transplant vessels, resulting in a hybrid vasculature within the transplant. In the case of free tumor cells transplanted subcutaneously or elsewhere in the body, vascularization is entirely dependent on infiltration of host blood vessels.

The cells contained in allografted tissues of these types are almost uniformly class I MHC-positive, and thus potentially recognizable by recipient CTL. However, in vitro studies had shown that naïve T cells (both CD4 and CD8) are essentially non-reactive to most parenchymal cells, whereas (as evidenced by MLC reactions) they respond vigorously to leukocytes. In vitro studies also demonstrated the need, within a stimulatory leukocyte population, for an adherent accessory cell to achieve full sensitization (Davidson, 1977). George Snell had proposed many years earlier, based on in vivo experiments, that leukocytes entrapped in graft vasculature ("passenger leukocytes") were likely to be the provoking cells in the rejection of skin allografts (Snell, 1957). Steinmuller showed that leukocytes alone could induce a state of allograft sensitization leading to subsequent accelerated rejection of solid-tissue allografts (Steinmuller, 1967). Some researchers thought passenger leukocytes would be necessary to provide a class II stimulus for the production of CD4 helper T cells (Bach et al., 1976), but others thought the situation would be more complex (Lafferty, 1983).

In fact we now know that the critical accessory cell for optimal T-cell activation, in vitro or in vivo, is the dendritic cell, which provides not only class I and class II MHC signals, but also numerous ligands interacting with co-stimulatory receptors that are part of the T-cell receptor complex (in the immunological synapse) discussed in Chapter 1. Dendritic cells (DC) are what is generally referred to when the term "professional APC" is used. DC encompass a loosely related and broadly distributed lineage of bone marrow-derived leukocytes that play a unique stimulating role in both innate and adaptive immune responses. Dendritic cells are found in virtually every tissue in the body, including skin (where they are known as Langerhan's cells) and lymphoid tissue, and they circulate in blood and particularly in lymph (Tew et al., 1982; Austyn and Larsen, 1990; Shortman and Liu, 2002). Some dendritic cells are not phagocytic, but all are vigorously endocytic and display elevated levels of both class I and class II MHC proteins. There is compelling evidence to support the notion that dendritic cells are the critical passenger leukocytes brought in with tissue transplants, and stimulate the host immune system to mount a rejection response. For example, if DC are destroyed in grafts prior to transplantation, rejection can be circumvented, but if donor DC are introduced into an animal onto which such an allograft has been placed, rejection is swift and certain (Lafferty et al., 1976; Lechler et al., 1982; Faustian et al., 1984; Benson et al., 1987; Iwai et al., 1989). Recently attention has been focused on the role of dendritic cells in inducing tolerance, and a possible role for this in managing allograft rejection (Thompson and Lu, 1999).

How T cells encounter DC during the course of sensitization to vascularized allografts has been a subject of intense investigation, driven by the possibility that if this interaction could be manipulated it might have profound effects on the course of transplant rejection. Key to this inquiry has been a study of trafficking patterns of both host and donor DC to and from lymphoid tissues (Austyn and Larsen, 1990). In general DC enter lymph nodes via primary afferent lymphatic vessels, which originate in open tissue spaces (lymphatic sinuses). Once inside the node, DC home to T-cell compartments such as the paracortex. Host dendritic cells are mobilized and attracted to inflammatory sites under the influence of locally produced inflammatory cytokines (Kaplan et al., 1987; Sallusto and Lanzavecchia, 1999). Activated DC drifting away from these sites and into the lymph fluid appear to have MHC-bound antigenic peptides on their surface. Antigen-pulsed DC tend to remain in regional lymph nodes for long periods of time, and only a few traffic onward to the blood stream. Those that do reach the blood stream in turn home into the spleen and liver (Kupiec-Weglinski et al., 1988).

In the specific case of allografts, the question of whether recipient DC circulate from the blood to implanted allografts and back to host lymphoid tissues is controversial and may depend on the grafted tissue (Larsen et al., 1990a; Saiki et al., 2001b). On the other hand, *donor* DC do migrate from allografts, both those revascularized by surgical anastamosis (cardiac, kidney; Larsen et al., 1990b), or by infiltration of host vasculature into skin or tissue fragments implanted e.g. under the kidney capsule (Hall, 1967; Tilney and Gowans, 1971). Both host and donor DC migrate into draining lymph nodes and to the spleen. Antigen-stimulated DC interact directly with and expand CD8 T cells (Tsunetsugu-Yokata et al., 2002). Draining lymph nodes are the major site for T-cell activation; little if any primary T-cell activation occurs at the graft site.

The involvement of DC in the activation of both CD4 and CD8 T cells poses interesting questions for allograft rejection that are not fully resolved. Do host T cells respond directly to allogeneic donor DC MHC alloantigens during a primary allograft response, or indirectly, to self (host) DC that have scavenged donor antigenic material? In vitro, in the mixed leukocyte culture (MLC) reaction, depletion of adherent accessory cells (now presumed to be dendritic cells) from the responding population has little effect on either the proliferative or cytotoxic phases of the reaction. Removal of adherent cells from the stimulating population, on the other hand, results in profound suppression of both phases. Moreover, as just discussed, when DC are depleted from allografts prior to transplantation, rejection is also greatly suppressed. These data have been interpreted to mean that responder T cells directly recognize and are activated by MHC and other cell surface products on allogeneic stimulator DC.

On the other hand, skin allografts taken from mice lacking functional class II genes are rejected just as rapidly as normal skin, implying that critical class II signals required for recipient CD4 T cell activation can be provided by host class II-positive cells (Auchincloss et al., 1993; Gould and Auchincloss, 1999). In such cases it is believed that damaged donor cells sloughed off after transplantation enter the host circulation, eventually reaching regional lymph nodes where they are phagocytosed by recipient interdigitating reticular cells, a form of dendritic cell that is phagocytic, which in turn present donor information to recipient T cells (Saikai et al., 2001a). Alternatively, donor material could be scavenged at the transplant site by recipient DC, and carried back to regional lymphoid tissues. The operation of an indirect activation pathway for CD4 T cells in allograft rejection has in fact been observed (Lee et al., 1997).

2.3 Initiation of allograft responses in free-cell suspensions

The rejection of allografts consisting of free-cell suspensions (usually allogeneic hematopoietic tumor cells introduced into the peritoneal cavity of an immunocompetent host), also presents some interesting problems in understanding both initiation of the response and ultimate rejection. In this situation there is no vascular connection between graft and host; oxygen and nutrients are obtained from ascites fluid accumulating in the peritoneum. The growth of ascites tumors in allogeneic hosts is initially similar to the unimpeded expansion observed in syngeneic hosts, but after several days a dramatic decrease in the number of tumor cells occurs, and abundant, highly potent peritoneal exudate CTL (PEL) specific for the allogeneic tumor are found within the peritoneal cavity (Chapter 3).

A major question in such cases is, what signals on the allografted cells control CTL generation? The vast majority of tumors used as allografts display neither class II antigens, nor important co-stimulatory ligands such as B7. PEL can be generated with tumor cells that have been maintained in vitro for many generations, eliminating potentially stimulatory contaminating cells such as dendritic cells or macrophages as the source of the stimulus. So how does the reaction get off the ground? How are naïve host CD4 cells, which are essential for CD8 cell maturation, stimulated to respond in the absence of stimulatory class II antigens, and the ligands for activation co-receptors on CD4 cells? The most likely possibility is the indirect route for sensitization described above for solid transplants. Host dendritic cells in the peritoneum could scavenge dead or dying graft cells, migrate to local lymph nodes, and present donor peptides in association with class II to host CD4 cells, which would then release the needed cytokines.

What is not entirely clear in this scheme is how CD8 cells receive required primary and costimulatory signals ordinarily obtained from CD4-activated dendritic cells. In allografts we think that CD8 cells are normally triggered by allogeneic class I plus peptide on donor DC, and fed costimulatory signals by the DC at the same time. There are no allogeneic class I molecules on self DC. It is possible that the responding CD8 cell could pick up this signal from the tumor cell itself *if* the tumor cell also displayed some of the required accessory molecules such as CD40. While in general CD8 cells are thought not to use their limited CD40L in activation sequences (Clarke, 2000), there have been reports that under some circumstances they may do so (Lefrancois et al., 1999; Eck and Turka, 1999). But clearly there are some things we do yet not fully understand about sensitization against allogeneic cell suspensions in vivo.

3 MECHANISMS OF ALLOGRAFT REJECTION IN VIVO

3.1 The histopathology of allograft rejection

Cellular and molecular mechanisms of allograft rejection worked out in vitro must ultimately be compatible with what we actually observe in vivo. Long before such explanations were even attempted, the classic and detailed histopathological analysis of allograft rejection undertaken by Kidd provided a solid framework for mechanistic interpretations of this process (Kidd and Toolan, 1950). He studied the fate of several types of tumor cell suspensions implanted subcutaneously into susceptible (syngeneic) and resistant (allogeneic) hosts. When implanted into syngeneic mice, the inbred strain in which they had originated, the cells of these cancers progressed and formed tumors that killed the animals within a few weeks. The cancer cells grew equally well for a time after implantation into allogeneic mice, often forming palpable nodules within a week to ten days. After that the tumors stopped growing and disappeared within about a week. The mice in which the tumor allografts had regressed were highly resistant to re-implantation with the same tumor cells, forming no visible or palpable nodules.

Microscopic analysis revealed that five-seven days after implantation of allogeneic tumor cells, lymphocytes began to accumulate in blood vessels serving the tumor nodules and surrounding tissues. Kidd observed that the number of lymphocytes increased rapidly, extravasated, and began to penetrate among the proliferating cancer cells, moving inward from the periphery of the nodules. The penetrating lymphocytes were highly pleomorphic, typical of activated lymphocytes observed in the presence of antigen in vitro. Not infrequently two or more of the lymphocytes were seen engaged with a single tumor cell. Kidd observed that once the lymphocytes had made contact with the tumor cells, the latter then began to die one after another in rapid succession. Importantly, he noted that the tumor cells died in a very different way from cells killed by heat or toxins. He was seeing lymphocyte-induced apoptosis long before anyone else.

The cancer cells died individually; in a given field, the microscope often disclosed lymphocyte-associated cells in the process of dying, whereas nearby tumor cells, not yet contacted by lymphocytes, remained unchanged; some were even in mitosis. Even when the bulk of the cells at the periphery had been overcome, tumor cells in the central parts of the regressing growths remained morphologically unchanged and viable.

When the same allogeneic tumor cells were reimplanted into animals that had previously overcome the tumor, lymphocytes infiltrated the tumor site much more promptly. They quickly surrounded the site of implantation and became intimately associated with the numerous tumor cells, which were nearly always overcome in the immune hosts before they had formed palpable nodules. Although the process of regression was greatly accelerated in the immune hosts compared with naive animals, it was otherwise the same.

Only lymphocytes seemed to be active in tumor allograft destruction. Neutrophils accumulated in abundance around tumor cells within the first eight hours after implantation into mice regardless of their immune status. However, they were notably less numerous at 24 hours and were essentially absent by 48 hours. A few macrophages were found in areas in which regression was taking place, but they were no more numerous here than in other areas of the tumor or in the surrounding tissue of the host, and they never had a conspicuous relationship to the necrobiotic cancer cells

Kidd mixed lymph node cells, harvested from mice immediately after rejection of a tumor allograft, with fresh allograft cells in vitro for one-two hours, and then injected the mixture into a naïve host. No tumors developed in such cases.

Kidd's results, which confirmed and extended elements of an even earlier study (Murphy, 1913), led him to postulate most of the rules of cell-mediated immunity that would be "rediscovered" over the next quarter century. For example, here is a quote from his closing argument:

> *"From the findings as a whole, it seems obvious that the "sensitized" lymphocytes participate actively in the process whereby cancer cells are overcome in resistant and sensitized hosts. Indeed, it might be assumed at present, as a working hypothesis not contradicted by any of the available facts, that the sensitized lymphocytes attach themselves to the individual cancer cells and actually kill them, thus inducing the sequence of necrobiotic changes already described."*

In this one passage, Kidd anticipates the work of Mitchison (1953), Govaerts (1960) and Kerr et al., (1972).

An ingenious experiment carried out in the mid-1960s provided further insight into the interaction of host immune cells and allograft cells (Strober and Gowans, 1965). "A" strain rats were connected by vascular anastamosis to a semi-allogeneic $(AxB)F_1$ kidney maintained outside the body (extracorporeal transplant). After varying periods of exposure to the semi-allogeneic kidney, the recipients were disconnected and a week later grafted with $(AxB)F_1$ skin. Six days later the skin graft showed typical characteristics, as judged by histological examination, of accelerated second-set rejection. The authors concluded that host lymphocytes had been sensitized to "B" strain transplantation antigens at the graft site, and traveled back to host lymph nodes and spleen where they completed their activation sequence. This would seem at odds with what we currently think about the sensitization process in vivo, and the role of dendritic cells. The extracorporeal kidneys were perfused prior to anastamosis with the host to remove donor blood elements, but it is possible that residual donor cells still migrated into the host circulation during the experiment. This experiment has never been refuted, and should be kept in mind by anyone thinking about donor-host interactions during allograft rejection.

Yet another approach to analyzing the involvement of cells in allograft rejection came later, through the use of "sponge allografts" (Roberts and Hayry, 1976). A synthetic, sponge-like material was used as a bed for growing adherent cells (fibroblasts or dissociated solid tumor cells). Sponges containing these cells were then implanted subcutaneously into allogeneic recipients, and after varying lengths of time the sponges were removed and their cellular contents analyzed. When implanted into a naïve animal CTL activity could be detected within the sponge matrix, peaking at day 8 – the same kinetics as development of cytotoxicity in the spleen and lymph nodes. The sponge also contained numerous macrophages and some granulocytes. The fact that activated CTL were found in regional lymph tissue suggests the graft had been infiltrated by blood vessels, although this was not directly determined. When sponges were implanted into a previously immunized mouse, typical accelerated kinetics of infiltration and appearance of reactivated CTL in local lymph nodes were observed (Roberts, 1977).

3.2 The role of vascular damage in allograft rejection

Initially it was imagined that allograft rejection by lymphocytes would involve a cell-by-cell destruction of the entire graft by infiltrating killer T cells, essentially as described by Kidd. However, another possibility was put forward early on. In those cases where the graft is implanted with its own vascular system intact, killer cells might only have to attack and destroy the vasculature itself; the rest of the graft would die quickly through ischemic infarction. Evidence that this might be the case was gathered from a number of different allografting situations in animals and in humans (Henry et al., 1962; Waksman, 1963; Kountz et al., 1963; Flax and Barnes, 1966).

One of the earliest detailed analyses of the interaction between host immune cells and donor blood vessels was carried out by Dvorak in allografted human skin. Using a highly sophisticated microphotometric technique, Dvorak had observed that in contact dermatitis (a form of DTH reaction), infiltrating lymphocytes tended to cluster around capillaries and other blood vessels, and that the vessels appeared to be seriously damaged. In most cases, however, the reparative phase of inflammation took over and prevented complete destruction of the vessels (Dvorak et al., 1976). Subsequently applying his technique to an analysis of events accompanying skin allograft rejection, he saw the same thing – lymphocytes clustering around blood vessels, followed in this case by actual destruction of the vessels. Destruction of blood vessels in every case could be observed before necrosis of dermal cells and other nonvascular graft elements, which he concluded died from infarction and subsequent necrosis (Dvorak et al., 1979). Interestingly, host blood vessels in the immediately adjacent graft bed were not injured, suggesting considerable specificity of the host immune response.

These basic observations were confirmed and extended using a SCID mouse system in which the interaction between human lymphocytes and allogeneic human skin were studied. Human skin was grafted onto a SCID mouse, and allowed to heal in before intravenous infusion of allogeneic human lymphocytes (Murray et al., 1994). One of the important contributions of this paper concerns graft vascularization. It was long thought that in skin allografts, donor vasculature died via necrosis and was replaced by infiltrating host blood vessels. However, Murray et al., showed there is also considerable spontaneous anastomosis between donor and host vessels, even across species barriers. He found that class II MHC molecules and adhesion ligands were quickly up-regulated on the human vascular endothelium, and human lymphocytes were seen clustering around the human vascular elements in the graft, but not mouse vascular elements. Many of the lymphocytes clustering about blood vessels were perforin-positive. The graft human blood vessels as well as the human/mouse hybrid vessels were ultimately destroyed, but the graft survived as mouse blood vessels replenished the graft bed. The fact that host vessels were also destroyed in the experiments of Dvorak et al., but not those of Murray et al., is likely related to the fact that the latter's observations were made on grafts that had already healed in prior to onset of the allograft reaction, whereas Dvorak's observations were made on allografts still in the process of healing.Dvorak's observations thus more closely approximate a normal allograft situation.

The destruction of graft blood vessels as a preliminary step in vascularized allograft rejection has been further documented in heart (Forbes et al., 1983), liver (Matsumoto et al., 1993), and kidney (Busch et al., 1971; Leszczynski et al., 1987) allografts. In these cases, host and donor blood vessels are surgically anastomosed, and vascular damage to the graft is rapid and lethal. Perivascular accumulation of lymphocytes and monocytes/macrophages is followed by vascular degeneration, which is in turn followed shortly by graft failure. Thus in all solid tissue and organ allografts, it seems clear that allograft rejection is not caused by cell-by-cell destruction of the graft parenchyma, but rather by vascular damage, ischemia, and global tissue necrosis.

Vascular endothelial cells can in fact be killed in vitro by appropriately sensitized CTL (Collins et al., 1984). There has also been a suggestion that endothelial cells may themselves be stimulatory for CD8 CTL, although this notion has come largely from a single laboratory. Although CTL can be generated that have a degree of endothelial cell specificity, the CTL are rather atypical and it is not obvious from the data presented that endothelial cells are a major target in primary allograft sensitization and rejection (Biederman and Pober, 1998; Dengler and Pober, 2000).

3.3 DTH reactions and allograft rejection

DTH reactions are inflammatory in nature, and as such are often accompanied by significant "collateral damage" – destruction of nearby healthy tissues antigenically unrelated to the provoking antigen. But inflammatory reactions are remarkably efficient in clearing a wide range of potentially lethal infections, and moreover have built-in mechanisms that limit collateral damage and initiate rapid healing. The establishment of a delayed hypersensitive state to biological or chemical antigens can be brought about in numerous ways, which are beyond the scope of the present discussion. Establishment of a DTH state in a naïve animal is absolutely T-cell dependent, and depending on the particular antigen, its mode of administration, and the species involved, either CD4 T cells, CD8 T cells, or both, may be involved. Memory T cells resulting from the primary response initiate an enhanced inflammatory reaction upon re-exposure of hypersensitive animals to the original antigen.

The role of T cells in DTH is generally thought to be restricted to antigen recognition and the release of numerous inflammatory cytokines. The antigen-presenting cells in DTH are the same as in other T-cell activation reactions, namely dendritic cells and probably macrophages. Among the cytokines important in the development of DTH are IL-3, GM-CSF, IFNγ, TNFβ, MAF and MIF. These cytokines play a role in attracting monocytes and other nonspecific cells to the site of the inflammation, and in inducing changes in local vascular endothelium that favor extravasation of these cells into surrounding tissue spaces. Perivenous accumulation of monocytes is one of the early hallmarks of DTH reactions. Although full development of a DTH lesion may take up to 48 hours, an accumulation of macrophages may be evident in as little as 4 hours. T-cell cytokines are also important in activation of inflammatory cells. Particularly important in this latter regard is the role of MAF (macrophage activating factor), which promotes maturation of monocytes into macrophages, and stimulates enzyme production and phagocytosis. Once activated, macrophages and other nonspecific cells release cytokines of their own which further promote the inflammatory response. The damage caused by activated macrophages, which are considered the principal inflammatory mediator in DTH, is due to release of numerous hydrolytic enzymes and other degradative molecules during phagocytosis of nearby damaged cells. This damage makes no distinction between the compromised cells provoking the initial reaction and surrounding normal cells. However, once the provoking antigen is removed, the reaction rapidly diminishes and the macrophages release a spectrum of mediators that promote healing of residual tissue.

It would seem not unreasonable that some portion of the tissue destruction that occurs during allograft rejection, and other in vivo immune phenomena we will discuss in subsequent chapters, could be caused by the inflammation accompanying DTH. In fact, for many years immunology textbooks have classified allograft rejection as a "Type IV" hypersensitivity reaction. Other Type IV (DTH) reactions include the tuberculin reaction, contact reactivity to certain substances such as poison ivy or picryl chloride, and inflammatory responses to various fungal substances (Gell and Coombs, 1968). The case for involvement of DTH in allograft rejection, or at least its concomitant occurrence, was first made by Brent et al., (1962) and Brent and Medawar (1966), who showed that allografting is accompanied by a hypersensitivity state demonstrable by intradermal injection of graft donor cell extracts. This results in a typical "wheal and flare" reaction (erythema and induration) at the injection site within 1-3 days. It gradually became accepted that transplantation immunity, along with previously defined DTH reactions, were different expressions of a newfound "cellular immunity", or "cell-mediated immune response" (Brent et al., 1962; Waksman, 1962a; Uhr, 1966; Turk and Stone, 1963; Cooper et al., 1968; Good et al., 1969).

Thus a major question that must be addressed is, to what extent is allograft rejection a result of inflammatory DTH reactivity, and to what extent is it caused by cell-mediated cytotoxicity – by killer lymphocytes? Although triggering of a DTH inflammatory response requires specific T-cell recognition of allografted cells, the subsequent attack mechanism could be independent of target cell recognition, and basically any cell in the region of the reaction may be killed. This is the in vivo equivalent of the "innocent bystander" issue raised with in vitro systems evaluating CTL function (Chapter 4). On the other hand, if allograft rejection in vivo is caused by CTL directly engaged with specifically recognized target cells, there should be little or no collateral damage to innocent bystanders.

Early insight into this question in vivo was provided by a study carried out in Sweden (Klein and Klein, 1972). In one variant of the experimental scheme, varying numbers of sarcoma cells syngeneic to the host were mixed with allogeneic tumor cells to see if the former were killed as part of a generalized response to the allogeneic tumor cells. In fact, even when the syngeneic tumor cells were mixed at a frequency of only 10^{-4} with the allogeneic tumor cells, they were not killed; they grew out as syngeneic tumors that eventually killed their host. In this subcutaneous environment the dissociated tumor cells very likely reassociate, but certainly remain closely associated once injected. Yet the syngeneic cells were clearly not killed. These results argue for a highly discriminatory immune attack on the implanted cells.

Additional insight into this question came from an ingenious experimental system using allophenic mice (Mintz and Silvers, 1970). Allophenic or tetraparental mice are produced by mixing cells of allogeneic embryos at preimplantation stages of development. Individual cells in allophenic mice express (for example) either H-2^a or H-2^b MHC antigens, with no cells co-expressing both H-2^a and H-2^b. Mintz and Silvers found that when allophenic skin was grafted onto one of the parental strains, only melanoblasts and hair follicle cells expressing the MHC of the opposite parental strain were rejected, whereas those expressing the host's MHC type did not suffer irreversible damage.

Rosenberg and Singer refined this basic experiment. When skin from H-$2^{a\text{-}b}$ allophenic mice was grafted onto immunoincompetent H-2^b nude mice, the resulting graft was accepted and produced patches of both white (from the H-2^a skin cells) and black (from the H-2^b skin cells) fur. Subsequently, T cells from a normal immunocompetent H-2^b mouse were infused into the engrafted H-2^b nude mouse. Initially, there was a vigorous inflammatory response at the graft site that caused extensive damage to epidermal cells of both H-2^a and H-2^b origin. However, once the inflammatory response subsided, the underlying dermal cells returned to a healthy state, and regenerated fur – but only black (H-2^b) fur. The H-2^a cells had been killed by the post-inflammatory immune response of the infused H-2^b T cells, but the H-2^b-expressing cells were not killed (Rosenberg and Singer, 1988). Other experiments illustrating the exquisite specificity of allograft rejection in vivo involved placing syngeneic skin grafts adjacent to allogeneic grafts on a single graft bed, with the rejection showing a clear demarcation at the division between the two grafts.

The researchers involved in all of these experiments drew two conclusions. Allograft reactions are indeed accompanied by an inflammatory response (DTH), and this response can cause considerable collateral damage that is independent of graft-cell MHC expression. However, this damage is not sufficient to lead to allograft rejection; only those graft cells specifically recognized by cytotoxic T cells appear to be killed. If graft rejection were due to an inflammatory process, they reasoned, we would expect both to be killed, since the inflammatory mediators would spread readily throughout the entire graft. The data strongly suggested that allograft rejection in vivo is caused by antigen-restricted CTL bound to specific target cells, and not by antigen-nonspecific inflammatory mediators. Moreover, as shown subsequently, in mice in which the genes for interferon, or for interferon plus IL-2 – the two principal inflammation-promoting cytokines – are deleted, allograft rejection is unimpeded (Saleem et al., 1996; Zand et al., 2000).

While the notion that inflammation is not an exclusive or even general mechanism of allograft rejection may be correct, it is based on assumptions about the functional radius of DTH reactions that are difficult to assess. One of the T-cell cytokines released at the site of DTH reactions is macrophage inhibitory factor (MIF), which exerts a potent cytostatic effect on macrophages and would be expected to inhibit their migration much beyond the immediate site of the initial response (David et al., 1964). Moreover, given that the principal cellular targets in allograft rejection in most cases are vascular endothelial cells, inflammatory damage to the blood vessels around which monocytes and macrophages are clustered could appear to be highly specific in the allophenic system described above, and sufficient for allograft rejection.

In the case of ascites tumor allografts, the early stages of rejection are marked by a massive influx of macrophages (and possibly dendritic cells), the number of peritoneal macrophages being proportional to the number of tumor cells injected. This correlation suggested initially that macrophages are responsible for the tumor rejection. At the peak of allogeneic ascites tumor growth, peritoneal macrophage populations consist primarily of small macrophages with compact cytoplasm and rounded nuclei; they also exhibit only slight phagocytic activity for the tumor cells. As the number of tumor cells decreases, larger macrophages, which contain irregular lipid granules and tumor cell fragments (Amos, 1962) and which adhere to the tumor cells in vivo (Baker et al., 1962), are observed. In the system studied by Baker and co-workers, the intraperitoneal injection of tumor cells elicited tumor cell-macrophage clusters within 30 minutes, and the generation of cellular debris and large macrophage-tumor cell clumps within several hours. Given the prime role of macrophages in mediating DTH reactions, it is hard to imagine that they are not causing inflammmatory damage to ascites tumor allografts. However, as we have seen, PEL invading the peritoneal cavity are very efficient CTL and specifically kill cognate tumor target cells in vitro. In addition, mixing experiments of the Klein and Klein type just discussed have also been carried out in the peritoneal cavity. Even at large allogeneic to syngeneic tumor cell ratios, only the allogeneic tumor is cleared from the peritoneum, leaving the syngeneic tumor to grow out and ultimately kill the host (G. Berke, unpublished observations). Still, given the rampant macrophage-mediated inflammatory reaction that develops in the peritoneum, it seems unlikely that some tumor cells – perhaps both syngeneic and allogeneic - are not killed as a result of inflammatory damage.

3.4 The role of CD8-mediated cytotoxicity in allograft rejection

Although antigen-specific lymphocytotoxicity was discovered in connection with allograft rejection, it must be admitted that some 40-plus years later the evidence for a role of cytotoxicity in allograft rejection remains indirect at best. That cytotoxic T cells are generated in allograft reactions is beyond question, nor is the primacy of CD8 T cells in causing allograft rejection at issue. If the development of CD8 cells is prevented in mice, by disruption of the gene for β-2 microglobulin, the ability to reject allografts is lost[5]. What is unclear is the extent to which CD8 T-cell-mediated *cytotoxicity* is a factor in allograft rejection in vivo. Antigen-specific, in vitro-cytotoxic T cells can be recovered from within rejecting allografts (Bradley et al., 1985). Immune CTL populations, as well as CTL clones, can cause accelerated graft rejection in immunoincompetent hosts. Perforin- and granzyme-containing cells can be identified by in situ hybridization at graft rejection sites, and local levels of these agents seem to correlate with the vigor of rejection (Griffiths et al., 1991; Clement et al., 1994). Given what we know about CTL, perforin and granzymes it is hard to imagine that the cytotoxic processes we observe in vitro with these cells are not taking place in vivo. But most of the data gathered over the past four decades, while *consistent* with a role for cytotoxicity in allograft rejection, do not *demand* that cytotoxicity be a factor in rejection.

Important insights into involvement of the two known mechanisms of CTL-mediated lysis have come from gene knockout mice (perforin) and the natural mutant mouse strains *lpr* and *gld* (Fas and FasL mutations, respectively). In non-vascularized islet cell or tumor allografts (Walsh et al., 1996; Ahmed et al., 1997), in skin allografts with mixed donor-recipient vascularization (Selvaggi et all., 1996), and in revascularized heart allografts (Schultz et al., 1995), the same striking results were obtained. When donor cells or tissues from an *lpr* mouse, which lacks functional Fas, are transplanted into an allogeneic perforin knockout mouse, the rate of rejection

[5] It should be noted that such mice are not absolutely class I-negative. Some class I, especially at the D locus in mice, does get through to the surface in the absence of β-2 microglobulin (Bix and Raulet, 1992).

of the allografts is not different from the rate of rejection of Fas-expressing tissues transplanted into allogeneic perforin-expressing mice. In the former cases, neither perforin nor Fas can be involved in allograft rejection, yet the rejection rate is unaffected. As expected, allograft rejection in such cases is not accompanied by generation of CTL capable of lysing target cells in vitro. These results – particularly those involving transplantation of hematopoietic tumor cells like those used as targets in vitro CTL assays – provide powerful evidence that cytotoxicity mediated by CTL is not an absolute requirement for allograft rejection in vivo.

In all the above instances, mononuclear and lymphocytic infiltration of the allografts was the same, whether functional CTL (as measured by in vitro assays) were generated or not. In the case of hematopoietic tumor allografts (Walsh et al., 1996), some of the Fas-negative tumors were resistant to TNF-mediated cytotoxicity, yet were rejected with the same kinetics as TNF-sensitive tumors, suggesting TNF-mediated killing (Chapter 5) is also not a factor in allograft rejection. Most (17/21 tested) Fas-deficient tumors placed in vivo undergo up-regulation of Fas expression (Rosen et al., 2000). Although it is technically possible that in the case of Fas-negative tumor allografts, some tumors may have expressed Fas after transplantation under the influence of inflammatory cytokines (Rosen et al., 2000), the results just cited with *lpr* tissue transplants reinforce a lack of requirement for Fas in allograft rejection.

Cytotoxicity may thus simply be a marker for the sensitization of CD8 T cells that utilize an as-yet unidentified, non-cytotoxicity-based mechanism to induce allograft rejection. There have been hints of the involvement of perforin or Fas in a few cases, under special circumstances. For example, in the study of Schultz et al. cited above, if the allograft donor differed from the recipient at only a single class I MHC gene (Sub-locus), then Fas-deficient hearts were rejected more slowly in perforin-negative recipients (88 days survival) than in perforin-positive recipients (31 days survival), suggesting a possible role in this situation for perforin. And when *gld* mice, which express mutated (non-functional) Fas ligand, are used as recipients for fully allogeneic cardiac transplants, rejection is prolonged compared to normal recipients. However, when cardiac tissue from *lpr* mice, which lack Fas antigen, is transplanted into fully allogeneic normal controls, there is no impact on the rate of rejection (Seino et al., 1996). Thus the general immune disorder in the *gld* mice may account for the observed differences in graft rejection, rather than a defect in the FasL/Fas pathway of lymphocyte-mediated cytotoxicity *per se*.

Interestingly, both perforin and Fas appear to be involved in an allograft-related situation in vivo, namely graft-vs.-host disease (GVHD). GVHD is a major complication of bone-marrow transplantation, wherein mature donor

T cells present in bone marrow preparations (in part from ruptured bone marrow blood vessels) react against host MHC antigens. In immunocompromised recipients, this can result in skin and gastrointestinal lesions, lymphadenopathy, wasting and death. In a mouse model wherein recipients are irradiated prior to infusion of donor T cells, CTL specific for recipient class I MHC antigens develop in lymphoid organs, and the recipients invariably die after 6-8 days. When either the perforin or Fas pathways are blocked in this model, mortality is significantly delayed; when both pathways are blocked, mortality is eliminated. (Braun et al., 1996; Baker et al., 1996; Baker et al., 1997). Whether disruption of these pathways affects other manifestations of GVHD was not addressed in these studies. A critical role for granzyme B has also been discerned in GVHD mediated by CD8 T cells (Graubert et al., 1996).

Another surprising situation in which the perforin and Fas pathways play a role involves cutaneous DTH. Contact skin sensitivity to the chemical dinitrofluorobenzene (DNFB) in mice is mediated by CD8 T cells that recognize DNFB-modified peptides on self class I MHC molecules. The resultant inflammatory reaction manifests itself as an obvious dermatosis upon secondary exposure to DNFB, and CTL specific for DNFB-modified target cells can be isolated from adjacent lymphoid tissues. When mice doubly deficient in the Fas and perforin lytic pathways were painted with DNFB, they developed neither DTH reactivity nor actively lytic CTL. Mice deficient in either pathway alone developed both DTH and CTL, suggesting that either pathway is sufficient to support both reactivities (Kehren et al., 1999).

4 CONCLUSIONS

So where does all this leave us in terms of defining the mechanism of allograft rejection in vivo? As discussed earlier, the exquisite specificity of in vivo rejection would seem to argue against a generalized inflammatory attack on allografts as a mechanism. The Klein and Klein tumor mixing experiments in particular are difficult to reconcile with inflammation as an exclusive means of allograft rejection. In that experiment, recall, the syngeneic and allogeneic tumor cells were confined in a physically constricted space. Yet at a concentration of one syngeneic tumor cell per thousand or less, only the allogeneic tumor cells, and not the syngeneic cells, were rejected.

The specificity of CTL killing (as measured in vitro) seems more consistent with the specificity of allograft rejection we observe in vivo, and has since its first description been the presumed effector mechanism of choice in explaining allograft rejection. Cytotoxic T cells arise in vivo with the onset of allograft rejection; they recognize, bind to and kill specific targets in vitro; and passive transfer of CTL to immunocompromised hosts results in rapid and specific graft rejection. Could we imagine that CTL do not conjugate with cognate target cells in vivo? And once conjugated, could we imagine that they do not release cytotoxic granule contents, or engage with Fas where it is expressed? Yet as we have just seen, mice deprived of the use of these pathways of CTL-mediated cytotoxicity, as well as membrane TNF, are unimpaired in their ability to reject allografts. As counter-intuitive as it seems, we may have to admit that the vigorous cell-mediated cytotoxicity we measure using ^{51}Cr release assays in vitro is not required for allograft rejection in vivo. Of course, failure to disrupt allograft rejection in the absence of any one, or even all, of these cytotoxic mechanisms does not mean that cytotoxicity has no hand in allograft rejection – only that we must assign it a more modest role. Other mechanisms are clearly able to compensate for its absence.

We are left then with the vague and somewhat unsatisfying view that multiple, overlapping effector mechanisms must exist for allograft rejection, some perhaps involving cytotoxicity and some clearly not, some with specificity for donor antigens, and others involving indirect effector processes. We began this chapter with a brief discussion of the historical entanglement of DTH reactions and allograft rejection, and took note of the fact that for many years cell-mediated cytotoxicity was listed as simply one of several manifestations of DTH. But the in vitro evidence for a swift, powerful, highly specific cytolytic mechanism that could cause the complete destruction of allogeneic cells in a matter of minutes led much of the field away from the view of CTL acting through a mechanism notable for its lack of discrimination. It seems that in the end we may have come full turn; it may be time to put CTL back where they started out, as one of the mechanisms of Type IV hypersensitivity.

Perhaps this should not be surprising: given that allograft rejection can never be regarded as anything other than an accident of nature, why should there be a distinct, highly focused mechanism to deal with it? The rejection of organ and tissue transplants is clearly secondary to some other, more natural process, and as best we understand it, that process is elimination of cells compromised by intracellular pathogens, and perhaps oncological transformation. We turn to these more orthodox immunological challenges in the following chapters.

Chapter 8

CYTOTOXICITY IN IMMUNE DEFENSES AGAINST INTRACELLULAR PARASITES

Although CTL were discovered in the course of trying to understand the immunological basis for allograft rejection, it was not until the 1970s that the true significance of CTL began to be understood: the detection and elimination of cells compromised by intracellular parasites, and possibly by oncological transformation. Viable intracellular parasites produce parasite-specific proteins during their sojourn in the host cell. Fragments of these peptides are displayed at the host cell surface in association with MHC class I molecules, which leads to their elimination by host CTL. This provides a critical adjunct to the role of antibodies, which are only able to effect the elimination of nonself antigens from the extracellular milieu. By far the heaviest intracellular parasite burden consists of the wide range of viruses that find vertebrate cells a good (and necessary) place to reproduce, but CTL responses are also provoked by intracellular bacteria and eukaryotic parasites with obligate host intracellular stages in their life cycles.

We will not, in this chapter, explore the full range of host defenses against intracellular parasites. In fact most parasites are susceptible to both innate and adaptive immune elements, both humoral and cellular, at some stage in their infectious life cycle. Nor will we examine the repertoire of ingenious mechanisms employed by these parasites to evade host immune responses (see e.g., Hengel and Koszinowski, 1997). Our primary focus will be on a single question, closely related to that raised in connection with allograft immunity: When CD8 T cells or other killer lymphocytes are involved in defenses against viral or microbial intracellular parasites, what is the evidence that it is their inherent cytotoxic function, rather than their contribution to lymphokine-mediated hypersensitivity and inflammation, that plays a role in parasite control? There is a second aspect of intracellular infection of which we will ask the same question. Not infrequently, infection by a virus or other microbe is not *a priori* harmful to a cell; rather, it is the ongoing immune system-mediated damage to infected and nearby uninfected cells that causes disease, particularly when the infection is not cleared quickly – when it becomes chronic rather than acute. We will ask to what extent lymphocytotoxicity contributes to this *immunopathology*, as well as to the curative immune response itself.

Unquestionably, inflammation accompanies essentially all infections by intracellular parasites, and is an element of both the curative response and any accompanying immunopathology. The release of proinflammatory lymphokines, particularly IFNγ, is not only involved in, but critical to, the responses of all of the cells involved in the response: α/β and γ/δ T cells, as well as NK and NKT cells. One of the most important roles for IFNγ is in the arming of macrophages which, once activated, attack and sweep away everything in their immediate vicinity. Against this powerful background, can we discern any role for cell-mediated cytotoxicity?

1 VIRUSES

1.1 CD8 T cells and resistance to viral infection

Studies of CTL generation in response to viruses both in vivo and in vitro have been complicated by the fact that the immune status of the animals or individuals studied, with respect to the virus in question, is often uncertain. Mice in most colonies harbor numerous active or latent viruses, and human exposures to viruses are frequent and varied. Thus one is never certain whether the reaction under study is primary, involving the initial activation of virgin CD8 cells, or a mix of virgin and memory CD8 cells. Unless one is working with mice in a germ-free colony, most often it is the latter; even then, there is the problem of vertically transmitted endogenous viruses.

As far as can be discerned, the generation of CD8 CTL in response to viruses appears to follow the rules gleaned from studying the generation of allospecific CTL. The realization that what would ultimately be identified as CD8 CTL are involved in immune responses to viruses in vivo emerged from studies of the involvement of T cells in tumor immunity in animal systems, where many of the tumors used were virally induced (Chapter 9). The first to show that in vivo-generated killer cells could destroy virally infected targets in vitro was Lundstedt (1969; see Wagner and Rollinghof, 1973 and Cerottini and Brunner, 1974, for reviews of early work).

A series of studies appearing in the mid-1970s established that the generation of CTL against tumor/viral antigens was highly similar to the generation of CTL in allograft reactions. The responding cells belonged to the same class of T lymphocytes, and the cellular dynamics, the kinetics of the response (both "primary" and "secondary"), and the kinetics and specificity of lysis of ^{51}Cr-labeled target cells in vitro were essentially indistinguishable in the two cases (Plata et al., 1975, 1976). The relationship between CTL allograft reactivity and viral reactivity was greatly strengthened by Zinkernagel and Doherty's landmark paper in 1974 showing that the interaction of CTL with virally infected cells was restricted by self class I MHC molecules. Very quickly thereafter, virus-specific cytotoxicity came to be viewed as an interaction of killer lymphocytes with self class I MHC modified by virally derived products (Townsend et al., 1984, 1986; Braciale et al., 1987). By the mid-1980s, CTL clones for most of the commonly studied murine (Andrew et al., 1985; Anderson et al., 1985) and human (Sterkers et al., 1985) viruses were readily available.

Early studies suggested that as with alloreactivity, the generation of virus-specific CTL in vitro requires an antigen-processing adherent accessory cell, initially assumed to be a macrophage (Finberg and Benacerraf, 1981), but now widely considered to be a dendritic cell. Virus-specific CTL are readily generated in vitro or in vivo against dendritic cells either infected with a given virus, or transfected with key viral antigenic proteins. A partial list of viruses against which CTL can be generated in this manner in vitro includes cytomegalovirus (Einsele et al., 2002), human papilloma virus (Davidson et al., 2001), HIV (Lisziewicz et al., 2001), and Epstein-Barr virus (Savoldo et al., 2002).

As mentioned in Chapter 2, infection of dendritic cells with an intracellular pathogen, or even contact of dendritic cells with circulating microbial products, can lead to a state of activation of dendritic cells which allows them to interact directly with CD8 cells, often reducing the need for involvement of CD4 cells in CD8 cell activation (Ridge et al., 1998; Bennett et al., 1998; Schoenberger et al., 1998). Interaction of virus-specific CTL with cognate stimulating (and target) cells occurs through the same kind of immunological synapse described in previous chapters for allospecific CTL; viruses able to interfere with formation of this synapse can evade a CTL response (Coscoy and Ganem, 2001).

One major piece of evidence supporting a role for CD8 CTL in controlling viruses comes from adoptive transfer studies. Adoptive transfer of EBV-specific CTL has been used successfully to treat human patients with EBV-induced lymphoproliferation and EBV lymphoma (Rooney et al., 1998; Comoli et al., 2002). Similarly, adoptive transfer of CTL clones has proven effective in treating opportunistic CMV infections (Walter et al., 1995).

1.2 NK and NKT cells in viral defense

As indicated in Chapter 6, NK cells play an important role in resisting viral infection. MICA and MICB proteins induced by microbial infection are recognized targets for NK activation. The exact degree of NK involvement varies among different viral groups, possibly depending on the initial cytokine pattern induced by infection. Cytokines released by virally infected cells, including dendritic cells, have major activating effects on NK cell activity. Interferons α and β increase cytotoxic function in NK cells, both in terms of per-cell cytotoxic activity, and the range of target cells killed. IL-12 is a powerful stimulant of IFNγ production in NK cells (Orange and Biron, 1996). These two signals can act antagonistically, significantly altering the effector pathways used by NK cells depending on the local cytokine environment (Nguyen et al., 2000). In addition to IFNγ, NK cells can also make TNFα and macrophage inflammatory protein-1α (Bluman et al., 1996).

The herpes group of viruses are among the more susceptible viral families to control by NK cells, although the signaling pathways involved are complex and not fully understood. Depletion of NK cells by treating mice with antibodies to asialo-GM1 greatly increased sensitivity to infection with the herpes virus CMV (Bukowski et al., 1984). However, it is not clear this marker is absolutely confined to NK cells, and it is possible other cell subsets may have been deleted as well. Like many other viruses, herpes viruses such as CMV interfere with peptide loading in the class I MHC pathway, greatly decreasing expression of classical class I molecules at the surface of the infected cell (Tortorella et al., 2000). While reducing their exposure to CD8 CTL, reduction of class I removes a potent inhibitory signal for NK cells. However, CMV can also up-regulate expression of HLA-E in humans and Qa-2 in mice. As discussed in Chapter 6, these non-classical class I molecules present peptides derived from degraded classical class I molecules, and they do so in a way that bypasses normal peptide loading. Both HLA-E and Qa-2 inhibit NK cell function. And finally, the viruses themselves encode "decoy" class I-like molecules that may be perceived as inhibitory signals by NK cells (Farrell et al., 1997; Tomasec et al., 2000).

A major activation receptor for NK cells during viral infection appears to be the DAP-12-associated Ly49H, although its exact mode of action is uncertain (Brown et al., 2001; Lee et al., 2001). In some strains of mice, we do know that Ly49H can be triggered by interaction with class I MHC-like proteins (e.g. m157) encoded by murine cytomegalovirus (Arase et al., 2002; Smith et al., 1998). However, in other strains the m157 protein acts as an inhibitor of activation, just as a class I molecule ordinarily does. Many viruses (papilloma viruses; poxviruses; most retroviruses; flaviviruses) have developed a sophisticated repertoire of mechanisms for evading NK cells (reviewed in Orange, 2002; Moser et al., 2000). In the case of viruses such as LCMV, NK cells may play a more important role in facilitating CTL responses than in directly controlling the virus (Biron et al., 2002; Liu et al., 2000). Although HIV infection of CD4 T cells decreases expression of MHC class I in humans, HIV-infected cells are not susceptible to attack by NK cells (Bonaparte and Barker, 2003). Thus while NK cells are still presumed to participate in viral control, aside from the herpes family in mice a great deal more needs to be learned before a definitive assessment of their direct contributions can be formulated.

NKT cells (Chapter 6) are a subset of T cells displaying NK-cell markers and receptors, and expressing a limited repertoire of α/β T-cell receptors restricted by the nonclassical class I structure CD1. NKT cells can be up to a quarter of resident hepatic lymphocytes (in mice), and they release IFNγ; together with the fact that liver cells are high expressers of CD1 molecules, NKT make cells credible candidates for a role in viral defense. Activated NKT cells have been shown to inhibit replication of hepatitis B virus in mice (Kakimi et al., 2000), and to protect against encephalomyocarditis virus (Exley et al., 2001). It is unclear at present whether NKT cells use their potential cytotoxic function against virally infected cells.

1.3 γ/δ T cells

Potentially cytotoxic γ/δ T cells are involved in numerous viral infections (see Dechanet et al., 1999, for review of early work). The Vγ9/Vδ2 subset proliferates in response to, and is cytotoxic in vitro against, HIV-infected cell lines (Martini et al., 2002). However, as is the case with CMV infection in humans (LaFarge et al., 2001), virus-specific γ/δ T cells tend to disappear in vivo when viremia is not quickly resolved. Nevertheless with most viral infections γ/δ T cells are among the earliest responders to infection, and there are fairly high levels of virus-reactive γ/δ T cells present in uninfected animals for most viruses, suggesting they respond to something other than viral peptides (e.g. Selin et al., 2001). γ/δ T cells have also been implicated in liver pathology during hepatitis infections (Tseng et al., 2001).

1.4 The role of cytotoxicity in clearing viral infections

Evidence for a role of classical CTL-mediated cytotoxicity in clearing viral infections in vivo comes from several directions. Initially, much of this evidence was indirect. For example, the level of perforin-positive lymphocytes at the site of viral infections correlates well with the kinetics of clearance of viral infections (Muller et al., 1989; Young et al., 1989), but this could be either CD8 CTL- or NK-cell-associated, and in any case is certainly not proof that cytotoxicity plays a role. The exquisite specificity of viral clearance in vivo has been taken as evidence in favor of direct CTL-mediated cytotoxicity. Insight into this question is provided by mixing experiments similar to those carried out in the study of allograft reactions (Chapter 7). Lymphocytic choriomeningitis virus (LCMV) and Pichinde virus (PV), both arenaviruses, infect many of the same host cells in mice; coinfection with the two viruses shows cells adjacent to each other in spleen infected with one or the other virus in a random fashion. When CTL specific for one or the other virus are transferred into immunoincompetent, doubly infected mice, only the virus specifically recognized by the transferred CTL is cleared (McIntyre et al., 1985). A similar experiment carried out with CTL specific for either of two strains of influenza A virus showed that again, when adoptively transferred into mice coinfected with the two strains, the CTL cleared infection only by the specifically recognized strain (Lukacher et al., 1984). In both cases, it was argued that the intimate proximity of cells infected by one or the other virus argues against a nonspecific clearance mechanism such as delayed-type hypersensitivity, and favored direct, MHC-guided, cell-mediated cytotoxicity.

However the most direct evidence bearing on the involvement of cytotoxicity in viral immunity comes from perforin knockout mice, and Fas-deficient mouse strains. The initial round of papers on perforin knockout mice showed that upon exposure to the noncytolytic murine lymphocytic choriomeningitis virus (LCMV), perforin-deficient mice could not clear the resulting infection even though CD8 T cells were clearly activated (Kägi et al., 1994; Walsh et al., 1994). NK cell activity was also greatly suppressed in these mice, which doubtless contributed to difficulty in clearing the virus. CD8-mediated immunopathologies associated with LCMV infection, such as hepatitis (measured by release of liver enzymes) and destruction of brain tissue (lethal endpoint) were greatly reduced or absent in perforin-deficient mice. These experiments provide strong evidence for a role of the degranulation pathway in both LCMV clearance and the accompanying immunopathologies (see below), a conclusion reinforced by a demonstration of the involvement of granzymes in some viral responses (Mullbacher et al., 1999a).

The Fas pathway plays no demonstrable role in clearance of LCMV, and neither the Fas nor perforin pathways are involved in clearance of cytopathic viruses such as vesicular stomatitis virus (VSV), influenza virus, or Semliki forest virus (SFV) (Kägi et al., 1995). The latter authors made the interesting suggestion that lymphocytotoxicity may be particularly important in the response to noncytopathic viruses. Because these viruses spend a smaller proportion of their life cycle in the extracellular milieu, they are less exposed to soluble mediators such as antibodies. In the case of cytopathic viruses, the virus may well destroy the infected cell before a CTL response can even be mounted. This proposal seems generally to be true. The granule exocytosis pathway has been shown to play a role (although rarely an exclusive role) in clearing infections with murine AIDS-like viruses (Tang et al., 1997), a murine neurotropic hepatitis virus (Lin et al., 1997), ectromelia, a cytopathic orthopoxvirus (Mullbacher et al., 1999), and murine CMV (Tay and Welsh, 1997; Riera et al., 2000). As with LCMV, it is difficult to know how much of the missing cytotoxic function would have been contributed by CTL, and how much by NK cells. Neither perforin nor Fas play a role in clearing noncytopathic murine rotavirus infections (Franco et al., 1997).

While both CD4 and CD8 cells are normally involved in the response to LCMV, acute clearance can be achieved in the absence of CD4 cells (Ahmed et al., 1988; Rahemtulla et al., 1991). The most likely explanation for this is that the key role for CD4 cells in such cases, which is activation of dendritic cell APC through the CD40-CD40L interaction, is rendered less important because the virus itself activates the DC. This is supported by the finding that CD40L-deficient mice also mount an acute response to LCMV (Thomsen et al., 1998). However, as we will discuss in Chapter 11, longterm maintenance of the CTL response, and development of T-cell memory, both require CD4 help.

Influenza is another good example of a noncytopathic virus in which cytotoxicity appears to play a role in viral clearance. Both CD4 and CD8 T cells are engaged in viral clearance, but CD8 cells – some of which are specific for the FluA NP-147 K^d-restricted nucleoprotein epitope (Townsend and Skehel, 1984) – are particularly crucial. However there are at least six addtional CD8 T cell epitopes for influenza in K^d mice including PB2-146, PB-2-185, PA, PB-1, PB1-RF2, which are as dominant as the NP epitope, and it is not clear which of the CD8 cells is required for viral clearance (Welsh, et al.,004).

Mice in which CD8 cells are eliminated cannot clear an influenza infection, suggesting that NK cells, alone or together with CD4 cells, do not provide enough protection (Topham, et al., 1997). Comparable levels of CTL develop in mice lacking CD4 cells (Belz et al., 2002), and cloned CD8 CTL alone can transfer protective immunity to flu, particularly if administered at the beginning of infection (Lin and Askonas, 1981). Transfer of CD8 cells is followed by perivascular accumulation of lymphocytes and macrophages, followed by destruction of nearby lung epithelial cells. After a short period the infection is cleared, the damaged cells are removed by phagocytes, and the cellular infiltrate recedes (MacKenzie et al., 1989).

Immunization in the absence of CD4 T-cell help results in defective CD8 T-cell memory, and diminished protective immunity. The stage during the immune response when CD4 T cells are essential in promoting functional CD8 T-cell memory has been studied by Bevan and colleagues (Sun et al., 2004). Memory CD8 T cell numbers decreased gradually in the absence of CD4 T cells, despite the presence of similar numbers of memory cell precursors at the peak of the effector phase. Adoptive transfer of effector or memory CD8 T-cells into wild-type or CD4 T cell-deficient mice demonstrated that the presence of CD4 T cells was important only after, not during, the early CD8 T-cell programming phase. In the absence of CD4 T cells, memory CD8 T cells became functionally impaired and decreased in quantity over time. In sum, CD4 T cells are required primarily during the maintenance phase long-lived CD8 memory, for which the IL-7 receptor is an important marker.

When influenza infection is studied in mice in which the perforin pathway, the Fas pathway, or both are disabled, it is clear that crippling of either pathway alone causes only a modest reduction in ability to clear the virus. When both are disabled, however, the mice cannot clear the infection (Topham et al., 1997). Interestingly, spleen cells from perforin-deficient mice after a week or two of influenza infection produce a hundred times more IFNγ, than control mice when presented with infected cells (Sambhara et al., 1998). The relative frequency of CD4 and CD8 T cells in the spleens of influenza-infected perforin knockout mice is the same as normal controls; the increased cytokine production is due to an estimated 10-fold expansion of flu-specific, hyper-producing CD8 cells in the T-cell pool. This could be due to continued antigen-driven expansion of T cells caused in turn by failure to clear the infection completely, or a failure to reduce the number of flu-specific CTL as the infection subsides. The latter may well be the case, as we will discuss in Chapter 11. But in spite of this greatly increased production of IFNγ, the perforin/Fas-deficient mice still cannot clear an influenza infection, underscoring the key role of cytotoxicity in defense against this virus.

An interesting problem associated with CTL killing of virally infected target cells is that target cell death, whether caused by CTL or NK cells, via either the degranulation or Fas pathway, is apoptotic, and apoptotic bodies released from dying cells can shelter viruses from soluble mediators as effectively as intact cells. These bodies are readily taken up by neighboring cells (Albert et al., 1998), and thus apoptotic cell death could be seen as promoting viral dissemination. In fact, chemical and radiation-induced apoptosis of virally infected cells has been shown to do just that (Holmgren et al., 2000; Mi et al., 2001; Sasaki et al., 2001), and as we will see below this approach has even been used to purposely deliver viral DNA into dendritic cells, which readily take up apoptotic bodies (Russo et al., 2000; Melero et al., 2000). However, in the case of CTL-induced apoptosis, we know that fragmentation of target-cell DNA begins within seconds of CTL-target conjugation (Russell, 1983). Thus, as suggested by Eric Martz (Martz and Howell, 1989; Martz and Gamble, 1992), the most important aspect of CTL interaction with virally infected cells may not be the destruction of the target cell *per se*, but rather degradation of intracellular viral DNA.

Although there is clear evidence for involvement of the degranulation pathway in clearing many viral infections, with the exception of influenza virus there is little data to support a role for the Fas pathway. This is not surprising, since not all cells are Fas-positive, and thus Fas-mediated killing is unlikely on its own to completely purge virus from an infected host; Fas-negative cells would provide a depot from which a constant supply of virus could emerge. While it is difficult to rule out completely an involvement of Fas in the clearance of many viruses, it seems that in the vast majority of cases where cytotoxicity is involved in controlling a viral infection, this is via the degranulation pathway. On the other hand, as we see in the following section, there is evidence supporting a role for both perforin- and Fas-mediated cytotoxicity in many immunopathologies associated with viral infection.

It is important to note that perforin knockout mice and Fas mutant mouse strains have no impairment in their ability to produce and secrete IFNγ upon stimulation with cognate antigen. If anything, IFNγ production is somewhat increased in such strains (Sad et al., 1996). These mice thus allow us to draw clear conclusions about the relative contributions of degranulation-based cytotoxicity, and production of IFNγ, which is often touted as the most important contribution of CD8 cells to the clearance of intracellular pathogens.

1.5 Cytotoxicity as an element of viral immunopathology

Viruses that are not cleared quickly from the system may go on to establish chronic infections, achieving a balance between destruction of virally infected cells and de novo infection of new cells. The chronic immune responses induced in such cases ultimately cause more problems for the host than the virus itself, which in many such cases is relatively harmless. This is different from a carrier state for a particular virus, in which the virus lies dormant and does not provoke a measurable immune response. The damage done to otherwise healthy tissues in chronic viral infections can seriously compromise the host, and can be lethal if untreated.

The immune-based, lethal neuropathology accompanying LCMV infections in mice is perhaps one of the most thoroughly studied immunopathologies in animal systems (Doherty and Zinkernagel, 1974). This pathology does not develop in mice without T cells (Buchmeier et al., 1980; Doherty et al., 1988), and involves infiltration into multiple brain tissues of both CD4 and CD8 T cells. In perforin knockout mice, this neuropathology does not develop (Kägi et al., 1994a), although infiltration of CD4 and CD8 cells is unimpaired, suggesting it is the perforin-based cytotoxic function of these cells rather than their ability to induce inflammation that is critical. LCMV also causes an immune-based viral hepatitis in mice (Zinkernagel et al., 1986), which can be measured by release of specific enzymes from damaged liver cells into the blood. This pathology appears to be absent in perforin knockout mice (Kägi et al., 1994a), suggesting the unaffected Fas pathway in these mice was not involved. However later studies have suggested that the Fas pathway is in fact at play (Nakamoto et al., 1997; Balkow et al., 2001), and can involve CD4 as well as CD8 cells (Zajac et al., 1996). Liver cells in mice are strong expressers of Fas; administration of whole Fas antibody to mice is lethal in a matter of hours (Ogasawara et al., 1993). The murine hepatitis B virus (HBV) mimics human hepatitis viruses in that mice failing to clear HBV infections efficiently develop a fulminant hepatitis. This condition is triggered by Fas-mediated killing of liver cells by CTL, which only secondarily develops into a full-blown inflammatory attack (Ando et al., 1993; Kondo et al., 1997). There is indirect but convincing evidence that Fas is involved in hepatitis B virus-induced chronic liver disease in humans as well as in promoting liver regeneration (Rivero, 2002). As discussed in Chapter 5, Fas-mediated damage to liver or other Fas-positive cells during the course of an infection involving FasL-positive host T cells could potentially be antigen non-specific, and cause wide-ranging damage. However, it seems likely that Fas-mediated damage occurs only to cells recognized via the T-cell antigen receptor, and therefore is likely to be relatively restricted (Li et al., 1998).

1.6 Killer lymphocytes and HIV/AIDS

Perhaps for more than any other virus, we would like to understand the role of killer lymphocytes during infection by HIV. A high frequency of HIV-specific CD8 CTL (up to 1% of peripheral blood T cells) has been observed in seroconverted but asymptomatic HIV-infected individuals

(Hoffenbach et al., 1989; Moss et al., 1995). Classical MHC class I-restricted, HIV-specific CTL have been shown to kill cells displaying HIV target peptides derived from core proteins (Gotch et al., 1990), reverse transcriptase (Lieberman et al., 1992), envelope proteins (Clerici et al., 1991), and various regulatory proteins (Koening et al., 1990). These CTL can kill HIV-infected CD4 cells in vitro (Sewell et al., 2000; McMichael and Rowland-Jones, 2001), and are key to controlling HIV infection during the following asymptomatic period (Borrow et al., 1994). Experimental reduction of CD8 cells in SIV infections results in increased viral loads (Schmitz et al., 1999); in humans, the variants of HIV that emerge in frank AIDS are CTL-resistant, implying some form of CTL selection (Borrow et al., 1997; Evans et al., 1999; Klenerman et al., 2002). There have been reports that CD8 CTL kill HIV-infected CD4 cells in a class I-independent manner (Grant et al., 1993, 1994; Bienzle et al., 1996), but this may have been due to contaminating γ/δ T cells (see below).

CD8 cytolytic function as manifested in vitro is impaired in HIV-infected individuals by the time full-blown AIDS sets in (Trimble et al., 1998; Appay et al., 2000). This impairment can be corrected in some cases by incubating the CTL overnight in IL-2, suggesting the defect may be secondary to loss of environmental IL-2, perhaps by the progressive loss of CD4 cells. Another impediment to CD8 CTL function is down-regulation of class I on HIV-infected cells (Kutsch et al., 2002). Finally, some of the drugs used after the onset of frank AIDS inhibit CTL function (Franck et al., 2002).

But after two decades of intense research on HIV-1-specific CD8 CTL, it must be admitted their precise role during the clinical course of HIV-1 infection in vivo is still unresolved. Without a suitable animal model that faithfully reproduces the human disease, we can say little about the contribution of cytotoxicity per se to whatever role CD8 T cells may play in the course of HIV infection in vivo. We know that the rise in HIV-specific cytolytic activity in vitro correlates with suppression of the initial viremia following HIV infection (Koup et al., 1994; Borrow et al., 1997), but whether this is achieved by production of proinflammatory cytokines or through lymphocytotoxicity is not known. Cytokines produced by CD8 cells play a number of roles that are known to seriously compromise HIV infection and reproduction (Copeland, 2000; Chun et al., 2001).

Lysis by CTL of HIV-infected targets in vitro is apparently exclusively via the degranulation pathway; Fas ligand is not expressed on human HIV-specific CTL (Shankar et al., 1991). A highly lytic CD4 CTL clone lysing Gag-peptide pulsed target cells in vitro is blocked by Concanamycin A, a potent inhibitor of degranulation (Kataoka et al, 1996), but not by Fas-blocking reagents (Norris et al., 2001). HIV-specific CTL from patients with progressive disease show decreased perforin content (Heintel et al., 2002;

Zhang et al., 2002), whereas the rare coterie of longterm non-progressing patients retain high levels of CD8 cell perforin (Migueles et al., 2002). Thus the potential for an important role of T-cell-mediated cytotoxicity is certainly there, but at present must remain simply an attractive possibility. Assuming cytotoxicity is a factor, we have no way of assessing the relative importance of cytolytic and non-cytolytic CD8 mechanisms in HIV control.

The role of NK cells in HIV infection is still largely undefined. NK cell lytic action is suppressed shortly after infection and remains low throughout progression to frank AIDS. This is true for lytic activity against both standard NK targets and HIV-infected primary CD4 T cells (Mansour et al., 1990; Bonaparte and Barker, 2003). This was somewhat surprising, since as just noted HIV is among those viruses that induce a down-regulation of class I MHC (Collins and Baltimore, 1999), which would be expected to increase susceptibility of HIV-infected cells to NK killing. The down-regulation is not complete, however, and only affects HLA-A and -B (Cohen et al., 1999). NK-associated cytotoxic activities, such as ADCC and LAK cells, are also suppressed (Tyler et al., 1990; Gryliss et al., 1990). Like CD8 cells, NK cells also control viral infections through soluble mediators. This function is also suppressed by HIV infection, although is apparently more sensitive to viremia than NK cytolytic function (Kottilil et al., 2003).

The minor subset of NK cells that is CD4-positive harbors HIV (Valentin et al., 2002), but the effect of this on overall NK activity is probably slight. NKT cells are rather severely depleted during HIV infection, which is perhaps not surprising since approximately half of this subset in humans is CD4-positive, and also displays HIV co-receptors (Fleuridor et al., 2003). No specific role for NKT cells in HIV infection has yet been elucidated.

The involvement of γ/δ T cells in HIV infection is intriguing, but not yet completely defined. First of all, in normal humans the Vδ2 subset accounts for roughly ninety percent of all γ/δ T cells, and the Vδ1 subset about ten percent. This ratio is essentially reversed in HIV-infected individuals, apparently caused by destruction of Vδ2 cells rather than expansion of Vδ1 cells (Hinz et al., 1994; Boullier et al., 1995). Vδ2 cells using the Jγ1.2 joining segment are particularly hard hit. Although there is no selective expansion of particular Vδ1 clonotypes, the overall number of γ/δ T cells in fact increases considerably in seropositive individuals (Enders et al., 2003). The loss of Vδ2 T cells is not understood at present; they may take up HIV (Lusso et al., 1995; Imlach et al., 2003), so direct killing by the virus is a possibility. Given the particular loss of the Jγ1.2 subset of Vδ2 cells, which suggests that the loss may be activation driven, it is also possible that activation-induced cell death is involved (Enders et al., 2003; see section below on immune homeostatis).

Vδ1 T cells lyse HIV-infected CD4 T cells in vitro, but they also lyse uninfected CD4 cells and other lymphoid targets in an MHC-independent fashion (Sindhu et al., 2003). Given what we know about γ/δ T cell target recognition generally, this cytolysis could well be due to the fact that activated CD4 cells, as well as many lymphoid cell lines maintained long-term in vitro, display stress molecules on their surfaces (Groh et al., 1999; Molinero et al., 2002). This could turn out to be an important factor in the overall immunopathology of HIV infection.

2 CELL-MEDIATED CYTOTOXICITY IN BACTERIAL INFECTIONS

Bacteria obviously pose a major threat to the survival (and hence ability to reproduce) of all animal species, so it is not surprising that virtually every arm of the vertebrate immune system is involved with their detection and elimination. The role of antibody in defense against bacteria was clear at the end of the nineteenth century, and was one of the major driving forces in the field of immunology. The role of neutrophils and macrophages was appreciated even earlier, but there was little reason to expect that the type of cell-mediated immunity apparent first in delayed hypersensitivity reactions, and then in allograft rejection, would have any role in the response to bacteria. But this proved to be the case for a limited number of bacteria, and we now appreciate that these are bacteria that spend at least a portion of their life cycle within cells. Macrophages are the cells most often found to harbor bacteria, since they take them in as a normal part of their function in phagocytosis. If the bacteria remain in phagolysosomes until their destruction, they will provoke only a CD4-mediated cellular response, with bacterial peptides being presented via the class II MHC pathway. However, if the bacteria escape from the phagolysosomes during their life history, fragments of their proteins can enter the class I pathway and activate CD8 cells. Cell-mediated destruction of cells harboring bacteria is an important component of defense against this type of infection (Kaufmann, 1993).

There is another, subtler but equally important way that bacteria mobilize the immune system, and that is through the collection of evolutionarily ancient receptors called, in vertebrates, Toll-like receptors (TLR) (Akira et al., 2001; Bendelac and Medzhitov, 2002). Bacteria produce a number of different products – flagellin lipopolysaccharides, unmethylated CpG sequences, among others – that interact with TLR on dendritic cells (DC), bringing them to a higher state of activation. Activated DC are highly efficient APC for T cells, but also interact with NK and NKT cells, both through direct contact and through production of activating cytokines

(Amakata et al., 2001). This highlights once again the role played by dendritic cells in the network of innate and adaptive immunities.

2.1 Listeria monocytogenes

Listeria infects primarily the spleen and liver in mice, and the heart and brain (meninges) in humans. Adoptive transfer of Listeria immunity in mice with lymphocytes was shown not long after cellular immunity to viruses had been established (Mackaness, 1969). Both CD4 and CD8 T cells play a role (Czuprynski et al., 1990; Ladel et al., 1994). CD8 cells appear to be the more critical; their neutralization or elimination results in a more marked reduction in Listeria protection than elimination of CD4 cells (Mielke et al., 1988; Roberts et al., 1993). Class I-restricted CD8 cells, highly similar to those arising from viral infection, develop during murine Listeria infections. They are able to kill both infected macrophages and infected hepatocytes in vitro (North, 1973; De Libero and Kaufmann, 1986; Jiang et al., 1997). Both CD4 cells and CD8 CTL produce IFNγ, which could activate macrophages or other innate responses. However, the protective role of cloned CD8 T cells was shown to be independent of IFNγ (Harty et al., 1992).

The role of NK cells in bacterial defenses generally is not well defined at present. In the case of Listeria, it is possible to get clearance in the absence of NK cells (Andersson et al., 1998). Mice lacking both α/β and γ/δ T cells, as well as NK cells, cannot clear an infection, but if α/β T cells (CD4 and CD8) are added back to such mice, they clear the infection at essentially normal rates, suggesting neither γ/δ T cells nor NK cells are essential for bacterial control. However, if the α/β T cells added back in this experiment were from IFNγ-deficient mice, Listeria could not be cleared (Bregenholt et al., 2001). This would appear to conflict with the results of Harty et al. just cited, but may be related to the fact Harty was testing a CD8 clone, whereas Bregenholt and colleagues were testing primary, nonimmune T cells. If NK cells are left intact in such mice, they could supply enough IFNγ to slow the infection considerably, even in the absence of perforin (Jin et al., 2001). What triggers NK cells to produce IFNγ in these cases is not clear – probably interaction with Listeria-activated dendritic cells. There are no data at present suggesting NK cells use cytotoxicity to kill cells harboring intracellular Listeria.

It was recognized some years ago that in some cases interaction of Listeria-primed CD8 cells with Listeria-infected cells in vivo or in vitro was not class I MHC restricted (Kaufmann et al., 1988; Lukacs et al., 1989). Further investigation of this apparent anomaly revealed that some Listeria peptides are presented to CD8 cells by non-classical class I molecules (Kurlander et al., 1992; Pamer et al., 1992). In vitro lysis of Listeria-infected hepatocytes by CD8 cells restricted by non-classical class I MHC has been demonstrated by Bouwer, et al. (1998).

As was the case for CTL-mediated protection against viral infections, insight into the question of whether cytotoxic function provided by these various cells plays a role in the host response to intracellular bacteria has come from gene disruption studies. In perforin-deficient mice, the primary response to *L. monocytogenes* in the liver was somewhat impaired, although bacterial clearance from spleen did not differ from normal controls. However the normal accelerated secondary response, and the ability of primed CTL to adoptively transfer protection, were greatly reduced in perforin-deficient mice (Kägi et al., 1994). It makes sense that degranulation-mediated processes would be more apparent in secondary responses. Primary CTL responses require 2-3 days to be effective, whereas memory CTL can be reactivated in a matter of hours after contact with antigen. Thus other mechanisms are likely to dominate in the primary response. CTL from Listeria-primed, perforin-deficient mice provide a limited amount of protection upon transfer to naïve hosts; this protection appears to be based on TNF-α, although whether secreted (inflammatory) or membrane-bound (cytotoxic) was not determined (White and Harty, 1998).

Mice lacking both the perforin and Fas pathways showed greatly increased Listeria burdens at each stage of infection, and clearance of the bacteria was significantly delayed compared to perforin-less or Fas-deficient mice (Jensen et al., 1998). Resistance to secondary infection was also decreased. In vitro experiments showed that CD8 CTL from perforinless mice killed Listeria-infected hepatocytes as well as CTL from normal mice, suggesting the Fas lytic pathway may be crucial in controlling Listeria infections in the liver. However, Fas-negative hepatocytes were still killed, albeit poorly, by normal CTL, indicating perforin may also play a role in liver protection. The ability of doubly deficient mice to eventually clear their infections presumably reflects the importance of IFNγ and/or TNFα in controlling Listeria; as discussed above, perforin-deficient mice are excellent IFNγ-producers. These results suggest that the apparent absence of Fas-mediated cytotoxicity in responses to intracellular bacteria based on absence of or interference with the Fas pathway alone may be misleading, since perforin and cytokine pathways also contribute to protection. In addition to its role in controlling infection, Fas has also been implicated in AICD control of Listeria-activated T cells (Fuse et al., 1997; see also Chapter 11).

2.2 Mycobacterium spp

Cell-mediated immunity also plays a role in the response to both *Mycobacterium tuberculosis* and *Mycobacterium leprae*. Mycobacteria reside mostly within vacules of macrophages, but the macrophages themselves do not act as APC for T-cell activation. Mycobacteria, or their peptides, are released from dying macrophages and picked up by DC (Schaible et al., 2003), which are key to activating both T cells and NK cells.

The response to *M. tuberculosis* involves both CD4 and CD8 T cells. CD4 cells were long thought to provide the most critical host response through release of proinflammatory cytokines. Class I-restricted CTL, however, also proved very important for bacterial clearance (Orme and Collins, 1983, 1984; Lalvani et al., 1998; Stenger et al., 2001), as well as class II-restricted CD4 CTL (Müller et al., 1987). IL-2-stimulated NK cells (LAK cells) have been found to lyse mycobacteria-infected monocytes in vitro (Blanchard et al., 1989). Non-classical MHC molecules (Chapter 1) also play a role in presenting bacterial peptides to both CD4 and CD8 T cells (Porcelli et al., 1992; Rosat et al., 1999; Seaman et al., 1999). These molecules, in addition to presenting bacteria-associated lipids, may also present formylated peptides common to most bacteria (Kurlander et al., 1992). γ/δ T cells, which can both produce IFNγ and kill *M. tuberculosis*-infected cells in vitro, are involved in in vivo defenses against this bacterium (Mombaerts et al., 1993; Tsukaguchi et al., 1995; Dieli et al., 2000). The response to *M. leprae* is essentially the same as that for *M. tuberculosis* (Chiplunkar et al., 1986; Silva et al., 1994).

The absence of perforin appeared to have no effect on the course of a primary infection by *M. tuberculosis* in mice (Cooper et al., 1997; Laochumroonvorapong et al., 1977). The latter authors also showed no apparent impact of the Fas pathway on infection. The possible impact of perforin or Fas on secondary infections was not assessed. The lack of Fas activity may reflect the fact that tubercle bacilli spend most of their life cycle in the phagolysosomes of macrophages rather than the highly Fas-positive hepatocytes. Also, to date resistance to mycobacteria has not been tested in mice doubly deficient for perforin and Fas. It is possible, as we have seen in Listeria, that the two pathways can complement one another, but only in the double-knockout mice would we see the effect of lack of cytotoxicity on the course of an infection. In humans, T cell lines have been developed that lyse *M. tuberculosis*-infected macrophages in vitro (Stenger et al., 1997). CD8 CTL lysed targets via the degranulation pathway; CD4-CD8-negative, α/β TCR T cells lysed targets via the Fas pathway. There is no information at present on the role of such cytotoxicity in vivo in humans. CD8 T cells also directly kill mycobacteria released from lysed target cells using exocytosed granulysin (see below; Stenger. et al., 1998).

2.3 Salmonella spp

Salmonella also reside and replicate inside macrophages. They are increasingly used as gene vectors to deliver potential antigenic determinants into cells (Beyer et al., 2001), but the strains used for human vaccination to Salmonella itself have proved only modestly successful (Levine et al., 1996). More recently strains have been engineered to enhance CD4 and CD8 responses to this bacterium (Salerno-Goncalves et al., 2002), based on the established effectiveness of T cells in defense against other intracellular bacteria. In mice, both T-cell subsets can be induced, including potent CD8 CTL that kill Salmonella-infected cells in vitro, adoptively transfer protection to naïve recipients, and which, when selectively depleted, reduce resistance to disease (Mastroeni et al., 1993; Lo et al., 1999; Mittrucker et al., 2000; Pasetti et al., 2002). γ/δ T cells also play a role in Salmonella infections (Jason et al., 2000), and nonclassical MHC molecules can be involved in presentation of salmonella peptides to classical T cells (Soloski and Metcalf, 2001). There is one report of possible Fas-mediated cytotoxic damage to liver cells in vivo, mediated by NKT cells (Shimizu et al., 2002), but in general there is little data on the role of cytotoxicity in clearing Salmonella infections.

2.4 Granulysin

In addition to killing cells that harbor bacteria, CTL also release granulysin, a member of the saponin-like protein (SAPLIP) family, which may represent a collection of fairly ancient antimicrobial defense molecules. Granulysin, which is stored in cytolytic granules and co-released with perforin and granzymes, can directly kill many bacteria that have been released as a result of cell lysis. It is effective against a wide range of microorganisms, including fungi and parasites as well as gram-positive and gram-negative bacteria (Stenger et al., 1998; Kumar et al., 2001; Ochoa et al., 2001). Absent a granulysin knockout mouse it is difficult to assess its exact importance overall in bacterial defense, but it seems likely that granulysin will be a major player. Granulysin may also help solve for bacterial defense the dilemma posed above for viruses: destruction of a pathogen-harboring cell could actually aid in release and dissemination of the pathogen.

3 EUKARYOTIC PARASITES

Only a limited number of single-cell eukaryotic parasites have become commensal with certain mammalian cells, mainly macrophages, for reasons of size as much as anything else. In some cases these intracellular parasites release proteins into the cytoplasm, where they are degraded and may enter the class I MHC pathway and provoke a CD8 cell response (Garg et al., 1997), as well as the expected CD4 response. Dendritic cells are particularly effective in promoting this kind of "cross-presentation" (Brossart and Bevan, 1997). Results of studies in a few parasites of the activation of CD8 cells, and potential cytotoxic activity, follow.

3.1 Plasmodium spp

This single-cell malarial sporozoan lives at various stages of its life cycle in mammals in liver cells (briefly), and red blood cells (long term). Red cells, because of their low expression of MHC proteins, do not interact measurably with T cells; most cellular reactivity to plasmodia takes place in the liver. CD8 T cells are actively involved in the response to malarial infection of liver cells. Depletion of CD8 cells in mice abolishes protection completely (Weiss, et al., 1990), and CD8 cells from sporozoite-immune mice can passively transfer immunity to naïve animals (Romero, et al., 1989; Rodriguez et al., 1991). Immunization strategies that elicit a CD8 response during the brief transit of sporozoites in liver cells are the most effective (Sedegah, et al., 1994; Tsuji, et al., 1998; Sano, et al., 2001). NKT cells responding to plasmodium antigens on CD1 molecules are also evoked during a malarial infection (Mannoor, et al., 2001), as are γ/δ T cells (Troye-Blomberg, et al., 1999). In a study with perforin-deficient and Fas-mutant mice, no involvement of either cytotoxic mechanism could be demonstrated in vivo (Renggli, et al., 1997).

3.2 Toxoplasma gondii

Most strains of *Toxoplasma gondii* cause a persistent but generally asymptomatic encephalitis in immunologically competent animals. This can become lethal in immunocompromised individuals, such as AIDS patients. A role for cell-mediated immunity in host responses to *T. gondii* has long been known. It involves both CD4 and CD8 T cells, as well as NK cells and NKT cells. The requirement for CD4 cells may be related mostly to their role in helping CD8 cells mature; NKT cells are also able to provide help to CD8 cells (Denkers et al., 1996). As with other intracellular parasites, IFNγ is absolutely key in the response to *T. gondii*, and is contributed by both CD4 and CD8 cells. IFNγ-knockout mice are completely unable to control infections (Scharton-Kersten et al., 1996). Passive transfer experiments suggest CD8 cells are particularly crucial in resisting infection (Suzuki and Remington, 1988), although mice that lack CD8 cells can overcome an infection if they have a sufficiently large supply of IFNγ-producing NK cells (Denkers, et al., 1993). CD8 cells (and CD4 cells) from *T. gondii*-infected mice do display cytotoxicity toward infected target cells in vitro (Kasper et al., 1992; Montoya, et al., 1996), raising the possibility that cytotoxicity may be a factor in host defense.

Two groups have addressed the question of a role for cell-mediated cytotoxicity in the response to *T. gondii.* Nakano et al., (2001) used Concanamycin A to show that blocking degranulation had a significant effect on the killing of *T. gondii*-infected macrophages by CD8 cells from *T. gondii*-infected mice. Blocking by Fas antibody had no effect. Using perforin knockout mice, Denkers, et al. (1997) found that when infected with an attenuated strain that normally confers protection against subsequent infection with a more virulent strain, perforin-deficient and control mice responded equally well. However with a different strain that induces chronic encephalitis, and ultimately death in a large proportion of mice, perforin-deficient mice developed more brain cysts and experienced accelerated mortality, suggesting that in this system cytotoxicity plays a significant role in controlling infection. Thus it seems likely that degranulation, and specifically perforin, plays at least some role in host resistance to *T. gondii.* Whether granulysin released during this process plays a role in preventing spread of parasite cells after target cell lysis has not been assessed. However Yamashita et al. (1998) reported that in vitro lysis of infected target cells did not result in significant death of intracellular parasites.

Toxoplasma gondii is unique in that there have been claims that CTL (Khan et al., 1988), LAK cells (Dannemann et al., 1989) and NK cells (Hauser and Tsai, 1986) can all kill parasite cells directly in vitro. Since this parasite does not pick up host cell markers at any stage of its development, it is not obvious how recognition by any of the putative effector cells would take place, and in fact subsequent work has failed to confirm these early findings (Eric Denkers, personal communication).

3.3 Leishmania major

This is another single-cell parasite that replicates within macrophages. Most of the parasite life cycle is restricted to the phagolysosome of the macrophages, allowing presentation of *Leishmania* peptides on class II MHC structures and activation of CD4 T cells. Initially it was thought that CD8 cells were not activated during Leishmania infection (Wang et al., 1993), but careful analysis showed that CD8 CTL are in fact generated, suggesting that some parasitic proteins may escape from the phagolysosomes and gain access to class I MHC trafficking compartments (Titus et al., 1987; Farrell et al., 1989; Conçeicão-Silva et al., 1998). CD8 cell involvement is particularly apparent when infection procedures similar to those occurring naturally – e.g., low dose infection of skin – are employed (Belkaid et al., 2002). γ/δ T cells are also expanded during Leishmania infections, but their significance is unclear (Rosat et al., 1995). Using the *lpr* and *gld* mutations bred onto susceptible mouse genetic backgrounds, it was found that the largely CD4-mediated host response to *Leishmania* infection was seriously compromised, and the mice were unable to resolve the leishmania-induced lesions. The response in perforin knockout mice, however, was equivalent to controls (Conçeicão-Silva et al., 1998).

3.4 Trypanosoma cruzi

An important role for CD8 cells in the response to *T. cruzi* is well established. *T. cruzi*, like L. monocytogenes, can escape from phagocytic granules and thrive in the cytoplasm of infected cells. Mice in which CD8 function is missing or inhibited respond very poorly to *T. cruzi* infections (Tarleton, 1990; Tarleton et al., 1992; Paiva et al., 1999). CD8 lines and clones derived from *T. cruzi*-infected mice are highly cytolytic to syngeneic *T. cruzi*-infected target cells in vitro (Nickell et al., 1993; Low et al., 1998; Wizel et al 1998). An analysis of CD8 function utilizing perforin knockout mice has shown that, as with other intracellular parasites, the degree of involvement of the perforin/degranulation pathway in CD8 protection depends on the *T. cruzi* strain used and the method of immunization. In various protocols tested, perforin/degranulation either was (Nickell and Sharma, 2000; Henrique-Pons et al., 2002) or was not (Kumar and Tarleton, 1998; Wrightsman et al., 2002) required for control of one or more parameters of infection. There has been one report that mice deficient in the Fas pathway are more susceptible to infection, and lose the ability to kill *T. cruzi*-infected cells (via CD4 effectors) in vitro (Lopes et al., 1999).

NK cells (Cardillo et al., 1996) and γ/δ T cells (Flynn and Sileghem, 1994; Himeno and Hiaeda, 1996) also play important roles in the host response to *T. cruzi*, but there have been no claims that this is through cytotoxic function.

4 CONCLUSIONS

The necessity to defend against a wide range of microorganisms capable of reproducing inside living animal cells is almost certainly the major rationale for the evolution of cell-mediated immunity. But adaptive cell-mediated cytotoxicity is, as we saw with allograft reactions, only one part of the story. Cell-mediated cytotoxicity is one of many often redundant mechanisms that have evolved to deal with this challenge. When perforin-deficient mice were first developed, researchers were extremely cautious in maintaining them, usually in barrier-sustained facilities, with handlers fully masked, gowned and gloved. No one was sure they would be able to survive in the absence of what hadbeen thought to be the major defense against viruses, for example. But in fact cautious testing showed that these animals were able to do just fine with no more than normal animal care protocols. As we have seen, special experimental designs are necessary to detect those situations in which perforin – or Fas, or TNF, or any other single defense mechanism – plays a major role.

Chapter 9

KILLER CELLS AND CANCER

The idea that the immune system could specifically recognize and respond to tumors was a major driving force in the early development of immunology. This hope arose in part from thoughts put forward by Paul Ehrlich (1909), and from the early observation that tumors passed from one outbred mouse to another did not survive. But Peter Gorer put the idea that rejection under these conditions had anything to do with the tumor nature of the transplant to rest in the 1930s, when he showed this was simply an example of allograft rejection. Subsequent research focused largely on the genetics and immunology of allograft reactions, and, together with advances in immunosuppression, would eventually lead to clinical transplantation of tissues and organs in humans. It also led, as we have seen, to the discovery of killer lymphocytes. Tumor immunology per se was pushed onto the back burner for nearly twenty years after Gorer's experiments. It was brought forward again by observations beginning in the 1950s that the immune system can indeed recognize something specific about tumors within the animal in which they arise, and mount a response that not only destroys the tumor, but renders the host resistant to future challenges with the same tumor.

One major step toward this realization, from an experimental point of view, was to focus exclusively on tumor immunity in syngeneic systems – the reaction to a mouse strain B tumor growing in inbred animals of strain B, and not in strain A. Indirect evidence had suggested that the immune system was likely involved in responses to syngeneic tumors, because treatments that suppressed immune reactions – radiation, certain immunosuppressive drugs – resulted in higher incidence or faster growth rates of tumors. Yet despite occasional claims to the contrary – which well may have been true in individual cases – attempts to show that animals had mounted even transient immune responses to their own spontaneously arising tumors have generally failed (see Hauschka, 1952 for a review of early work). Investigators had found in the vast majority of cases that a spontaneously arising tumor, when extirpated and passed through several syngeneic hosts, did not immunize them - secondary and even tertiary implants of the same tumor grew at the same rate and to the same extent as the primary tumor. In retrospect this is not surprising: any cells in a clinically detectable tumor capable of provoking an immune response would have done so, and been detected and destroyed by the immune system. Spontaneous tumors that grow out are the end products of a rigorous natural selection[6]. But repeated failures to demonstrate immune detection of spontaneously arising tumors raised concerns that tumors were, by their very nature as a form of self, somehow invisible to the immune system.

The breakthrough in the rebirth of tumor immunology came with a shift from the study of spontaneously arising – and by definition already established – syngeneic tumors, to the study of de novo-induced tumors. The initial observations on induced tumors in a syngeneic system were made by Foley, who used the polycyclic hydrocarbon methylcholanthrene to induce skin fibrosarcomas in mice (Foley, 1953). When Foley induced sarcomas with methylcholanthrene, and used the resulting tumor cells in an attempt to immunize syngeneic mice, he was successful. Mice that had been exposed to the tumor, and then had it surgically removed, were much more resistant to implantation of that same tumor than were naïve mice. This showed that tumors do have distinct antigens, and that the mice had indeed responded immunologically to the tumor. The fact that the tumor grew in the first place suggested this immunity was not sufficient to prevent primary tumor growth. But by immunizing an animal by allowing the tumor to grow for a limited time before resecting it surgically, animals could be made much more resistant to future attempts to implant the same – but not other – tumors. The animal had developed tumor-specific immunological memory.

Remarkably, over the next several years after publication of this seminal report, only two other papers were published following up on its findings (Baldwin, 1955; Prehn and Main, 1957). Neither group had difficulty reproducing Foley's results; surely others in this interval must have tried and succeeded. Why was nothing published? The main concerns seem to have been about the relevance of chemically induced tumors to cancer as a spontaneous disease, and perhaps more importantly, about whether syngeneic mice were really 100 percent genetically homogeneous. Gorer's imprint on the field remained large; could the apparent immune response to a

[6] This is clearly illustrated by the fact that spontaneously occuring tumors in immunodeficient mice are rejected with much higher frequency when implanted into syngeneic normal mice than are tumors arising in animals that are not immunodeficient (Svane et al., 1996; Engel et al., 1997; Street et al., 2002).

transplanted "syngeneic" tumor in fact be due to residual genetic heterogeneity, perhaps at some equivalent of a minor histocompatibility locus? But Foley and Prehn had both rightly argued that if this were the case, then one should also see immune responses to transplanted spontaneous tumors, and this had not happened. The possibility that the observed rejection might be due to a slight genetic heterogeneity between mice within the same inbred strain was finally laid to rest by George Klein, who carried out his experiments entirely within a single animal, with the same results (Klein et al., 1960). By that time the field of tumor immunology was well on its way.

To everyone's initial surprise, all of the early studies on chemically induced tumors showed that the resulting immunity was non-crossreactive; the same chemical inducing the same tumor (usually a sarcoma) in the same tissue in two syngeneic animals, and even two tumors induced in one animal provoked two non-crossreactive immune responses (reviewed in Old et al., 1962). Non-crossreactivity was also noted in immunogenic tumors induced by radiation (Kripke and Fisher, 1976). Once it was understood that the basis for tumor induction was random chemical- or radiation-induced damage to DNA, it made perfect sense that each tumor would be different. The likelihood that a physical carcinogen would hit exactly the same gene in separate attempts, while not zero, had to be small indeed.

The situation with virally induced tumors was quite different. That viruses can induce tumors had been known since early in the century, mainly through studies with the Rous sarcoma virus (see e.g. Banting and Gairns, 1934). But unlike chemically induced tumors, the same virus would always induce the same immune response in every test animal of a given strain (Habel, 1961; Sjögren et al., 1961). In this case the information encoding whatever it is the immune system recognizes is embedded in the viral genome, and is the same in each infection by the same virus. We will talk about the molecular nature of tumor antigens in more detail when we discuss vaccines that induce tumor-specific CD8 T-cell responses. And of course, all of the early thinking about tumor immunity focused on antigen-specific (adaptive) immunity; the major role played by NK cells and other innate cellular mechanisms would not be appreciated for another dozen years or so.

1 HOST CELLULAR IMMUNE RESPONSES TO TUMORS

Unlike immune responses to intracellular parasites, where humoral mechanisms can play a major role in host defense (since most such microbial parasites are exposed to the extracellular milieu at some point in their life history), the effective immune response to cancer is almost entirely cellular. The cellular response consists of both innate (NK/NKT cell and macrophage) and adaptive (both CD4 and CD8 T cell, and possibly γ/δ T cell) components. These cells can exert direct cytotoxic effects against tumors, and release inflammatory cytokines that further compromise tumor survival.

The first studies suggesting the response to chemically induced syngeneic tumors are largely cell-mediated were carried out by George Klein and his associates. They found that mixing these tumor cells with lymphocytes from syngeneic tumor-immune animals prior to implantation in a new syngeneic host greatly reduced or abolished the ability of the tumor cells to form a new tumor, whereas pretreatment with immune serum had no such effect (Klein and Sjögren, 1960; Klein et al., 1960). Cell-mediated immunity against virally induced tumors was established shortly thereafter (Habel, 1962; Sjögren, 1964). Within a short time most work on tumors in animals involved virally induced tumors (Habel, 1965).

1.1 CD8 T cells

That the cells in mice responsible for syngeneic tumor rejection, whether chemically or virally induced, are at least in part T cells was suggested by studies showing such rejection is seriously compromised by neonatal thymectomy (Miller et al., 1963; Ting and Law, 1965). The first demonstration of an in vitro effect of lymphocytes sensitized to syngeneic tumor antigens in vivo involved colony inhibition of target cells (Hellstrom and Hellstrom, 1968), but was followed soon thereafter with demonstrations of antigen-specific cytotoxicity as defined in allograft systems (Ortiz de Landazuri and Herberman, 1972; Ting et al., 1974). Finally, in vitro primary and secondary generation of CTL recognizing tumor-specific antigens was demonstrated using virus-induced syngeneic tumors as stimulator cells in mixed lymphocyte-tumor reactions (Rollinghoff, 1974; Plata et al., 1975).

The specificity and activity of CD8 CTL raised against syngeneic tumors are indistinguishable from CTL generated in response to alloantigens (Plata et al., 1976a,b; Fishelson and Berke 1981). The case for CTL immunity to tumor-induced viruses is fairly straightforward (Moss et al., 1977; Thorley-Lawson et al., 1977), and is not different from that discussed in Chapter 9 for immune responses to viral infections generally. The antigens detected by CTL are for the most part viral antigens. For chemically induced tumors, as just discussed each tumor is antigenically unique, and the cognate antigen in each case is presumed to be a mutated self-protein, probably but not necessarily one involved in cell-cycle control. That such antigens include mutated products of cell cycle regulatory genes has recently been established; however, the number of such defined antigens remains small (Ikeda et al., 1997; Klein, 1997). The situation for spontaneously arising tumors is more ambiguous, and the case for CTL involvement in the response is not always clear. In those cases where viruses are involved in primary oncological transformation, the CTL response is essentially anti-viral. For other spontaneous tumors, the cognate antigen, if there is one, could be derived from a normal or a mutated molecule involved in cell-cycle control, or one of the proteins involved in tumor suppression such as an oncogene, p53 or telomerase.

CTL specific for a number of different syngeneic human tumors have been isolated from patients and cloned (Vose and Bonnard, 1982; Slovin et al., 1986; Itoh et al., 1988; Mukherji et al., 1989). These CTL can be shown to lyse cognate tumor cells in vitro with varying degrees of efficiency. In vivo, tumor antigen-specific CD8 cells can be identified with peptide-MHC tetramers at the tumor site in concentrations significantly greater than in normal tissues (Romero et al., 1998). One convincing proof that CTL can be effective in syngeneic tumor control would be to show that tumor-specific CTL, when adoptively transferred to a tumor-bearing animal, cause regression of the tumor. A variant of this approach in humans using LAK cells was discussed in Chapter 7. In that case the sensitization event occurred in vivo, the effector cells were usually nonspecific, and re-injection was accompanied by administration of high levels of IL-2 which proved harmful to the patient.

Preclinical studies showed that in a few cases in mice, especially when the tumor is of viral origin, and often when combined with chemotherapy (Greenberg, 1991), adoptive transfer of tumor-specific CTL could be effective in causing tumor regression, and even cure (Melief, 1992). Clinical trials in humans showed occasional successes (Bakker et al., 1994; Rivoltini et al., 1995), particularly in EBV tumors (Rooney et al, 1998). But in the case of non-virally induced tumors, adoptive transfer of CTL generated in vitro in mixed lymphocyte-tumor cultures, particularly when expanded with repeated stimulations in the presence of IL-2, had little or no effect on growth of the tumor recognized by the CTL, even though their in vitro cytolytic effect was strong (Hammond-McKibben et al., 1995; Yee et al., 1997).

Even the use of highly lytic CTL clones specific for tumor antigens gave at best transient tumor regressions in patients. However recent advances in our understanding of what happens in adoptive transfer of tumor-specific CTL offers hope that this approach may yet yield important benefits (Dudley and Rosenberg, 2000). For example, Shimizu et al. (1999) described a CD4 cell (bearing the CD25 marker) whose removal greatly increased tumor immune responses in mice. The CD4/CD25 cell probably acts on CD4 cells needed to promote tumor responses, and its removal enhances both CTL and NK cell function. Patient treatments aimed at reducing this cell population, even transiently, improved development of cellular responses to their tumors, and in combination with adoptive transfer of tumor-specific CTL (TIL; see Chapter 6) showed the strongest tumor regressions yet seen (Dudley et al., 2002; Rosenberg et al., 2003).

And there may yet also be hope for use of potent CTL clones in treating tumors. It had been proposed that at least part of the reason clones have proved ineffective is that their continued expansion in the presence of high levels of IL-2 alters them, particularly in terms of surface information controlling migration patterns in vivo and extravasation at the tumor site (Byrne et al., 1986; Klarnet et al., 1987). The occasional case where adoptively transferred CTL to non-viral tumors seemed to be effective in fact involved CTL that had undergone little or no expansion in the presence of IL-2 (Keyaki et al., 1985; Klarnet et al., 1987).

This was examined in detail in a study involving transfer of CTL generated in vitro in a mixed lymphocyte-tumor reaction and expanded in the complete absence of exogenous IL-2 (Iwashiro et al., 2002). Only about ten percent of the clones generated were able to survive five or more rounds of stimulation in the absence of IL-2, presumably by selection of CTL able to secrete enough of their own IL-2 to support proliferation in vitro. When these clones were injected intravenously into syngeneic mice carrying the stimulating tumor in a subcutaneous site, complete eradication of the tumor could be achieved with as few as 3 million CTL. Although most emphasis on new approaches to cancer treatment has been focused on cancer vaccines (see Chapter 11), this study may clear the way for renewed interest in adoptive immunotherapy for human cancer (Yee et al., 1997).

1.2 A word about "immunosurveillance"

As mentioned above, Paul Ehrlich was the first to suggest that the immune system, in addition to surveying the body for intruders from without, might also function to guard against dangers from within – specifically, from cancer. This idea was not really developed further from a theoretical point of view until the 1950s when, largely in response to the experiments of Foley and Prehn, it was restated separately by Burnet and by Thomas (Burnet, 1957a,b, 1970; Thomas, 1959). Both Burnet and Thomas even predicted that the biological *raison d'être* for the cellular arm of the immune system, which all agreed could have nothing to do with transplant rejection, was likely to be, at least in part, tumor surveillance. At the time, they were thinking entirely in terms of adaptive CTL responses.

While conceptually attractive – and probably true for CTL – this hypothesis, which guided much of tumor immunology in its formative years – often seemed to run afoul of its own predictions (Möller and Möller, 1976). For example, the nude mouse, which has no thymus and thus should be compromised in the T-cell arm of immune responsiveness, seemed to have no greater incidence of spontaneous tumors than wild-type mice (Rygaard and Povlsen, 1974; Stutman, 1974). And although it was clear by the early 1970s that immunosuppressed transplant patients had a higher than normal incidence of cancer (Penn and Starzl, 1972), these were predominantly hematopoietic and usually viral in origin, suggesting that either the immunosuppressive drugs were themselves carcinogenic, or the increased microbial infections experienced as a side effect of immunosuppression were oncogenic, or were driving immune system cells to higher rates of cell division, with accompanying opportunities for cancer-causing mutations. Robert Good had reported a higher frequency of cancer in naturally occurring immune deficiency states (Good, 1970), but again with a predominance of lymphoid malignancies. These findings in both mice and humans suggested to many that the concept of immune surveillance for cancer, at least as then formulated, had serious shortcomings.

Progressive refinements during the 1980s of our understanding of cell-mediated immunity suggested several mitigating explanations of the foregoing data. Nude mice, we now know, have intact NK and NKT cell function, as well as γ/δ T cells that can develop extrathymically, and all of these can and do participate in tumor control. Moreover, nude mice actually have small numbers of functional α/β T cells (Ikehara et al., 1984; Maleckar and Sherman, 1987), although not enough to cause allograft rejection. A much cleaner system in which to analyze the role of adaptive immunity in tumor surveillance is offered by the RAG-1 and RAG-2 knockout mice. These mice are unable to rearrange either immunoglobulin or TCR genes, and thus lack α/β and γ/δ T cells, as well as B cells and NKT (but not NK) cells. In these mice, the response to both induced and spontaneous tumors is in fact measurably decreased, suggesting that T cells do play a role in guarding the body against oncogenic aberrancies (Shankaran et al., 2001).

Equally importantly, perhaps, we now appreciate the powerful role played by cytokines such as IFNγ (Dighe et al., 1994; Kaplan et al., 1998; Shankaran et al., 2001) and TNFα in tumor control. Thus in the body's response to cancer, as in its response to infection by intracellular pathogens – and indeed to allografts – there are multiple, overlapping mechanisms brought to bear, and the absence of any one mechanism may or may not cause an obvious perturbation in the overall response of the host animal. While immune surveillance for tumors was initially formulated largely in terms of cell-mediated cytotoxicity, its near-failure as a working hypothesis may simply reflect the larger complexity of cell-mediated immune responses.

The obvious failure of immune surveillance in the case of clinically emerging spontaneous tumors also presented special problems. Were these tumors simply not seen by the immune system? Or had they been seen, but somehow managed to induce a state of tolerance in the host? A strong case for the immunological neutrality of spontaneously arising tumors was made by Hewitt et al. (1976) who published an exhaustive study of over 20,000 transplants of 27 histologically distinct tumors in an attempt to detect an immune response in a syngeneic or autochthonous host. They were unable to do so, suggesting that under the conditions they used the tumors appeared to have lost the ability to induce an effective immune response.

However, that spontaneously arising tumors may still be potentially capable of stimulating an immune response was suggested by data showing that if secondarily mutagenized, a proportion of the cells making up a tumor could become immunogenic (Boon and Kellerman, 1977). This showed that the cells themselves were not in some a priori fashion immunologically invisible. But interestingly, animals that had acquired immunity to these mutagenized tumors in some cases now also displayed immunity to the original unmutagenized tumor (Van Pel et al., 1983), although not to other tumors. This breaking of a tolerance-like state in vivo was also reflected by in vitro activity; mice in which tolerance had been broken could now generate CTL capable of lysing target cells from the original tumor in ^{51}Cr release assays. Thus it may be that some tumors do induce a state of something like tolerance, which can be broken by cross priming with a different antigen (see Boon et al., 1997 for discussion).

1.3 NK and NKT cells

As we saw in Chapter 6, the ability to detect and cause the elimination of tumor cells in vivo without prior sensitization was the founding definition of NK cells. The evidence that they do so is overwhelming, but until recently only indirect (Trinchieri, 1989). It was observed at the very beginning, for example, that tumors sensitive to NK killing in vitro grew more slowly, or were rejected, in vivo (Kiessling et al., 1975a,b). Because many of the features of NK cells are shared with other lymphocyte subsets, it has been difficult to make absolute correlations between NK cells and their putative in vivo functions. In attempting to deplete NK cells in vivo with antibodies to NK-associated surface markers, investigators also depleted other subsets to varying degrees. NK1.1, for example, is also found on NKT cells (Chapter 7), so the NK1.1 antibody will delete these cells as well, making assignment of lost functions to NK cells unclear. The beige mouse is often taken as a model for deletion of NK function, but the underlying mutation affects degranulation in many cell types, again obscuring the involvement of NK cells in lost functions.

Nevertheless, there is persuasive indirect evidence for the role of NK and NKT cells in controlling tumors. For example, IL-12 has been shown to be a potent stimulator of defenses against many tumors in mice. This effect is equally strong in wild type and RAG2-deficient mice (Kodama et al., 1999); the latter have no α/β or γ/δ T cells, but do have NK cells. The protective effect of IL-12 in both types of mice is lost when NK cells are depleted. However, in line with the ambiguities pointed out above, this effect may have been at least partly due to NKT cells (Smyth et al., 2000a). Both NK and NKT cells are involved in the host response to MCA-induced tumors (Smyth et al., 2001a).

The closest we have to an NK "knockout" mouse is a mouse strain made transgenic for Ly49A under a Granzyme A promoter (Kim et al., 2000). Ly49A delivers a negative signal to NK cells when it encounters murine class I molecules such as $H\text{-}2D^d$ (Chapter 6; Roland and Cazenave, 1992). Spleen cells from these mice are unable to kill standard murine NK targets such as YAC cells in vitro. But they are also unable to kill MHC class I-negative targets such as RMA-S. These mice in fact have very few "classical" NK cells in any lymphoid tissues. On the other hand, NKT cells were present and appeared functionally normal. Ly 49A was expressed on about half of the T cells in these mice, but these T cells showed normal reactivities against all cells except those displaying $H\text{-}2D^d$, the ligand recognized by Ly49A. Thus the presence of Ly49A in lymphoid cells does not cause a generalized immune paralysis. In vivo, these mice were unable to clear YAC and RMA-S tumors, which were quickly rejected by wild-type mice. Most importantly, these tumors were rejected in the transgenic mice if the mice were reconstituted with spleen cells from syngeneic SCID mice, which have NK cells but no T or B cells. Using other tumor systems, it was shown that these NK cells are also involved in suppression of metastases. These data are the most direct evidence we have to date that NK cells do in fact play a major role in tumor control in vivo.

1.4 γ/δ T cells

Both subsets of γ/δ T cells discussed in Chapter 2 may be involved in tumor control. In some situations γ/δ cells may be at least as important as α/β cells (Girardi et al., 2001). The Vγ9/Vδ2 subset is associated with hematological malignancies (Kunzmann et al., 2000), while the subset using Vδ1 seems to be involved with solid tissue tumors (Fisch et al., 2000), although these distinctions are relative, not absolute. The Vδ2 subsets are particularly effective against class I-deficient tumor targets (Rothenfusser et al., 2002). Like α/β T cells, γ/δ cells are attracted to reaction sites by chemokines (Ferrarini et al., 2002). As discussed in Chapter 2, γ/δ T cells interact with a variety of peptide and non-peptide structures on cell surfaces, not always in the context of classical or even non-classical MHC structures, and in some cases it is not even clear that recognition is via the TCR. Perhaps most important for tumor control, γ/δ T cells recognize stress-related proteins, such as MICA and MICB and/or heat shock peptides, which signal that something is wrong inside a cell. The same may be true for phosphorylated nucleotide products (Constant et al., 1994). Tumor cells are under a great deal of metabolic stress, driven to divide constantly and often in the face of insufficient food and oxygen, and they display many of the same stress signals as parasitically infected cells.

Both γ/δ subsets express perforin (Smyth and Trapani, 2001), up-regulate FasL upon activation (Li et al., 1998), and can lyse autologous tumor targets in vitro (Zocchi et al., 1990; Ferrarini et al., 2002). There appears to be a good correlation between the ability to mobilize γ/δ T cells and tumor progression or regression (Lowdell et al., 2001), but there is no direct data suggesting that it is the cytotoxic function of these cells, rather than their production of cytokines, that is critical. Advantage has been taken of the ability of γ/δ T cells to respond to certain microbial products (Kato et al., 2000), and to natural products such as bisphosphonates (Kunzmann et al., 2000), to sensitize otherwise unrecognized tumor cells for lysis. As more information is gathered on the biology of γ/δ T cells, it is entirely possible they may turn out to be an important defense against tumor growth in vivo, and one that can be manipulated for therapeutic benefit.

2 THE ROLE OF CYTOTOXICITY IN HOST RESPONSES TO SYNGENEIC TUMORS.

One of the first comprehensive looks at the role of perforin in rejection of syngeneic tumors was carried out by van den Broek et al. (1995, 1996). They looked at rejection of a wide range of lymphoid and non-lymphoid tumors in perforin knockout mice, compared to normal controls. The results varied widely, with no obvious relation to tumor type. A melanoma was equally lethal in mice with or without perforin. A fibrosarcoma grew about ten times more readily in perforin-deficient mice, although both the knockout mice and normal controls developed stable tumors. A modified adenovirus-induced tumor was rapidly rejected by both mice. In general, among lymphoid tumors, perforinless mice were 100-1000 times more susceptible to some, and totally unable to control others. At least one tumor, RMA, was controlled equally well in both types of mice. Among some of the tumors controlled by perforin-deficient mice, the presence of Fas was shown to be critical. On the other hand, RMA-S, a class I-deficient variant of RMA whose rejection is presumably under control of NK cells, had previously been shown to grow without restriction in perforin knockout mice, whereas it is adequately controlled in normal mice. Finally, MCA-induced tumors were found to develop sooner and grow faster in perforin knockout mice (van den Broek et al., 1995, 1996).

The impact of perforin on spontaneous tumor growth was examined in another study. Researchers maintaining perforin knockout mouse colonies had not observed an unusual incidence of spontaneous tumors, but the numbers under observation were generally small, and few mice were kept more than a year or so. When observed in large numbers and over a longer time span, perforinless mice developed a high incidence of tumors of lymphoid cells (lymphomas), particularly of T, NKT and B cells. These tumors were also highly metastatic. Because mutations of p53 are commonly found in tumors of both mice and humans, it was decided to look at the incidence of tumors in p53-mutant mice that were also perforin-deficient. Both the incidence and time of onset of lymphomas were greatly accelerated in the doubly mutant mice. Surprisingly, a much lower proportion of the lymphomas were metastatic. Other types of tumors, principally sarcomas, began showing up in the doubly deficient mice as well, but this was likely due to the lack of p53 rather than perforin. Control of primary tumor growth appeared to be under control of CD8 cells, and the impact of perforin suggests it is the cytotoxic function of these cells that is critical (Smyth et al., 2000). In a separate study, these same authors had shown that the key cell type involved in perforin-associated control of metastasis is the NK cell (Smyth et al., 1999).

The role of perforin in syngeneic tumor control must be evaluated carefully. In the BALB/c kidney tumor system Renca, where tumor control is absolutely dependent on CD8 cells, disruption of the perforin gene resulted in a major decrease in killing of tumor cells in vitro, but had little impact on tumor control in vivo (Seki et al., 2002). Although Fas appeared to be used for in vitro killing in this system, neither Fas nor TRAIL appeared to have any effect on tumor growth in vivo.

As discussed in Chapter 5, CTL can use perforin alone to kill target cells in vitro (as evidenced by ^{51}Cr release), but the target cell DNA fragmentation normally accompanying CTL killing requires the participation of granzymes (Simon et al., 1997). The granzymes studied most intensely in this regard are granzymes A and B. In GrzA knockout mice, DNA fragmentation is unaffected, while in GrzB-deficient mice, DNA fragmentation is slowed considerably. In doubly granzyme-deficient mice, there is no DNA fragmentation. In situations where perforin appears to be critical for tumor control in vivo, the absence of granzymes in one report appeared to make no difference, whether the putative effector cells were CTL or NK cells (Davis et al., 2001). However, a subsequent report using some of the same tumor cells lines and essentially the same knockout mice, reached the opposite conclusion (Pardo et al., 2002). At present this discrepancy is unresolved.

Perforin also plays a role in innate cellular responses to syngeneic tumors, most likely mediated by NKT cells (Street et al., 2001). In RAG2-deficient mice, where tumor control was exerted largely by NK and/or NKT cells (Kodama et al., 1999), the efficacy of these cells was shown to be dependent on perforin, but not on Fas. However numerous other studies have shown that NK and NKT cells also utilize the FasL-related molecule TRAIL (TNF-related apoptosis-inducing ligand) in controlling tumors (Vujanovic, 2001). TRAIL in soluble form appears to be highly selective for tumor cells in in vitro cytotoxicity tests and in vivo (Walczak et al., 1999). IFNγ-stimulated NK cells, especially those found in liver, are strong expressers of TRAIL, and blocking of TRAIL in some cases can profoundly inhibit the ability of NK cells to block tumor development (Takeda et al., 2002). TRAIL knockout mice were found to be more susceptible to both transplanted and spontaneous tumors (Cretney et al., 2002; Sedger et al., 2002). TRAIL is now thought to be a major component of tumor resistance in both mice and humans, and trials to test the efficacy of the TRAIL ligand alone as an antitumor agent in humans are underway (Norris et al., 2001; de Jong et al., 2001).

It is also possible that granulysin could play a role in tumor control. Granulysin has been shown to kill tumor cells in vitro (Pena et al. 1997; Sekiya et al., 2002). Its expression in tumor-infiltrating lymphocytes correlates with progression or regression more closely than that of perforin (Kishi et al., 2002). However at present, absent a granulysin knockout mouse, there is no definitive evidence for a role for granulysin in tumor suppression in vivo, and little insight into how granulysin might selectively target tumor cells. Taken together, the data in perforin and granzyme knockout mice confirm that cytotoxicity plays an important role in tumor surveillance and control, but as in other immune defense systems, it is only one mechanism at the disposal of host animals to guard against aberrant and potentially life-threatening tumor cells.

3 TUMOR ESCAPE FROM IMMUNE ATTACK

It is now generally accepted that tumor growth in vivo elicits similar, if not identical, forms of immune resistance as do intracellular parasites and allografts, and that cancer cells do, at least at their inception, display characteristic antigens. These may be small peptides bound to MHC I and II membrane proteins (Boon et al., 1997; Rosenberg 1997); stress molecules at tumor-cell surfaces can also induce immune activity. So if cancer cells are immunogenic, why are they not destroyed by the cellular immune response, in particular CTL and/or NK cells, as is the case with virus-infected cells or allografts? The answer to this question is highly relevant to clinical oncology; despite rapidly accumulating new immunotherapeutic approaches to cancer, a breakthrough is still not at hand.

Obviously cancer, like some intracellular parasites, may evade immune attack if growth rate exceeds the destructive capacity of the immune system. This is likely a major mechanism whereby certain infectious agents as well as some tumors evade the immune response. In addition, factors such as aging, diet, certain drugs, or tumor cell products could impair immune responsiveness and thus fail to prevent the appearance of a neoplastic cell or halt its progression to a frank tumor.

Aberrant T-cell receptor signaling (Deeths et al., 1999) appears to be another important factor accounting for the failure of immune attack against a tumor. Immune tolerance or anergy to cancer antigens, for example, due to exposure to potential cancer-associated antigens in very early life, may account in some cases for the failure to respond positively against neoplastic growth in adult life. Immune paralysis might occur if a large amount of tumor antigen is released by large or disseminated tumors. In animals with frank tumors induced by either chemicals or viruses, removal of the tumor often renders the animal resistant to re-implantation of the same tumor. This indicates that an intrinsic failure of the immune response is not directly responsible for the original lack of immune reaction against at least some tumors.

The failure of the immune response to control tumor growth may not necessarily involve just the onset phase of the anti-tumor response. It may well involve active, tumor-induced effects which counteract cytocidal (NK or CTL-based) responses against the tumor. For example, splenocytes from mastocytoma-bearing DBA/2 mice restimulated with mastocytoma cells in vitro exhibit considerable cytocidal activity against the tumor. However, this is time-dependent. Anti-tumor activity is manifested by splenocytes obtained as early as 5 days after tumor inoculation, peaks 9-12 days post inoculation, and then declines rapidly (Figure 9.1). However, unlike operationally defined 'memory' responses in alloreactivity and in virus-specific immunity, virtually no (memory) CTL activity is detected 3 weeks after tumor inoculation. Thus, cytocidal lymphocytes appear to be primed in vivo, as measured by the ability to restimulate them in vitro, during the early phase of tumor growth. Later, the primed CTL may become functionally 'anergic/tolerant' or possibly deleted altogether. Deletion may occur as a result of contact with the increasing tumor mass, or products thereof, by a process called activation-induced cell death (AICD; Chapter 11), or through active annihilation of the CTL by a Fas-L-expressing tumor (tumor "counterattack") (Li et al., 2002).

FasL was reported to maintain a state of 'immune privilege' in mouse testes (Bellgrau et al., 1995) and in the anterior chamber of the eye (Griffith et al., 1995). FasL expressed in these tissues appears to induce apoptosis of activated lymphocytes that infiltrate the sites, thus preventing rejection of implanted histoincompatible normal or malignant tissue implanted at these sites, as well as inflammatory responses that could occlude vision. Based largely on the finding that FasL-positive tumors can kill Fas-expressing cells, it has been suggested that FasL-expressing tumors can "counterattack" tumor-reactive CTL and NK cells, which themselves express Fas (and Fas-L), thereby evading anti-tumor responses (O'Connel et al., 1996; Hahne et al., 1996; Strand et al., 1996). These experiments and conclusions (reviewed in Green and Ware, 1997) were challenged (Zaks et al., 1999; Restifo, 2000), on the basis that that Fas-L expressed by the T lymphocytes, rather than tumor cells, upon activation by the cognate tumor, could cause the activated lymphocytes to kill themselves and each other.

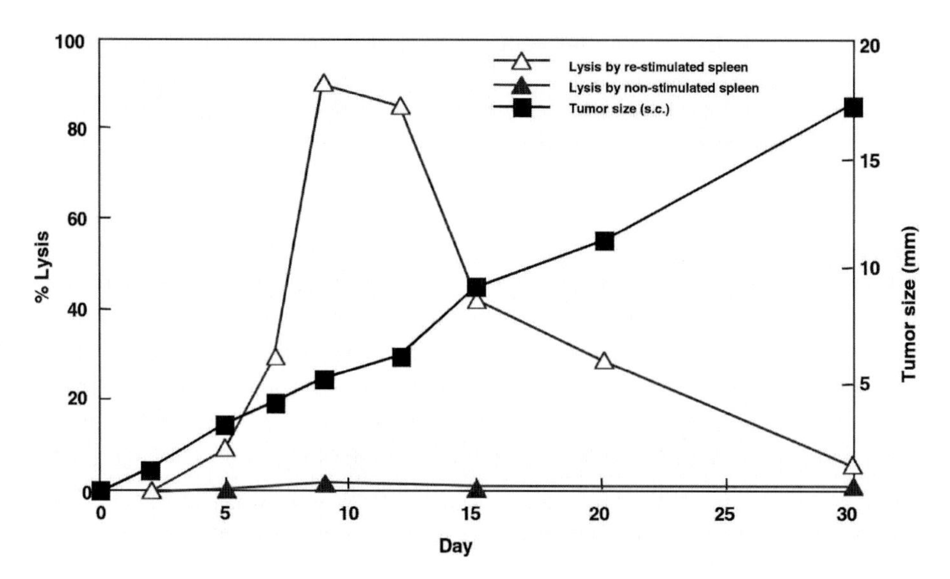

Figure 9-1. Induction of cytocidal lymphocytes in tumor-bearing mice

Exposure of CTL to FasL-expressing cognate targets does appear to cause inhibition of CTL function through at least two mechanisms (Restifo, 2000; Li et al., 2002) (Figure 9.2). First, FasL expressed on tumor target cells induces some direct inactivation (or lysis) of the CTL through interaction with Fas on the CTL surface. Second, cognate recognition of antigens on the target cell, independent of its FasL expression, brings about FasL-Fas-mediated activation-induced cell death (AICD) of CTL, mediated by other antigen-activated CTL in the population. Bystander CTL inactivation also occurs, which may account for the more generalized immune suppression seen in advanced cancer. Regardless by which mechanism, CTL inactivation will occur (Li et al., 2002).

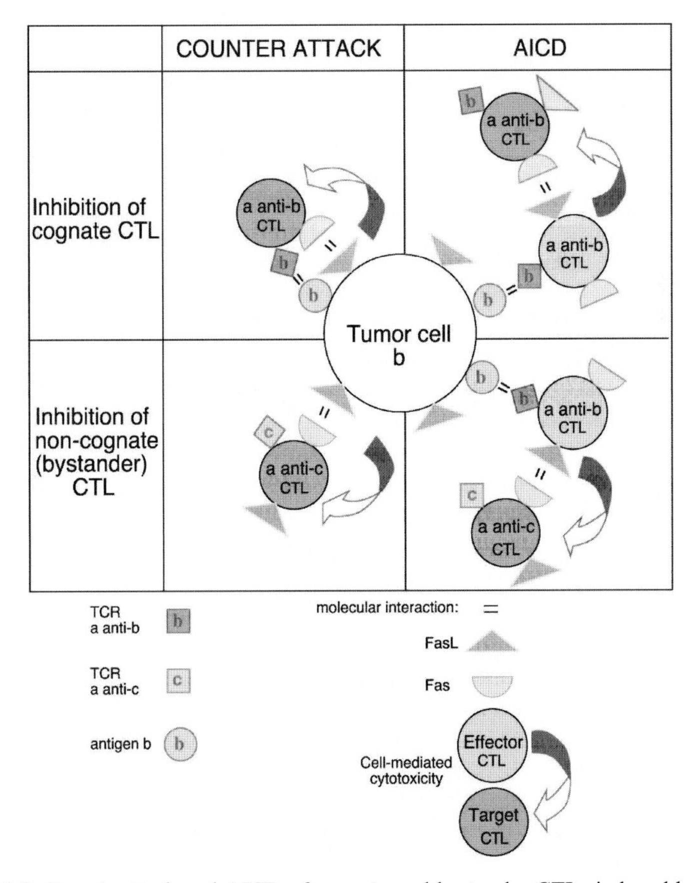

Figure 9-2. Counterattack and AICD of cognate and bystander CTLs induced by interaction with FasL expressing tumors. Specifically conjugated CTL express more Fas and FasL (as well as auxiliary molecules) and become susceptible to AICD by homicide or suicide. They kill bystander effector CTL (not conjugated to the cognate targets) since their Fas- FasL-based cytocidal potential is facilitated upon cognate TCR-engagement (from Li et al., 2002).

Chapter 10

AUTOIMMUNITY

The subject of autoimmunity brings into focus one of the central issues in immunology: How does an individual organism distinguish self from nonself? The molecules of which human "self" is constructed are qualitatively the same as those used in the construction of any other biological organism, including those that could seriously damage the host. Yet clearly the immune system must be able to make this distinction, or else it could very well self-destruct. Paul Ehrlich recognized this conundrum in the early years of immunology in his formulation of the concept of *horror autotoxicus* (Ehrlich and Morgenroth, 1901).

There are a number of generalizations about autoimmune disease that seem to hold true. Females often have much higher rates of a given autoimmune disease than do males, with the major exception of Type 1 diabetes. Although there is some evidence of a slight increase of autoimmunity in older individuals, in fact the most devastating forms of these diseases usually manifest during the reproductive years, which presents a particular biological challenge. There is almost always a genetic predisposition toward autoimmunity, usually centered around MHC genes; concordance for most autoimmune diseases in genetically identical twins typically runs between twenty and fifty percent. Finally, virtually every arm of the adaptive and innate immune systems is involved at some level in autoimmunity, and every such disease studied to date is characterized by a strong initial inflammatory reaction. The innate immune system may play a role in initiation of autoimmunity. For example, infectious microorganisms interfacing with the immune system via Toll-like receptors (Chapter 8) can stimulate production of cytokines that may divert the adaptive immune system away from a delicately balanced state of tolerance toward self antigens (Klinman, 2003).

The question of autoimmunity thus intersects closely with the subject of immunological tolerance, both central (clonal deletion) and peripheral (clonal regulation). These subjects are enormously complex and beyond the scope of the present chapter; the reader is advised to consult a recent advanced immunology text for the current status of the field. Here we address ourselves to the question of what role, if any, killer lymphocytes play in the pathology of autoimmune diseases. The question of involvement of cytotoxic function can only really be answered clearly in those diseases for which a good animal model is available. We begin with a disease in which the autoimmune damage observed is in fact almost entirely mediated by killer T cells – Type 1 diabetes.

1 DIABETES

Type 1A insulin-dependent diabetes mellitus (IDDM)[7] is a chronic autoimmune disease (Eisenbarth, 1986; Devendra and Eisenbarth, 2003) with a strong genetic component (Vyse and Todd, 1996). However, concordance for IDDM in mutually susceptible identical twins rarely exceeds fifty percent, suggesting the involvement of one or more environmental "triggers" in the onset of disease. One such environmental factor could well be viral infection. Some viruses cause IDDM as a direct result of infection of pancreatic tissues, either by virus-induced cell death or as a result of immune attack on virally infected β cells. Other mechanisms have been proposed as well (reviewed in Jaeckel et al., 2002). Serotype B Coxsackie virus in particular has been implicated in triggering IDDM in susceptible individuals (Barrett-Connor, 1985; Hyoty et al., 1995). Very often newly diagnosed patients will have recently experienced a Coxsackie virus infection, or display Coxsackie antibodies in their serum. However, the only virus definitively associated with IDDM is congenital rubella, which predisposes to a number of autoimmune disorders (Szopa et al., 1993; Robles and Eisenbarth, 2001). To the extent that viruses play a role in onset of IDDM, we may expect similarities in the immunopathologies of IDDM and certain chronic viral diseases.

The antigens identified as provoking true β-cell autoimmune responses are diverse, but at present the major antigen identified in both mice and humans in eliciting cell-mediated responses is glutamic acid decarboxylase

[7] The terms "juvenile onset" diabetes, and "IDDM" have been replaced with a more specific terminology, namely Type 1A. However for convenience sake we use IDDM herein to denote Type 1A diabetes. For current classification of diabetes syndromes, see the American Diabetes Association Expert Committee recommendations in Diabetes Care 20:1183 (1997).

(GAD; Baekkeskov et al., 1990; Kim et al., 1993). Although autoantibodies to unmodified GAD protein and other normal islet-associated antigens are almost always present in IDDM, the onset of this disease in an individual with X-linked agammaglobulinemia suggests antibodies may be correlative but are not causative (Martin et al., 2001). While GAD is not islet-specific, the damage produced during diabetes appears relatively islet restricted. Failure of GAD tolerance in T cells of NOD mice has been proposed as a major factor in onset of IDDM (Kaufman et al., 1993), and experiments with GAD-islet-specific transgenic NOD mice showed that the severity of the disease correlated directly with the level of GAD expression (Yoon et al., 1999). Intriguingly, Coxsackie virus infection causes increased GAD expression in the pancreas (Hou et al., 1993); moreover, one of the Coxsackie-encoded proteins contains a peptide region with strong homology to a region of the GAD protein (Kaufman et al., 1992).

Insulin itself is also an important antigen in IDDM, and not just for antibodies. A diabetogenic CD8 T cell clone has been reported that recognizes an insulin-derived peptide (Wong et al., 1999). Events that interfere with natural tolerance to insulin peptides (see below) could thus be important in IDDM. How important a factor insulin autoreactivity is in the overall cellular response during IDDM, however, is still uncertain. Analysis of the T-cell response to an unidentified "islet cell antigen" suggested that T-cell activation is triggered in the pancreatic lymph nodes well before the first measurable insulitis (Höglund et al., 1999). Thus if activation is via "molecular mimicry", secondary to microbial infection, the mimicry event must originate in the pancreas itself.

A wide range of immune components are mobilized in IDDM, but the damage leading to loss of insulin production is attributable mostly to CD4 and CD8 T cells (Yoon and Jun, 2001). Our current understanding of immune pathways involved in the development of IDDM comes to a large extent from studying the NOD (non-obese diabetic) mouse, and to a lesser extent the BB (BioBreeding) rat. Both are reasonably good models for spontaneous onset IDDM in humans. Early in the disease in NOD mice, which occurs almost exclusively in females, macrophages and dendritic cells infiltrate the pancreas, followed by T cells, B cells and NK cells. The T cells cluster around and physically penetrate the pancreatic islets. Eventually β cells within the islets are selectively killed, and insulin production is compromised. Islet cell death in NOD mice is apoptotic (O'Brien et al., 1997).

Athymic and SCID NOD mice do not develop diabetes (cited in Yoon and Jun, 2001), and ablation of either CD4 cells (Koike et al., 1987; or CD8 cells in NOD mice (Wicker et al., 1994; Sumida et al., 1994) prevents full development of diabetes. CD8 T cells are critical to disease onset, and are more apparent in early infiltrates of the islets than CD4 cells in NOD mice (Jarpe et al., 1990). The same may be true in humans (Bottazzo, et al., 1985). Both T cell subsets are able to transfer at least some aspect of the disease (Bendelac et al., 1987; Miller, et al., 1988; Haskins and McDuffie, 1990; Wonget al., 1996). While immune CD4 cells, like immune CD8 cells, can accelerate development of disease upon transfer to young, pre-disease NOD mice (both male and female), only immune CD8 cells are able to kill islet cells in vitro (Nagata, et al., 1992).

Perforin knockout mice and Fas-deficient mice have been used to analyze possible cytotoxic mechanisms employed by T cells in generating the diabetic state. In normal mice, infection with LCMV induces insulitis and diabetes within two weeks. Perforin-deficient mice develop the insulitis, but not diabetes, suggesting perforin is crucial for the islet cell destruction phase of virally induced diabetes (Kägi et al., 1996). In NOD mice with disrupted perforin genes, the incidence of spontaneous diabetes is both delayed and greatly reduced in severity, again highlighting an important role for perforin in islet destruction, but suggesting other mechanisms may be involved as well (Kägi et al., 1997).

NOD mice crossed to the *lpr* and *gld* mutations have also been used to assess the role of the Fas pathway in diabetes. Fas is not expressed on pancreatic β-cells of pre-diabetic NOD mice, but begins to be expressed with the early wave of inflammatory infiltrates, presumably under the influence of cytokines such as IFNγ (Chervonsky et al., 1997; Stassi et al., 1997; Suarez-Pinzon, 1999). A similar situation appears to hold in humans (Stassi et al., 1995; Loweth et al., 1998). NOD.*lpr* mice, which lack expression of surface Fas at all stages, fail to develop either insulitis or diabetes, and transfer of immune NOD splenocytes into NOD.*lpr* mice failed to cause disease (Itoh et al., 1997; Chervonsky et al., 1997). NOD.*gld* mice, which do not express the Fas ligand, also fail to develop diabetes (Su et al., 2000). Fresh NOD islets implanted into NOD mice with ongoing disease were protected to a substantial degree by Fas-ligand antibody (Suarez-Pinzon et al., 2000).

On the other hand, islet cells from NOD.*lpr* mice were only slightly (Allison and Strasser, 1998) or not at all (Kim et al., 1999) protected from destruction when transplanted into diabetic NOD. Fas ligand antibody also failed to block the action of immune NOD splenocytes transferred into pre-diabetic NOD mice (Kim et al., 1999). Given that perforin also plays a role in islet destruction in IDDM, it is not surprising that neutralization of the Fas pathway does not uniformly interfere with IDDM. A close study of the involvement of Fas suggests it is important in the early stages of the disease, with perforin perhaps more important in the final destructive stages (Nakayama et al., 2002; Savinov et al., 2003).

Membrane TNFα has also been implicated as a possible effector mechanism in murine IDDM. In mice expressing the B7 co-stimulatory molecule under an insulin promoter, but lacking perforin or Fas, onset of islet cell destruction appeared to be mediated by TNFα (Herrera et al., 2000).

The role of other potential killer lymphocytes in IDDM is complex. Most current evidence actually points to a protective rather than a destructive role. NKT cells, for example, increase in the pancreas during the pre-destructive insulitis stage in NOD mice, but then disappear from the pancreas and decrease in the circulation generally. They are also hypofunctional (Falcone et al., 1999). Moreover, the diabetic state in NOD mice can be markedly suppressed by regimens that activate NKT cells (reviewed in Sharif et al., 2002; Beyan et al., 2003). Exactly how NKT cells exert an effect on CD4 and CD8 effectors is unclear, but presumably involves manipulation of local cytokine environments. Interaction of regulatory NKT cells with CD4 T cells in IDDM renders the CD4 cells essentially anergic, without killing them (Beaudoin et al., 2002). NKT cells may also perturb the dendritic cell compartment, altering presentation of self-peptides (Naumov et al., 2001). The nature of internal or environmental signals that perturb NKT function in the first place is also uncertain, but could involve stress signals such as heat shock proteins. How the immune system would distinguish between stress signals generated by true danger to the host (e.g. during infection) and autoimmune-induced stress has yet to be elucidated. A similar case has been made for a role of NK cells in IDDM, although the evidence is less compelling. It should also be noted that a claim of skewing of NKT cells in human IDDM subjects has recently been challenged (Lee et al., 2002).

γ/δ T cells are also potentially cytotoxic, but there is no evidence that they are involved in damage to islet cells in IDDM. Rather, like NKT and possibly NK cells, they appear to play a role in regulation of peripheral self-tolerance (reviewed in Hänninen and Harrison, 2000). IDDM can be suppressed in NOD mice by intranasal aerosol (mucosal) immunization with insulin or GAD peptides, a procedure that stimulates γ/δ T cell production. As with NKT cells, it is not yet entirely clear how γ/δ T cells are induced, or exactly how they interact with CD8 effector cells to alter disease.

2 MULTIPLE SCLEROSIS

Multiple sclerosis (MS) is a chronic, initially relapsing disease of the central nervous system (CNS) accompanied by a complex immunopathology that is presumed to be autoimmune in nature. It is characterized by intense inflammatory infiltration of CNS tissues, beginning with post-capillary exudates but rapidly extending into the surrounding parenchyma. The most obvious damage is to myelin sheaths, and to the cells that produce them, the oligodendrocytes. However, long-term clinical sequelae generally include damage to neurons as well. While damage to the sheath is repairable to some degree, neuronal loss, and the accompanying neurological deficits, are presumably irreversible. An experimental model of the disease (experimental autoimmune encephalomyelitis; EAE) can be induced in animals by immunization with myelin sheath material or even an isolated sheath component, myelin basic protein (MBP).

For many years the tissue damage accompanying MS was presumed to be essentially if not entirely inflammatory in nature. Patients with ongoing disease develop numerous plaques along nerve tracts that are filled with macrophages and T cells, as well as cellular debris. CD4 T cells were thought to interact with some component of myelin sheath material, driving macrophage activation, which was responsible for the ensuing damage. CD4 activation may also be important in up-regulating expression of MHC molecules on CNS cells, many of which normally express low to undetectable levels of these structures (Vass and Lassmann, 1990; Neumann, et al., 1995).

While myelin sheath damage may in fact be largely inflammatory, more recently it has been recognized that this may be only part of the story. Although CD4 T cells are unquestionably key to the initiation of MS, it is now clear that CD8 cells are also involved, especially in progression of the disease. In fact CD8 cells outnumber CD4 cells in demyelinating lesions, with CD4 cells predominating only in perivascular exudates (Booss et al., 1983; Sobel, 1989). Freshly isolated, MBP-specific CD4 cells do not themselves cause direct, cell-mediated damage to oligodendrocytes (Lee and Raine, 1989). The CD8 cells in MS and EAE express a highly restricted repertoire of T-cell receptors, highlighting the antigenic specificity of the CD8 response (Tsuchida et al., 1993; Monteiro et al., 1996; Babbe et al., 2000). CTL specific for CNS cells displaying myelin-associated proteins have been demonstrated in MS patients (Tsuchida et al., 1994; Jurewiecz et al., 1998), and most neural cells (including oligodendrocytes) are, under some conditions, susceptible to CTL killing (reviewed in Neumann et al., 2002). EAE can be induced in naïve mice by the transfer of myelin-specific CD8 T cell clones (Sun et al., 2001; Huseby et al., 2001).

Unlike IDDM, NK cells may play a role in MS damage, particularly if activated by IL-2. Such cells, depleted of CD3 cells, can kill oligodendrocytes in an MHC-unrestricted fashion (Antel et al., 1998; Morse et al., 2001). The blood levels of NK cells drop in MS patients during periods of myelin destruction; it has been proposed this reflects a movement of NK cells from the circulation to the CNS where they attack oligodendrocytes (Kastrukoff et al., 1998), but other interpretations are possible.

Similar to IDDM, NKT cells have been implicated in regulation of MS. Disappearance of NKT cells from the circulation is a hallmark of MS and EAE (Sumida et al., 1995; Yoshimoto et al., 1995), although they do not show up in significant numbers in myelin sheath lesions (Illés et al., 2000). Stimulation of NKT cells has been reported to significantly reduce the pathology in EAE (Miyamoto et al., 2001). Simultaneous administration of the NKT-activating molecule α-GalCer and MBP actually exacerbates EAE, while activation of NKT prior to exposure to MBP significantly decreases EAE (Jahng et al., 2001). NKT cells are thought to exercise their effect in EAE by determining the types of cytokines released by CD4 cells (Miyamoto et al., 2001).

The role of γ/δ T cells in MS and EAE is not entirely clear. They are found in higher than normal concentrations in demyelinated plaques and in the cerebrospinal fluid of MS patients (Wucherpfennig et al., 1992; Shimonkevitz et al., 1993; Hvas et al., 1993). γ/δ T cells from either blood or CSF can kill oligodendrocytes in vitro, raising the possibility they could be involved in the myelin damage seen in vivo (Zeine et al., 1998). In mice infected with murine hepatitis virus, one of the pathologies seen is a demyelination not unlike MS, and this appears to be mediated at least in part by γ/δ T cells (Dandekar and Perlman, 2002). On the other hand, they have also been proposed to play an immunoregulatory role similar to that proposed for IDDM (Stinissen et al., 1998). But γ/δ knockout mice undergo the same disease process as normal mice upon transfer of EAE T cells (Clark and Lingenheld, 1998), and EAE cannot be induced in mice lacking α/β T cells but with an intact γ/δ T cell compartment (Elliot et al., 1996). Further work is required to clarify the precise role (or roles) of γ/δ T cells in MS.

Both Fas- and perforin-mediated killing have been demonstrated using various CNS cells as targets. In a demyelinating disease caused by infection with Theiler's virus, which shows many similarities to MS and EAE, CD8 T cells were shown to be critical for neuronal cell death, and appeared to use the perforin rather than the Fas pathway (Murray et al., 1998). Neurons can be destroyed by cytotoxic granules or by purified perforin (Rensing-Ehl et al., 1996). In a more recent series of studies, it was shown that neuronal cell bodies, after MHC class I-induction and pulsing with a CTL-recognized peptide, suffered membrane damage as measured by Ca^{+2} flow into the cell. This damage was slow, was absent in Fas-negative target cells, and could be mediated by perforin-deficient CTL (Medana et al., 2000). When damage to neurites was examined, it was much more rapid and appeared to be perforin-mediated (Medana et al., 2001). It may be of interest that in the former case the cells were treated with tetrodotoxin, whereas in the latter study they were not.

3 SYSTEMIC SCLEROSIS

Systemic sclerosis (scleroderma) is a progressive fibrosis of the skin and various internal organs such as the lungs, heart and GI tract (LeRoy et al., 1988). At the cellular level, apoptosis of endothelial cells (EC) appears to be a key pathogenic event (Sgonc et al., 1996). Long considered an inflammatory response to an undefined external agent, scleroderma is increasingly viewed as an autoimmune disease. T cells, both CD4 and CD8, dominate the inflammatory infiltrate, which occurs in the earliest stages of the disease (White, 1996).

No direct role for α/β T cells in tissue damage has been discerned. However, γ/δ T cells have been implicated as possible effectors in scleroderma. They are elevated in the circulation, accumulate at sites of inflammation, and display a limited junctional diversity, as well as activation and adhesion markers (Yurovsky et al., 1994; Giacomelli et al., 1998). γ/δ cells isolated from scleroderma patients displayed cytotoxicity toward EC in vitro that was not blocked by TCR or class I antibody, but was reportedly blocked by granzyme A antibody (Kahaleh et al., 1999; Giacomelli et al., 2001). γ/δ cells likely play a role in regulating inflammation through cytokines as well.

NK cell number and activity are also elevated in scleroderma, and apoptotic death of EC can be induced by activated NK cells in the presence of EC antibody; this lysis was blocked by Fas reagents, but not by concanamycin A, suggesting degranulation is not a lytic mechanism (Grazia-Cifone et al., 1990; Sgonc et al., 2000).

4 MUSCULAR DYSTROPHY

Duchenne muscular dystrophy (DMD) in humans is a lethal muscle wasting disease having as a root cause absence of functional dystrophin protein. Dystrophin connects the cytoskeleton of sarcolemma cells to the extracellular matrix through a multimeric transmembrane structure called the dystrophin glycoprotein complex (Koenig et al., 1987; Spence et al., 2002). There is a mouse model of the disease – the mdx mouse – that displays the essential features of at least the early stages of DMD in humans (Bulfield et al., 1984; Partridge, 1991). Although the muscle-wasting syndrome in both humans and mice is accompanied by marked leukocyte infiltration of damaged tissues, this was long presumed to be an *a posteriori* phenomenon of inflammation in response to dead and dying cells.

However, more recent data suggest a more active role of immune system cells, both innate and antigen-induced, in the progression, if not the initiation, of the disease. The leukocytic infiltration of muscle tissue in DMD is closer to that seen in autoimmune disease than in simple mechanical tissue damage (Giorno et al., 1984), and DMD patients show a high degree of conservation of a particular α/β T-cell receptor in damaged muscle (Gusssoni et al., 1994), suggesting that infiltrating T cells may be responding to something more than generalized inflammatory signals (reviewed in Spencer and Tidball, 2001).

A high proportion of both CD4 and CD8 cells in the infiltrates display activation markers consistent with an ongoing, antigen-driven immune response. Depletion of either of these subsets reduces the pathology developing in mdx mice, and adoptive transfer of these subsets to healthy recipients results in development of muscle pathology (Spencer et al., 2001). Humans respond favorably to treatment with prednisone (Fenichel et al., 1991), which dramatically reduced CD8 cells in the tissue infiltrates (Kissel et al., 1991).

The possible involvement of cytotoxic pathways in dystrophic disease was examined in the mdx mouse. While these mice develop a wasting disease similar to human DMD in its initial phases, they go on to transient recovery and regeneration of damaged muscle tissue, which does not occur in humans (Bulfield et al., 1984). Cell death in the early stages of mdx disease is apoptotic (Tidball et al., 1995). These two processes – wasting and regeneration – continue in parallel until about 15 months of age, when wasting overcomes regeneration and the mice die (Lefaucheur et al., 1995).

Mdx mice depleted of CD8 cells before the onset of initial pathology show a delay and a reduction in severity of disease; mdx mice with the perforin gene knocked out showed essentially the same thing (Spencer et al., 1997). In the latter case, the cell death that did occur was by necrosis rather than apoptosis, and its cause is not clear. These results suggest CTL-mediated apoptotic cell death may be an early and important factor in development of dystrophin myopathies. The fact that muscle cells are Fas-negative suggests Fas-mediated killing is unlikely to be a major contributor to the overall pathology, although this has not been formally tested.

DMD patients show an increase in peripheral T cells bearing the Vγ9/δ2 receptor (Hasui et al., 1995). γ/δ T cells have also been proposed as a possible effector cell in mdx mice, and are found in significant numbers among the inflammatory cells infiltrating the muscles of mdx mice as well as DMD patients (Montecino-Rodriguez, et al., 2003 submitted). These cells bear surface markers suggesting a high degree of activation. Whether they are involved as effectors of muscle damage, or as regulators of the autoimmune response, is not known at present.

5 AUTOIMMUNE THYROID DISORDERS

Chronic autoimmune thyroid disease occurs in several forms in humans. Hashimoto's thyroiditis is characterized by leukocytic infiltration, dominated by T cells and macrophages. There is initially a hypertrophy of the gland, followed by gradual fibrosis and atrophy and reduced ability to produce thyroid hormone (hypothyroidism). A number of viruses have been implicated in the onset of Hashimoto's thyroiditis (e.g. Mine, et al., 1996). Chronic atrophic thyroiditis is similar, but does not pass through the hypertrophic (goiter) phase that characterizes Hashimoto's disease. Grave's disease is also autoimmune, but has minimal inflammatory infiltration and goiter, and results in hyperthyroidism accompanied by exophthalmia. Grave's disease is not classified as a thyroiditis. In all three cases there is a loss of thyroid cells through apoptotic death, which presumably is the basis for compromised function. The clinical features of these diseases have recently been reviewed (Dayan and Daniels, 1996).

As with most autoimmune diseases, the triggering immune event is unclear, and the target antigens, at least for the cellular portion of the response, are unknown. Autoantibodies produced during these diseases recognize thyroglobulin, thyroid peroxidase and the thyrotropin receptor, among other antigens. Each syndrome begins with the activation of CD4 T cells (Dayan et al., 1991), followed in the case of the thyroiditises by an influx of CD8 cells, macrophages, and some B cells (Del Prete et al., 1986). Infiltrating CD8 cells are perforin-positive (Wu et al., 1994), and it has generally been assumed they are responsible for the cellular destruction accompanying these diseases, but the evidence for this is admittedly scant and indirect. There have been reports that NK cells may mediate ADCC in Hashimoto's thyroiditis (Bogner et al., 1984), but in general NK cells are not thought to be major effectors in the thyroid autoimmunities. A major impediment in working out details of immune mechanisms has been the lack of a good spontaneous-onset animal model for any of the recognized human diseases.

Analysis of the role of cell-mediated cytotoxicity in autoimmune thyroid disease has been complicated by several factors. Thyrocytes normally display low to undetectable levels of Fas, but this is up-regulated substantially under the influence of inflammatory cytokines such as IL-1. Moreover, it has been claimed (although this has been challenged) that even normal thymocytes express Fas ligand and that this, too is up-regulated during autoimmunity, particularly in Hashimoto's thyroiditis. Thyrocytes from autoimmune patients appear to be able to kill Fas-positive target cells in vitro (Giordano et al., 1997; Mitsiades et al., 1998; Stokes et al., 1998; Fiedler et al., 1998). This raised the possibility that thyrocytes might actually be killing themselves, via fratricide or suicide, in thyroid autoimmunity. And in fact, normal thyrocytes cultured in the presence of IL-1 do appear to die by apoptosis (Giordano et al., 1997). While this model for autoimmunity has provoked great interest, at present it requires further validation.

The question of cytotoxic function in thyrocytes from glands of patients with thyroid autoimmunity has been examined by Hammond et al. (2001). Previous studies used primary thyrocytes, which are a mixture of true thyrocytes and various infiltrating leukocytes, the proportion of monocytes being considerably greater (up to 50 percent) in Hashimoto's thyroiditis. Thus it is difficult to determine just which cells are acting as effectors in any apparent CMC. Hammond et al. tried to sort this out by fractionation of primary cell populations. They found that lymphocytes isolated from Grave's disease glands displayed reasonably good cytotoxicity toward Fas-positive targets, whereas lymphocytes from Hashimoto glands were modestly cytotoxic at best. Purified thyrocytes from the Grave's disease glands also gave reasonable cytotoxicty, while those from Hashimoto glands gave none at all. However, the effect of Fas and FasL blocking reagents were ambiguous, and exocytosis and TRAIL blocking reagents had no effect at all. It has been suggested that the primary role of Fas-FasL interactions may be in controlling lymphocytic infiltration into the thyroid, rather than self-destruction (de Maria et al., 1998; Salmaso et al., 2002), particularly in Grave's disease (Mitsiades et al., 2001; Giordano et al., 2001).

There are numerous reports of self-reactive γ/δ T cells in the various autoimmune thyroidopathies, but no role for these cells either in cytotoxicity or in autoimmune regulation has yet been either established or excluded (see e.g. Paolieri et al., 1995; Catalfamo et al., 1996). As noted earlier, NK cells do not appear to play a role either. Thus at the present time, it must be admitted that a role for killer lymphocytes in causing cellular damage in autoimmune throiditis, while certainly possible, has yet to be convincingly demonstrated.

6 MYOCARDITIS

Myocarditis is an inflammatory process affecting the heart, resulting in injury to the myocardium in the absence of ischemia. It is an incompletely understood syndrome of multiple etiologies, and a major suspected precursor of idiopathic dilated cardiomyopathy, accounting for 25% of heart failure cases (Brown and O'Connell, 1995). Although myocarditis may resolve within weeks or months of onset, in some patients the disease becomes chronic, with the persistence of interstitial fibrosis and inflammation, myocardial hypertrophy, chamber dilatation, and congestive heart failure (dilated cardiomyopathy), which may require a heart transplant (Kopechy and Gersh, 1987).

Like Type 1A diabetes, autoimmune myocarditis is often associated with viral infection. Various viruses are capable of inducing myocarditis, but the picornavirus family predominates. The clinical manifestations of acute viral myocarditis vary from that of a benign disease to a fulminant, life-threatening illness, particularly in children (Wenger et al., 1982).Ventricular enlargement may be present, and left ventricular regional wall motion abnormalities have been demonstrated. Ventricular dysfunction can be severe enough to result in congestive heart failure, leading to severe clinical implications. An understanding of the pathophysiology of myocarditis in humans has been advanced by the development of murine models of viral myocarditis, mostly Coxsackievirus B3 (CVB3; Hershkowitz et al., 1987; Brown and O'Connell, 1995), which have provided insight into the complex interactions between the virus and the host, culminating in tissue injury. Indeed, the murine model of CVB3-induced heart disease bears a striking resemblance to human myocarditis.

An important paradigm of experimental myocarditis is myosin-induced autoimmune myocarditis, developed by Neu et al, (1987). This model is based on the hypothesis that in susceptible mice, the late phase of CVB3-induced myocarditis is mediated by a cross-reactive autoimmune response to cardiac myosin. The restriction of the disease to the cardiac form of myosin, as well as the genetic parallel between susceptibility to the late phase of viral myocarditis and myosin-induced myocarditis, supports the view that myosin is one of the prominent antigens capable of inducing autoimmune myocarditis (Rose et al., 1993). The availability of a specific myocarditis-inducing autoantigen in genetically defined mouse strains provides a useful tool to explore the immune effector mechanisms causing auto-immune myocarditis, as well as to characterize the pathophysiology of affected cells resulting in cardiac dysfunction. By using adoptive transfer and immunodepletion techniques, both forms of myocarditis have been shown to be T-cell-dependent diseases.

Despite increased evidence for an autoimmune component of myocarditis, as well as the increasing recognition of myocarditis as an important precursor of dilated cardiomyopathy and heart failure, the specific effectors responsible for impaired cardiac function at the cellular and molecular levels have not yet been determined. At least three mechanisms that cause cardiac damage during the chronic phase are possible: 1) infitration and direct action of CTL and NK cells; 2) antibodies to viral and cardiac antigens; and 3) locally or systemically produced cytokines such as IL-1, IL-6 and TNF-α.

Chapter 11

HOMEOSTASIS, MEMORY AND CTL VACCINES

1 THE ROLE OF CYTOTOXICITY IN IMMUNE HOMEOSTASIS

We have long known that one of the consequences of T-cell activation is vigorous clonal expansion of responding T cells. CD8 T cells specific for a particular viral epitope, for example, can expand a few orders of magnitude over a roughly one-week period (Murali-Krishna et al., 1998; Butz and Bevan, 1998). But what becomes of these greatly expanded pools of potentially dangerous cells once a pathogen is cleared? Given the frequency with which T-cell activation must occur, the body obviously cannot maintain all the cells produced from each round of infection. Some will go on to become memory T cells, by a process that is still not well understood, but the vast majority of cells must be cleared to prepare the way for response to other infections. The way in which these cells are removed has become a topic of great interest during the past decade.

This important element of immune system homeostasis is mediated at least in part through an apoptotic process called "activation-induced cell death" (AICD; Kabelitz et al., 1993; Kabelitz and Janssen, 1997; Krammer, 2000). AICD may be the default pathway for activated T cells; activation primes them to die unless rescued by other signals. Awareness that activation via the TCR could sometimes result in a lethal signal to T cells arose in the late 1980s using antibodies to the TCR (Nau et al., 1987; Breitmeyer et al., 1987; Webb and Sprent, 1987), but was soon extended to activation by antigen as well (Russell et al., 1988; Suzuki et al., 1988;

Brocke et al., 1990). And while initial studies were focused on CD4 T cells, CD8 cells were found to undergo a similar down-regulation (Walden and Eisen, 1990; Pemberton et al., 1990; Russell et al., 1991). AICD occurs in responses to intracellular parasites as well as to alloantigens (Mamalaki et al., 1993; Christensen et al., 1996) and may be key in the body's response to tumors (Li et al., 2002); we will discuss each of these in more detail below.

AICD is mediated by several apoptosis-inducing pathways, depending on the cell type to be eliminated, nature of the activating signal, and availability or absence of cytokines that promote survival of activated cells (Gorak-Stolinska et al., 2001). It could be triggered simply by decreased availability of key cytokines as the offending pathogen is cleared from the system (Vella et al., 1998; Marrack et al., 1999; Tan et al., 2001), or it could represent a more active process. One AICD pathway, prominent particularly in CD4 cells, uses the Fas-FasL system (Dhein et al., 1995; Brunner et al., 1995; Nagata, 1997; Krammer, 1999). Activated T cells under the influence of IL-2 (signal 2) up-regulate surface expression of both Fas and Fas ligand, and down-regulate intracellular apoptosis inhibitors (Algeciras-Schimnich et al., 1999; Van Parijs et al., 1999). In this sense, physiologically "correct" activation of T cells is necessary to set them up for ultimate self-destruction. Engagement of costimulatory molecules such as CD28 with B7 during activation can delay, at least transiently, the susceptibility to AICD, for which the activated T cell is poised (Boise et al., 1995; Kirchhhoff et al., 2000; Borthwick et al., 2000). But as discussed in Chapter 1, after activation T cells begin to express CTLA-4, which out-competes CD28 for the APC ligand B7 and delivers a negative, rather than an activating, signal to the T cell. Within a day or two after initial triggering, subsequent and repeated engagement of activated T cells with antigen normally results in cell death.

An interesting question is whether AICD proceeds via "fratricide" or "suicide". That is, do activated T- cells kill each other, or do individual activated cells kill themselves? Two of the above studies showed that isolated single T cells, upon activation through the TCR complex, died via apoptosis, and that this cell death was inhibited by reagents blocking the Fas-FasL interaction (Dhein et al, 1995; Brunner et al, 1995). Whether this occurs by release from the activated T cells of soluble FasL, or by interaction of different regions of the membrane of single cells expressing Fas and FasL, is not clear. The fact that T cells can die by self-inflicted Fas damage does not, of course, mean that fratricidal cell death is not involved when T cells can interact (see also Figure 9.2).

The importance of the Fas pathway in immune homeostasis can be seen in *lpr* and *gld* mice, which suffer beginning shortly after birth from uncontrolled lymphocyte proliferation (lymphadenopathy) with widespread inflammatory, autoimmune-like disease (Andrews et al., 1978; Roths et al.,

1984; Nagata and Suda, 1995). Mice with the *lpr* mutation are impaired in their ability to delete T cells after activation by a wide range of signals (Russell et al., 1993; Mogil et al., 1995; Fuse et al., 1997; Singer and Abbas, 199). However, *lpr* mice are not completely impaired in AICD, suggesting other pathways may be involved as well. There have been suggestions that at least in mice TNFα (Zheng et al., 1995; Sytwu et al., 1996) and lymphotoxin (Sarin et al., 1995) may also be involved in AICD. In the liver, Fas may play an important role in down-regulation of CD8 as well as CD4 T cells (Dennert, 2002), and death in this case appears to involve fratricide. Removal of effector immune cells when they are no longer needed could be mediated by newly expressed FasL on non-lymphoid cells at the site of immune response. Normally not expressed on quiescent tissues, it has been shown that FasL is promptly up-regulated during a local immune response; IL-10 appears to play a role in this process (Barreiro et al., 2004).

Perforin knockout mice have been used to assess the role of the degranulation pathway in AICD in recently activated T cells triggered by TCR antibodies. The absence of perforin resulted in a reduced susceptibility of such cells to AICD, mainly in CD8 T cells (Spaner et al., 1998; Badovinac et al., 2000). In these experiments, AICD appeared to be by suicide rather than fratricide; that is, activated CD8 cells died autonomously, rather than by killing one another. The basis for such a role of perforin is unknown at present, given the apparent resistance of CTL to perforin-mediated damage. A study with human T cells, using a granzyme B inhibitor, also suggested a role for the degranulation pathway in AICD of CD8 T cells, and showed both CD4 and CD8 cells to be susceptible to Fas-mediated AICD (Gorak-Stolinska et al., 2001). It has also been proposed that perforin may play more of a cytostatic role in AICD, and that much of the actual CTL killing is via IFNγ; IFNγ knockout mice are seriously impaired in their ability to clear CTL at the end of an infectious cycle, as are perforin knockout mice. But only the double knockout (perforin/IFNγ) mice are totally crippled in the ability to carry out AICD (Badovinac et al., 2000).

Perforin-mediated AICD in normal mice can be seen in the kinetics of the CD8 response to LCMV infection. As discussed earlier, one result of infection is an enormous expansion of LCMV-specific CD8 T cells. However, once the virus titer is significantly reduced, the level of these cells greatly decreases. Perforin-deficient mice chronically infected with LCMV contain much larger numbers of anti-viral T cells late in infection than do normal mice. In fact, perforin-deficient mice usually die within a month or so from non-specific immune damage, probably mediated by excessive production of IFNγ, and possibly by Fas- or TNFα-mediated damage to normal tissues. Mice deficient in both perforin and Fas develop even more pronounced inflammatory problems (Matloubian et al., 1999; Kägi et al.,

1999). However the Fas-mediated immune regulation probably affects mostly CD4 cells, as neither Fas nor TNFα appear to be involved in down-regulating CD8 cells during LCMV infection (Lohman et al., 1996; Reich et al., 2000; although see Zhou et al., 2002). A similar pathology is evident in perforin-deficient mice infected with Theiler's virus (Rossi et al., 1998) and Herpes simplex (Ghiasi et al., 1999).

Interesting insights into the role of perforin in immune homeostasis is provided by human diseases stemming from a loss of perforin. Absence of a functional perforin gene in humans due to a missense or nonsense mutation results in familial hemophagocytic lymphohistiocytosis (FHL) (Stepp et al., 1999). Patients with FHL over-respond to many infections, developing harmful, long-lasting systemic inflammation. Although not formally proved at present, by analogy with mouse models it would seem likely that FHL patients have an impaired ability to control CD8 responses through AICD (Stepp et al., 2000).

Another interesting question relating to AICD as manifested in vitro is how highly activated CTL clones can be maintained for long periods of time in culture with repeated stimulation by cognate antigen. As discussed in Chapter 4, long-term CTL clones are resistant to perforin-mediated lysis, presumably as a result of an as yet undefined selective mechanism. An examination of the presence of Fas and TNF receptors on cultured CD8 CTL clones showed that these receptors have been down-regulated, providing a ready explanation for insensitivity to these potential mediators of AICD (Sugawa et al., 2002). The fact that only CTL that have developed resistance to these lytic mechanisms can be maintained long-term in culture with repeated stimulation reinforces the significance of AICD.

2 THE ENIGMA OF CTL MEMORY

The question of homeostasis is tied in closely with the subject of immunological memory, particularly in CTL. Memory, along with diversity and specificity, is one of the cardinal characteristics of adaptive immune responses. NK cells, which play a key role in defense against viral infection, do not develop a memory state. But in CTL, initial exposure to a given antigen leads to a state wherein subsequent exposure to the same antigen elicits a response that is more rapid and of greater amplitude than the first (primary) response. It is clear that one key to memory is expansion during the primary response of cells displaying receptors recognizing the antigen under consideration. But equally important is the question of what happens at the end of the response; some activated cells survive to become memory cells, but most die. How are these decisions made?

Several general questions pertain to immunological memory, whether we are considering B cell, T helper cell or CTL memory. Are memory cells the direct lineal descendents of the cells responding in the primary response, or a separate population costimulated with the responding cells? Is memory quantitative or qualitative? That is, can the faster-stronger secondary response be accounted for entirely by the fact that there are more responding cells around, or is the qualitative nature of the secondary response in some way different from the primary response? And finally, what are the requirements for maintenance of memory? In particular, what role – if any – does antigen play in the long-term maintenance of the memory state? These questions are interrelated, and only partly resolved for all forms of immunological memory (Sprent and Surh, 2001).

The precise lineage of memory CTL is still unclear. While the preponderance of evidence favors direct lineal descent of memory cells from CTL effectors (Jacob and Baltimore, 1999; Opferman et al., 1999, 2001; Lauvau et al., 2001), there are data consistent with distinct effector and memory cell precursors (Manjunath et al., 2001). The vast majority – 90 percent or more – of CTL die off at the end of the primary response through AICD, and probably by withdrawal of cytokines crucial to T-cell survival as well. Nevertheless, some CTL do escape and go on to become memory cells. Were these cells ever fully activated, or did they shunt off from the main activation pathway at some early point, thereby gaining immunity to apoptotic cell death at a later stage? We do not know, although in line with the direct lineal descent model, there is evidence that CTL that become memory cells were in fact at one time expressing intracellular molecules associated with a fully activated state, such as perforin (Opferman et al., 1999), and granzyme B (Jacob and Baltimore, 1999). CD8 memory cells continue to express high levels of the surface activation marker CD44 for some time, and are often designated $CD44^{hi}$.

One factor in selection of memory cells from the CTL generated in a primary response may be related to the strength and duration of the activating signal (Lanzavecchia and Sallusto, 2002). Full activation, driven by sufficient contact with both peptide/class I and costimulatory molecules on dendritic APC, appear to render cells more resistant to AICD-induced apoptosis, possibly through induction of anti-apoptotic molecules such as BCL-2 and $BCL\text{-}X_L$, and expression of receptors for cytokines such as IL-7 and IL-15.

As discussed in earlier chapters, it is possible to generate CD8 CTL in the absence of CD4 help if the antigen-presenting dendritic cell is activated by direct infection, by previous contact with a CD4 cell, or by interaction with a range of soluble or pathogen-associated products (Buller et al., 1987; Liu et al., 1989; Rahemtulla et al., 1991). However, chronic lack of T-cell help greatly impairs the development and maintenance of long-term memory (Matloubian et al., 1994; von Herrath et al., 1996). This is likely related to the need for CD40L activation of dendritic cell APC, and is critical early in the development of the CD8 response (Borrow et al., 1998). The need for T cell help in development of memory can be replaced by IL-2 (Yang and Wilson, 1996).

There clearly are differences in the requirements for activation of virgin and memory T cells, implying some sort of qualitative difference in these two cell types. As pointed out in Chapter 1, naïve T cell activation is accompanied by expression of new, function-related genes (Agarwal and Rao, 1998; Grayson et al., 2001). Although these genes may not be actively transcribed in resting memory CTL, they often have elevated levels of mRNA that allow rapid resumption of protein production (Bachmann et al., 1999; Veiga-Fernandez et al., 2000). The reactivation of memory CTL, at least as studied in vitro, is less strict than activation of virgin pre-CTL; the requirement for both APC-embedded costimulatory signals and soluble help factors is markedly less. The reappearance of cytotoxic function after restimulation of memory cells is very rapid (Kedl and Mescher, 1998); in fact, memory CTL can be reactivated under conditions where cell division is completely blocked (MacDonald et al., 1975). However, once reactivated, there is no apparent difference in the mode, specificity, rate or extent of killing in CTL derived from virgin or memory CD8 cells.

Memory CTL in vivo also have different circulatory and residential patterns compared with virgin CTL and activated CTL. It has been useful from this point of view to recognize two distinct subsets of CTL memory cells (Salusto et al., 1999; Weninger et al., 2001). *Peripheral memory cells* are CTL that express reduced or even no effector function, but remain at the original site of infection. Peripheral memory CTL lack the surface markers required to gain re-entry to lymph tissues, and thus upon leaving the infection site are forced to remain for a time in the lymph and blood circulation. A great many of them take up residence in non-lymphoid tissues, and this in fact may be a major portion of the memory pool at any given time (Masopust et al, 2001). These cells require mainly IL-2 for maintenance. *Central CTL memory cells* represent a later stage of memory development. They have acquired the surface information necessary to guide them back into lymphoid tissue scattered throughout the body; some also settle into bone marrow (Moser and Loetscher, 2001; Feurer et al., 2001a,b). Central memory cells are highly dependent on IL-15 and IL-7 for maintenance (Tan, et al., 2002). IL-15 may promote CD8 memory cell survival by up-regulation of anti-apoptotic molecules such as Bcl-2 (Grayson, et al., 2000; Sprent and Surh, 2002). Memory CTL can be stimulated to proliferate, albeit slowly, by

IL-15 (Zhang et al., 1998), and IL-15-deficient mice have greatly reduced CD8 memory T cell function (Kennedy et al., 2000).

Closely tied in to the concepts of AICD and memory cell survival is the notion of clonal anergy/exhaustion. In chronic infections, and perhaps in cancer where CTL are repeatedly exposed to antigen over long periods of time (see below), some of the responding clones become anergic – they do not physically disappear, but lose their ability to express effector function. In some cases, it has been reported that the CTL are essentially tolerized, and are completely unable to respond to antigen (Moskophidis et al., 1993). In other cases, the CTL persist in an activated state, as suggested by the presence of the marker $CD44^{hi}$, and are even able to proliferate in response to antigen, but they cannot secrete IFNγ and cannot kill specific target cells (Zajac et al., 1998). In other cases, IFNγ production may be intact, but the ability to kill is lost (Appay et al., 2000). Several studies, particularly with HIV infection in humans, have suggested the degree of CTL anergization may be a function of viral load (Kostense et al., 2001; Edwards et al., 2002). The relationship among clonal anergy, clonal deletion (AICD) and clonal memory, though doubtless important, is not understood at present.

The issue of CTL memory maintenance is also not completely resolved. A major question is whether continuous contact with the initial antigen is required for maintenance of memory. This question is made difficult by the fact that failure to detect persisting antigen does not mean it is not present in amounts too small to detect, but sufficient to drive memory cells. In humans, CTL memory to viruses or to foreign MHC (allogeneic) can persist for years, even decades, in the absence of detectable antigen (Demkowicz et al., 1996; Van Epps et al., 2002). Rafi Ahmed has dissected this problem with great care in the LCMV/mouse system. He proposed some years ago that virus-specific CTL memory cells can persist for very long times in the absence of viral antigenic stimulation (Jamieson and Ahmed, 1989; Lau et al., 1994), a point that has been confirmed by others (Mullbacher, 1994; Hou et al., 1994). More recently, Ahmed has shown that such memory cells upon transfer to class I MHC-negative mice persist for a very long time, proliferating slowly if at all, suggesting that not only is antigen not required, neither is continuous stimulation through the T-cell receptor (Murali-Krishna et al., 1999). The same has been shown for CD4 memory cells (Swain et al., 1999). This may be one reason that the TCR repertoire of CD8 memory cells does not change over time (Busch et al., 1998; Sourdive et al., 1998). It should be kept in mind that although antigen may not be required for maintenance of CTL memory, in most cases antigen will be present, even if at very low levels (Ciurea et al., 1999), and may well have some influence on the shape of memory responses. However, we would expect that memory CTL in that case have somehow become resistant to AICD induced by TCR activation (Inaba et al., 1999). An important point that has not yet been clarified is the need for antigen in the maintenance of peripheral vs. central tolerance.

3 CTL VACCINES

One of the most important applications of our evolving understanding of CTL activation, homeostatsis and memory is the production of vaccines designed specifically to activate CTL and facilitate the emergence of CTL memory. This new approach will have important implications for infectious diseases such as HIV/AIDS, and for cancer.

From our understanding of antigen processing, it is clear why "classical" vaccines of killed or disabled viruses, bacteria and eukaryotic parasites generally do not work. While able to facilitate antibody formation, through the class II/CD4 helper pathway, these potential immunogens only rarely insert themselves into the class I/CD8 antigen-processing pathway, which in most cases requires active intracellular synthesis of parasite proteins for CTL induction. Thus many current strategies for vaccines against intracellular parasites focus on means for introducing antigenic information associated with these parasites into the class I pathway, and enhancing their presentation to both naïve and memory CTL. These strategies have been extensively reviewed in the recent literature (Gherardi et al., 2001; Schirmbeck et al., 2001; Kaech et al., 2002; Robinson, 2002), and we present here only some of the highlights of this current research.

The evidence that CTL play an important role in defense against tumors has also led to interest in the development of vaccines that selectively enhance CD8 T cell responses to tumor antigens. In humans, tumor-associated antigens are produced within the tumor cell, and should enter the class I processing pathway. There is reason to believe that in some cases this process fails, so vaccine strategies for cancer are aimed at overcoming this deficit. Obviously such vaccines cannot be used to prevent cancer in the human setting, but they could conceivably be effective, alone or in combination with other therapies, in helping to control a tumor that has already become established.

3.1 Vaccines for intracellular parasites

Dendritic cells are by far the most effective antigen-presenting cells for both naïve and memory CD8 T cells, and play a key role in the response to every virus studied (Klagge and Schneider-Schaulies, 1999). Dendritic cell presentation of antigen is increasingly at the heart of vaccine strategies developed in the past several years, so a complete understanding of the life cycle of each virus within DC is absolutely essential (Blauvelt et al., 1995; Bhardwaj, 1997; Salio et al., 1999; Raftery et al., 2001). Dendritic cells are able to present high concentrations of MHC-associated antigen at their surface, where they also display a wide range of activating co-receptor ligands, and they can provide cytokines needed for T-cell maturation (Banchereau and Steinman, 1998). Thus increasing the presentation of antigen through dendritic cells, and ensuring that the DC itself is optimally activated (Romani et al., 1996; Maraskovsky et al., 1996), will help assure optimal activation of CTL. DC, once activated (often as a direct result of parasite infection) can activate CD8 cells with a minimum of CD4 help.

There are many ways of delivering antigen to DC. Selected peptides known to be a focus of CTL attack can be "pulsed" (Chapter 1) directly onto DC class I MHC molecules (Ludewig et al., 1998). This method has worked well for induction of CTL immunity in inbred strains of mice (Schulz et al., 1991; Kast et al., 1991). As shown clearly in the case of influenza virus in mice, DC can pick up antigen very effectively by ingesting apoptotic bodies in their environment, and present antigens contained therein via the class I pathway through cross-priming (Chapter 1; Albert et al., 1998; Fontenau et al., 2002; Zinkernagel, 2002). Transfection of DC with genes encoding potent microbial antigenic peptides, which are transcribed and ultimately expressed on DC class I molecules, has also proved effective (Operschall et al., 2000; Lewinsohn et al., 2002). Another approach takes advantage of DC receptors for heat shock proteins; antigenic peptides are fused to hsp 70 or hsp 110, which deliver them very effectively into dendritic cells (Sato and Srivastata, 1995).

The use of viral peptides, or genes for peptides, in creating vaccines has one inherent limitation. Because of the tremendous polymorphism of MHC proteins, each individual allele of a class I protein will have a different array of binding residues in the antigen binding site. Each different class I allele will thus likely recognize a different portion of a particular viral protein, or may not bind any of the derivative peptides from a given viral protein. The number of different possible combinations of alleles in an individual heterozygous at just two class I loci is enormous, making the preparation of individually tailored vaccines economically unfeasible. One partial way around this might be to prepare vaccines that combine large numbers of peptides, or genes for peptides, determined empirically to bind to a limited array of the most common class I alleles in a given population.

In addition to loading dendritic cells with various forms of microbial antigen, some vaccine strategies attempt to provide the DC with highly focused exposure to activating cytokines. GMCSF has proved particularly useful in this respect, and has significantly enhanced the response to vaccines for hepatitis C (Ou-Yang et al., 2002), HIV (Barouch et al., 2002), and respiratory syncitial virus (Bukreyev et al., 2001) in mice, among others. Dendritic cell affinity for GMCSF is sufficiently great that fusion proteins between GMCSF and antigenic peptides have been generated for delivery of the peptide to DC (Tao and Levy, 1993).

Another approach to promoting effective responses to vaccines intended to induce a killer cell response is to amplify signals passing through the T-cell receptor and co-activation molecules in the immunological synapse. We have learned a great deal about the differences between CD8 and CD4 T-cell activation, and advantage can be taken of these new insights to selectively drive a CTL response. There are both quantitative and qualitative aspects to this approach. First of all, it is clear that the size of the memory cell pool is related to the "burst size" of the initial immunizing event (Kaech et al., 2002), so it is desirable to stimulate as many naïve CD8 cells as possible. This is relatively easy to do for CD8 cells, which require less antigen and briefer exposures than CD4 cells, as long as the activation and coactivation signals are provided on fully activated dendritic cells.

One way of increasing DC presentation of a desired peptide supplied exogenously is to slightly alter the peptide at those positions known to interact with the DC class I molecule in order to increase binding affinity without altering the residues perceived by responding CD8 cells. Care must be taken not to disrupt residues in the peptide that interact with the TCR, however. Proper modification can both increase DC surface density of antigen, and assure that it does not get competed away by other peptides (Slansky et al., 2000). Another approach, useful when the antigen is supplied in the form of a gene for an antigenic protein or peptide, is to build into the immunizing construct genes for costimulatory molecules such as B7, the ligand for CD28 (Kim et al., 1997; Agadjanyan et al., 1999). There have also been attempts to build cytokine genes into vaccinating constructs. Because cytokines such as IL-2, IL-12 and IL-15 can profoundly modify the fate of reacting T cells, it may be possible to affect not only the strength of a response but its quality (Ahlers et al., 1997).

3.2 Cancer vaccines that elicit CTL

The strategies for development of cancer vaccines have much in common with those used with viral vaccines, but address some unique challenges as well (Finn, 2003). In the case of viruses, the viral peptides known to be antigenic can be defined, and are relatively constant for a given virus in the context of defined class I haplotypes. For tumors the situation is more complex. A limited number of tumors may arise as a result of viral infection, in which case the strategies for vaccine development are essentially those for viral vaccines. But in humans in particular, the vast majority of tumors arise sporadically, and the associated antigens are highly variable and difficult to predict. Early attempts to define tumor antigens based on antibody reactivity to tumor surface structures seemed of limited use, particularly once it was realized that CTL do not recognize native antigen at-cell surfaces.

One of the more efficient ways to identify tumor-associated antigens and peptides is by CTL screening (Van der Bruggen et al., 1991; Traversari et al., 1992). CTL (bulk or cloned; from peripheral blood or isolated from the tumor) known to lyse a particular tumor are used as effector cells, and ^{51}Cr-labeled, MHC-compatible normal cells, transfected with pooled cDNA from the cognate tumor, are used as targets. A target cell that is lysed is assumed to express one or more peptides recognized by the CTL, and candidate peptides are eluted from target MHC molecules and analyzed until the triggering peptide (and parent protein) is found. In many instances, the frequency of tumor-peptide-specific CTL is considerably higher in patients bearing the tumor, reinforcing the notion that such peptides were recognized in vivo and served to expand the corresponding T-cell population (Marincola et al., 1996; Anichini et al., 1999; Nagorsen et al., 2000).

Using these and other approaches, a large number of tumor-associated peptides have been identified (Renkvist et al., 2001; Dermime et al., 2002; Parmiani et al., 2002). Some are derived from proteins normally expressed at low levels in a limited number of cells, but over-expressed in tumor cells. These may be of limited value because of the possibility of cross-reactivity with normal cells and tissues, inducing a state of autoimmunity (e.g., Yee et al., 2000; Dudley et al., 2002; Overwijk et al., 2003). Some represent proteins (differentiation antigens) normally present only at restricted stages of development, but expressed – again, often at very high levels - in tumor cells. The use of such antigens also carries the possibility of cross-reaction with normal cells, but in fact differentiation-related peptides have already been used with some success for vaccination, as we will see. In the long run, however, so-called "tumor-unique antigens" are probably the most promising targets for vaccine development. These are proteins unique to the oncogenic state, such as oncogenes or other mutated cell-cycle proteins, or in fact any mutated protein associated with, even if not causative of, oncogenesis (Robbins et al., 1996; Lupetti et al., 1998; Nottage and Siu, 2002).

Like viral antigens associated with oncogenesis, over-expressed normal proteins and differentiation antigens in tumors are fairly defined and constant. But for spontaneous tumors lacking expression of such proteins, only tumor-unique antigens – the mutations in cell-cycle proteins thought to exist in every cancer – may be feasible as targets for vaccine production. These mutations occur randomly and unpredictably, either in genes or in their associated introns (Guilloux et al., 1996) or control sequences. (Mutations in the latter will be useful for vaccines only if they lead to aberrant protein production.) Thus the number of unique peptides that could be associated with any given cancer is potentially huge. Moreover, as discussed in connection with viral vaccines, the tremendous heterogeneity in class I structures within a population makes it difficult, even with non-mutated tumor proteins, to predict which peptides will interact well with individual MHC structures, or to prepare peptide "cocktails" that will cross-react with sufficiently large numbers of alleles to make this approach economically feasible. On top of all that, selected peptides must find CTL with cognate receptors to be effective as vaccines. In spite of these difficulties, some striking results have been obtained with this approach, and are discussed briefly herewith.

In humans, extensive work has been done to define peptides associated with malignant melanoma (reviewed in Parmiani et al., 2002; Weber, 2002), and clinical trials have been underway for the past several years. Typical of the antigens defined are MART-1 and gp100, both non-mutated differentiation antigens associated with melanocytes. These were identified using TIL specific for melanoma cells from patients with the HLA-A2 class I allele (Kawakami et al., 1994a,b). The genes for these antigens were cloned, and cognate peptides identified for use in vaccines. Clinical trials using a modified gp100 peptide administered together with IL-2 yielded promising results. Objective tumor regressions were obtained in 42 percent of HLA-A2-positive patients with advanced melanoma (Rosenberg et al., 1998). IL-2 appeared to promote sequestration of responding CTL at the tumor site (Rosenberg, 1999). In an unrelated trial a selected peptide from MART-1 was shown to induce significant CTL responses in melanoma patients, which correlated with a prolonged time to relapse (Wang, 1999).

Vaccine trials with peptides representing other cancers including breast (Disis et al., 1999), cervical (Muderspach et al., 2000), and pancreatic (Gjertsen et al., 2001), are underway, and have given broadly similar results where reported. As with the melanoma trials, these are all Phase I/II studies, restricted to patients with advanced cancer who have failed conventional treatments. However it must be admitted that at present, although CTL and NK cell responses can often be elicited by these vaccination procedures, success has been considerably less than with melanoma. As more peptides are identified, and modified where appropriate to broaden class I affinities, or to enhance responses within individuals, and as treatment is extended to patients with less advanced disease, it is reasonable to expect the present encouraging results will improve still further.

Dendritic cells are key to induction of CD8 T-cell responses to tumor antigens as well as to intracellular parasites. DC are attracted to tumor sites by cytokines released either by the tumor itself, or nearby DC, NK cells or other elements of the innate immune system. After ingestion of potentially antigenic material, the DC migrate to nearby lymph nodes where they in turn activate CD4 and CD8 T cells to peptides displayed on MHC molecules (Huang et al., 1994; Dallal and Lotze, 2000). Peptide-presenting DC can be prepared in the laboratory either by peptide pulsing or by transfection with cDNA encoding key peptides, as described above for viral vaccines. It is also possible to present larger antigenic protein structures to DC, and allow them to process these naturally into peptides for MHC loading. A promising recent approach involves introducing tumor antigen genes into hematopoietic stem cells, and then infusing these back into mice where they travel to bone marrow and produce antigen-laden DC. The advantage is better homing of such cells to lymph tissues where they can interact with T cells (Cui et al., 2003). Key to many current DC-based vaccines is inclusion of substances to activate the DC, such as microbial products. The route of administration of peptide-bearing DC may be critical to the effectiveness of this approach (Fong et al., 2001). And as with viral vaccines, inclusion of activating cytokines such as GMCSF as part of the vaccine protocol may further enhance the response (Zitvogel et al., 1996; Grohmann et al., 1997). Indeed, provision of GMCSF alone to tumors greatly increases CTL activation and tumor reduction (Neumanaitis et al., 2004).

These approaches to tumor vaccination have produced encouraging results in animal models (e.g., Noguchi et al., 1994). Clinical trials are now underway with pulsed DC as vaccines to immunize patients against their own tumors. Again, most such trials involve melanoma patients (Thurner et al., 1999; Lotze et al., 2000; Scadendorf and Nestle, 2001; Banchereau et al., 2001; Dudley et al., 2002). Other trials using DC are aimed at prostate and bladder cancer (Murphy et al., 1999; Small et al., 2000; Nishiyama et al., 2001). The prostate trials have been disappointing, but a trial using DC pulsed with B-cell lymphoma antigen achieved a response rate of 75 percent (Hsu et al., 1996). Vaccines based on cell-cycle proteins are typified by Her-2, an oncogene expressed in a wide range of human and animal tumors (Kiessling et al., 2002). A recent Phase I trial using DC pulsed with Her-2 peptides resulted in production of CTL able to lyse Her-2-transfected cells in vitro, and positive DTH responses in about a third of patients (Kono et al., 2002). A major drawback with the use of DC for vaccination procedures is the difficulty in harvesting large numbers of these cells from patients to use in their own vaccines. This limitation may be overcome by using non-dendritic cells geneticaly modified to function as DC (Maus et al., 2002; Sili et al., 2003).

REFERENCES

Abrams, S. I., and Z. Brahmi. 1986. The functional loss of human natural killer cell activity induced by K562 is reversible via an interleukin-2-dependent mechanism. *Cell Immunol 101:558.*

Abrams, S. I., D. E. McCulley, P. Meleedy-Rey, and J. H. Russell. 1989. Cytotoxic T lymphocyte-induced loss of target cell adhesion and lysis involve common and separate signaling pathways. *J Immunol 142:1789.*

Abrams, S. I., and J. H. Russell. 1991. CD4+ T lymphocyte-induced target cell detachment. A model for T cell-mediated lytic and nonlytic inflammatory processes. *J Immunol 146:405.*

Acha-Orbea, H., L. Scarpellino, S. Hertig, M. Dupuis, and J. Tschopp. 1990. Inhibition of lymphocyte mediated cytotoxicity by perforin antisense oligonucleotides. *Embo J 9:3815.*

Agadjanyan, M. G., J. J. Kim, N. Trivedi, D. M. Wilson, B. Monzavi-Karbassi, L. D. Morrison, L. K. Nottingham, T. Dentchev, A. Tsai, K. Dang, A. A. Chalian, M. A. Maldonado, W. V. Williams, and D. B. Weiner. 1999. CD86 (B7-2) can function to drive MHC-restricted antigen-specific CTL responses in vivo. *J Immunol 162:3417.*

Agarwal, S., and A. Rao. 1998. Modulation of chromatin structure regulates cytokine gene expression during T cell differentiation. *Immunity 9:765.*

Ahlers, J. D., N. Dunlop, D. W. Alling, P. L. Nara, and J. A. Berzofsky. 1997. Cytokine-in-adjuvant steering of the immune response phenotype to HIV-1 vaccine constructs: granulocyte-macrophage colony-stimulating factor and TNF-alpha synergize with IL-12 to enhance induction of cytotoxic T lymphocytes. *J Immunol 158:3947.*

Ahmed, R., L. D. Butler, and L. Bhatti. 1988. T4+ T helper cell function in vivo: differential requirement for induction of antiviral cytotoxic T-cell and antibody responses. *J Virol 62:2102.*

Ahmed, K. R., T. B. Guo, and K. K. Gaal. 1997. Islet rejection in perforin-deficient mice: the role of perforin and Fas. *Transplantation 63:951.*

Akira, S., K. Takeda, and T. Kaisho. 2001. Toll-like receptors: critical proteins linking innate and acquired immunity. *Nat Immunol 2:675.*

Albert, M. L., B. Sauter, and N. Bhardwaj. 1998. Dendritic cells acquire antigen from apoptotic cells and induce class I-restricted CTLs. *Nature 392:86.*

Albrecht-Buehler, G., and A. Bushnell. 1979. The orientation of centrioles in migrating 3T3 cells. *Exp Cell Res 120:111.*

Algeciras-Schimnich, A., T. S. Griffith, D. H. Lynch, and C. V. Paya. 1999. Cell cycle-dependent regulation of FLIP levels and susceptibility to Fas-mediated apoptosis. *J Immunol 162:5205.*

Allavena, P., G. Bianchi, P. Giardina, N. Polentarutti, D. Zhou, M. Introna, S. Sozzani, and A. Mantovani. 1996. Migratory Response of Human NK Cells to Monocyte-Chemotactic Proteins. *Methods 10:145.*

Allbritton, N. L., C. R. Verret, R. C. Wolley, and H. N. Eisen. 1988. Calcium ion concentrations and DNA fragmentation in target cell destruction by murine cloned cytotoxic T lymphocytes. *J Exp Med 167:514.*

Allen, P. M., and E. R. Unanue. 1987. Antigen processing and presentation at a molecular level. *Adv Exp Med Biol 225:147.*

Allison, J., and A. Strasser. 1998. Mechanisms of beta cell death in diabetes: a minor role for CD95. *Proc Natl Acad Sci U S A 95:13818.*

Altman J. D., P. A. Moss, P. J. Goulder, D. H. Barouch, M. G. McHeyzer-Williams, J. I. Bell, A. J. McMichael and M. M Davis 1996. Phenotypic analysis of antigen-specific T lymphocytes. Science. 274:94-6.

Amakata, Y., Y. Fujiyama, A. Andoh, K. Hodohara, and T. Bamba. 2001. Mechanism of NK cell activation induced by coculture with dendritic cells derived from peripheral blood monocytes. *Clin Exp Immunol 124:214.*

Amos, D. B. 1962. The use of simplified systems as an aid to the interpretation of mechanisms of graft rejection. *Progr Allergy 6:648.*

Anderson, J., J. A. Byrne, R. Schreiber, S. Patterson, and M. B. Oldstone. 1985. Biology of cloned cytotoxic T lymphocytes specific for lymphocytic choriomeningitis virus: clearance of virus and in vitro properties. *J Virol 53:552.*

Andersson, A., W. J. Dai, J. P. Di Santo, and F. Brombacher. 1998. Early IFN-gamma production and innate immunity during Listeria monocytogenes infection in the absence of NK cells. *J Immunol 161:5600.*

Ando, K., T. Moriyama, L. G. Guidotti, S. Wirth, R. D. Schreiber, H. J. Schlicht, S. N. Huang, and F. V. Chisari. 1993. Mechanisms of class I restricted immunopathology. A transgenic mouse model of fulminant hepatitis. *J Exp Med 178:1541.*

Ando, K., K. Hiroishi, T. Kaneko, T. Moriyama, Y. Muto, N. Kayagaki, H. Yagita, K. Okumura, and M. Imawari. 1997. Perforin, Fas/Fas ligand, and TNF-alpha pathways as specific and bystander killing mechanisms of hepatitis C virus-specific human CTL. *J Immunol 158:5283.*

Andrew, M. E., V. L. Braciale, T. J. Henkel, J. A. Hatch, and T. J. Braciale. 1985. Stable expression of cytolytic activity by influenza virus-specific cytotoxic T lymphocytes. *J Immunol 135:3520.*

Andrews, B. S., R. A. Eisenberg, A. N. Theofilopoulos, S. Izui, C. B. Wilson, P. J. McConahey, E. D. Murphy, J. B. Roths, and F. J. Dixon. 1978. Spontaneous murine lupus-like syndromes. Clinical and

immunopathological manifestations in several strains. *J Exp Med 148:1198.*
Andrin, C., M. J. Pinkoski, K. Burns, E. A. Atkinson, O. Krahenbuhl, D. Hudig, S. A. Fraser, U. Winkler, J. Tschopp, M. Opas, R. C. Bleackley, and M. Michalak. 1998. Interaction between a Ca2+-binding protein calreticulin and perforin, a component of the cytotoxic T-cell granules. *Biochemistry 37:10386.*
Anel, A., A. M. O'Rourke, A. M. Kleinfeld, and M. F. Mescher. 1996. T cell receptor and CD8-dependent tyrosine phosphorylation events in cytotoxic T lymphocytes: activation of p56lck by CD8 binding to class I protein. *Eur J Immunol 26:2310.*
Anichini, A., A. Molla, R. Mortarini, G. Tragni, I. Bersani, M. Di Nicola, A. M. Gianni, S. Pilotti, R. Dunbar, V. Cerundolo, and G. Parmiani. 1999. An expanded peripheral T cell population to a cytotoxic T lymphocyte (CTL)-defined, melanocyte-specific antigen in metastatic melanoma patients impacts on generation of peptide-specific CTLs but does not overcome tumor escape from immune surveillance in metastatic lesions. *J Exp Med 190:651.*
Antel, J. P., E. McCrea, U. Ladiwala, Y. F. Qin, and B. Becher. 1998. Non-MHC-restricted cell-mediated lysis of human oligodendrocytes in vitro: relation with CD56 expression. *J Immunol 160:1606.*
Appay, V., D. F. Nixon, S. M. Donahoe, G. M. Gillespie, T. Dong, A. King, G. S. Ogg, H. M. Spiegel, C. Conlon, C. A. Spina, D. V. Havlir, D. D. Richman, A. Waters, P. Easterbrook, A. J. McMichael, and S. L. Rowland-Jones. 2000. HIV-specific CD8(+) T cells produce antiviral cytokines but are impaired in cytolytic function. *J Exp Med 192:63.*
Arase, H., N. Arase, K. Ogasawara, R. A. Good, and K. Onoe. 1992. An NK1.1+ CD4+8- single-positive thymocyte subpopulation that expresses a highly skewed T-cell antigen receptor V beta family. *Proc Natl Acad Sci U S A 89:6506.*
Arase, H., N. Arase, Y. Kobayashi, Y. Nishimura, S. Yonehara, and K. Onoe. 1994. Cytotoxicity of fresh NK1.1+ T cell receptor alpha/beta+ thymocytes against a CD4+8+ thymocyte population associated with intact Fas antigen expression on the target. *J Exp Med 180:423.*
Arase, H., N. Arase, and T. Saito. 1995. Fas-mediated cytotoxicity by freshly isolated natural killer cells. *J Exp Med 181:1235.*
Arase, H., T. Suenaga, N. Arase, Y. Kimura, K. Ito, R. Shiina, H. Ohno, and T. Saito. 2001. Negative regulation of expression and function of Fc gamma RIII by CD3 zeta in murine NK cells. *J Immunol 166:21.*
Arase, H., E. S. Mocarski, A. E. Campbell, A. B. Hill, and L. L. Lanier. 2002. Direct recognition of cytomegalovirus by activating and inhibitory NK cell receptors. *Science 296:1323.*
Armitage, R. J. 1994. Tumor necrosis factor receptor superfamily members and their ligands. *Curr Opin Immunol 6:407.*
Auchincloss, H., Jr., R. Lee, S. Shea, J. S. Markowitz, M. J. Grusby, and L. H. Glimcher. 1993. The role of "indirect" recognition in initiating rejection of skin grafts from major histocompatibility complex class II-deficient mice. *Proc Natl Acad Sci U S A 90:3373.*
Austyn, J. M., and C. P. Larsen. 1990. Migration patterns of dendritic leukocytes. Implications for transplantation. *Transplantation 49:1.*

Avery, R. K., K. J. Bleier, and M. S. Pasternack. 1992. Differences between ATP-mediated cytotoxicity and cell-mediated cytotoxicity. *J Immunol 149:1265.*

Babbe, H., A. Roers, A. Waisman, H. Lassmann, N. Goebels, R. Hohlfeld, M. Friese, R. Schroder, M. Deckert, S. Schmidt, R. Ravid, and K. Rajewsky. 2000. Clonal expansions of CD8(+) T cells dominate the T cell infiltrate in active multiple sclerosis lesions as shown by micromanipulation and single cell polymerase chain reaction. *J Exp Med 192:393.*

Bach, F. H., and N. K. Voynow. 1966. One-way stimulation in mixed leukocyte cultures. *Science 153:545.*

Bach, M. L., M. B. Widmer, F. H. Bach, and J. Klein. 1973. Mixed leukocyte cultures and immune response region disparity. *Transplant Proc 5:369.*

Bach, F. H., M. L. Bach, and P. M. Sondel. 1976. Differential function of major histocompatibility complex antigens in T-lymphocyte activation. *Nature 259:273.*

Bachmann, M. F., M. Barner, A. Viola, and M. Kopf. 1999. Distinct kinetics of cytokine production and cytolysis in effector and memory T cells after viral infection. *Eur J Immunol 29:291.*

Badovinac, V. P., A. R. Tvinnereim, and J. T. Harty. 2000. Regulation of antigen-specific CD8+ T cell homeostasis by perforin and interferon-gamma. *Science 290:1354.*

Baekkeskov, S., H. J. Aanstoot, S. Christgau, A. Reetz, M. Solimena, M. Cascalho, F. Folli, H. Richter-Olesen, P. De Camilli, and P. D. Camilli. 1990. Identification of the 64K autoantigen in insulin-dependent diabetes as the GABA-synthesizing enzyme glutamic acid decarboxylase. *Nature 347:151.*

Bain, B., and L. Lowenstein. 1964. Genetic studies on the mixed leukocyte reaction. *Science 145:1315.*

Bajpai, A., B. S. Kwon, and Z. Brahmi. 1991. Rapid loss of perforin and serine protease RNA in cytotoxic lymphocytes exposed to sensitive targets. *Immunology 74:258.*

Baker, P., R. Weiser, and J. Jutila. 1962. Mechanisms of tumor homograft rejection: the behavior of sarcoma I ascites tumor in the A/J and C57BL/6K mouse. *Ann N Y Acad Sci 101:46.*

Baker, M. B., N. H. Altman, E. R. Podack, and R. B. Levy. 1996. The role of cell-mediated cytotoxicity in acute GVHD after MHC-matched allogeneic bone marrow transplantation in mice. *J Exp Med 183:2645.*

Baker, M. B., R. L. Riley, E. R. Podack, and R. B. Levy. 1997. Graft-versus-host-disease-associated lymphoid hypoplasia and B cell dysfunction is dependent upon donor T cell-mediated Fas-ligand function, but not perforin function. *Proc Natl Acad Sci U S A 94:1366.*

Bakker, A. B., M. W. Schreurs, A. J. de Boer, Y. Kawakami, S. A. Rosenberg, G. J. Adema, and C. G. Figdor. 1994. Melanocyte lineage-specific antigen gp100 is recognized by melanoma-derived tumor-infiltrating lymphocytes. *J Exp Med 179:1005.*

Balaji, K. N., N. Schaschke, W. Machleidt, M. Catalfamo, and P. A. Henkart. 2002. Surface cathepsin B protects cytotoxic lymphocytes from self-destruction after degranulation. *J Exp Med 196:493.*

Balbi, B., M. T. Valle, S. Oddera, D. Giunti, F. Manca, G. A. Rossi, and L. Allegra. 1993. T-lymphocytes with gamma delta+ V delta 2+ antigen receptors are present in increased proportions in a fraction of patients with tuberculosis or with sarcoidosis. *Am Rev Respir Dis 148:1685.*

Baldwin, R. 1955. Immunity to methylcholanthrene-induced tumors in inbred rats following atrophy and regression of the implanted tumors. *Brit J Cancer 9:652.*

Baldwin, R. W., M. J. Embleton, J. S. Jones, and M. J. Langman. 1973. Cell-mediated and humoral immune reactions to human tumours. *Int J Cancer 12:73.*

Baliga, P., K. D. Chavin, L. Qin, J. Woodward, J. Lin, P. S. Linsley, and J. S. Bromberg. 1994. CTLA4Ig prolongs allograft survival while suppressing cell-mediated immunity. *Transplantation 58:1082.*

Balk, S. P., and M. F. Mescher. 1981. Specific reversal of cytolytic T cell-target cell functional binding is induced by free target cells. *J Immunol 127:51.*

Balk, S. P., J. Walker, and M. F. Mescher. 1981. Kinetics of cytolytic T lymphocyte binding to target cells in suspension. *J Immunol 126:2177.*

Balkow, S., A. Kersten, T. T. Tran, T. Stehle, P. Grosse, C. Museteanu, O. Utermohlen, H. Pircher, F. von Weizsacker, R. Wallich, A. Mullbacher, and M. M. Simon. 2001. Concerted action of the FasL/Fas and perforin/granzyme A and B pathways is mandatory for the development of early viral hepatitis but not for recovery from viral infection. *J Virol 75:8781.*

Ballas, Z. K., and W. Rasmussen. 1987. Lymphokine-activated killer (LAK) cells. III. Characterization of LAK precursors and susceptible target cells within the murine thymus. *J Immunol 139:3542.*

Ballas, Z. K., and W. Rasmussen. 1990. NK1.1+ thymocytes. Adult murine CD4-, CD8- thymocytes contain an NK1.1+, CD3+, CD5hi, CD44hi, TCR-V beta 8+ subset. *J Immunol 145:1039.*

Banchereau, J., and R. M. Steinman. 1998. Dendritic cells and the control of immunity. *Nature 392:245.*

Banchereau, J., A. K. Palucka, M. Dhodapkar, S. Burkeholder, N. Taquet, A. Rolland, S. Taquet, S. Coquery, K. M. Wittkowski, N. Bhardwaj, L. Pineiro, R. Steinman, and J. Fay. 2001. Immune and clinical responses in patients with metastatic melanoma to CD34(+) progenitor-derived dendritic cell vaccine. *Cancer Res 61:6451.*

Banner, D. W., A. D'Arcy, W. Janes, R. Gentz, H. J. Schoenfeld, C. Broger, H. Loetscher, and W. Lesslauer. 1993. Crystal structure of the soluble human 55 kd TNF receptor-human TNF beta complex: implications for TNF receptor activation. *Cell 73:431.*

Banting, F., and S. Gairns. 1934. A study of serum of chickens resistant to Rous sarcoma. *Amer J Cancer 22:611.*

Barker, E., J. A. Wise, S. Dray, and M. B. Mokyr. 1989. Lysis of antigenically unrelated tumor cells mediated by Lyt 2+ splenic T-cells from melphalan-cured MOPC-315 tumor bearers. *Cancer Res 49:5007.*

Barouch, D. H., S. Santra, K. Tenner-Racz, P. Racz, M. J. Kuroda, J. E. Schmitz, S. S. Jackson, M. A. Lifton, D. C. Freed, H. C. Perry, M. E. Davies, J. W. Shiver, and N. L. Letvin. 2002. Potent CD4+ T cell responses elicited by a bicistronic HIV-1 DNA vaccine expressing gp120 and GM-CSF. *J Immunol 168:562.*

Barreiro R,. G. Luker, J. Herndon and T. A. Ferguson 2004. Termination of antigen-specific immunity by CD95 ligand (Fas ligand) and IL-10. J Immunol. 173:1519-25.

Barrett-Connor, E. 1985. Is insulin-dependent diabetes mellitus caused by coxsackievirus B infection? A review of the epidemiologic evidence. *Rev Infect Dis 7:207.*

Barry, W. H. 1994. Mechanisms of immune-mediated myocyte injury. *Circulation 89:2421.*

Barry, M., Bleackley, R. C. 2002. Cytotoxic T lymphocytes: all roads lead to death. *Nature Reviews Immunology 2:401.*

Bauer, S., V. Groh, J. Wu, A. Steinle, J. H. Phillips, L. L. Lanier, and T. Spies. 1999. Activation of NK cells and T cells by NKG2D, a receptor for stress-inducible MICA. *Science 285:727.*

Baume, D. M., M. J. Robertson, H. Levine, T. J. Manley, P. W. Schow, and J. Ritz. 1992. Differential responses to interleukin 2 define functionally distinct subsets of human natural killer cells. *Eur J Immunol 22:1.*

Beaudoin, L., V. Laloux, J. Novak, B. Lucas, and A. Lehuen. 2002. NKT cells inhibit the onset of diabetes by impairing the development of pathogenic T cells specific for pancreatic beta cells. *Immunity 17:725.*

Beckman, E. M., S. A. Porcelli, C. T. Morita, S. M. Behar, S. T. Furlong, and M. B. Brenner. 1994. Recognition of a lipid antigen by CD1-restricted alpha beta+ T cells. *Nature 372:691.*

Belkaid, Y., E. Von Stebut, S. Mendez, R. Lira, E. Caler, S. Bertholet, M. C. Udey, and D. Sacks. 2002. CD8+ T cells are required for primary immunity in C57BL/6 mice following low-dose, intradermal challenge with Leishmania major. *J Immunol 168:3992.*

Bellgrau, D., D. Gold, H. Selawry, J. Moore, A. Franzusoff, and R. C. Duke. 1995. A role for CD95 ligand in preventing graft rejection. *Nature 377:630.*

Bellone, G., M. Aste-Amezaga, G. Trinchieri, and U. Rodeck. 1995. Regulation of NK cell functions by TGF-beta 1. *J Immunol 155:1066.*

Belz, G. T., D. Wodarz, G. Diaz, M. A. Nowak, and P. C. Doherty. 2002. Compromised influenza virus-specific CD8(+)-T-cell memory in CD4(+)-T-cell-deficient mice. *J Virol 76:12388.*

Bendelac, A., C. Carnaud, C. Boitard, and J. F. Bach. 1987. Syngeneic transfer of autoimmune diabetes from diabetic NOD mice to healthy neonates. Requirement for both L3T4+ and Lyt-2+ T cells. *J Exp Med 166:823.*

Bendelac, A. 1995a. Positive selection of mouse NK1+ T cells by CD1-expressing cortical thymocytes. *J Exp Med 182:2091.*

Bendelac, A., O. Lantz, M. E. Quimby, J. W. Yewdell, J. R. Bennink, and R. R. Brutkiewicz. 1995b. CD1 recognition by mouse NK1+ T lymphocytes. *Science 268:863.*

Bendelac, A., M. N. Rivera, S. H. Park, and J. H. Roark. 1997. Mouse CD1-specific NK1 T cells: development, specificity, and function. *Annu Rev Immunol 15:535.*

Bendelac, A., M. Bonneville, and J. F. Kearney. 2001. Autoreactivity by design: innate B and T lymphocytes. *Nature Rev Immunol 1:177.*

Bendelac, A., and R. Medzhitov. 2002. Adjuvants of immunity: harnessing innate immunity to promote adaptive immunity. *J Exp Med 195:F19.*

Benlagha, K., T. Kyin, A. Beavis, L. Teyton, and A. Bendelac. 2002. A thymic precursor to the NK T cell lineage. *Science 296:553.*

Bennett, S. R., F. R. Carbone, F. Karamalis, R. A. Flavell, J. F. Miller, and W. R. Heath. 1998. Help for cytotoxic-T-cell responses is mediated by CD40 signalling. *Nature 393:478.*

Benson, M. T., G. Buckley, E. J. Jenkinson, and J. J. Owen. 1987. Survival of deoxyguanosine-treated fetal thymus allografts is prevented by priming with dendritic cells. *Immunology 60:593.*

Benvenuto, R., A. Bachetoni, P. Cinti, F. Sallusto, A. Franco, M. E., V. Barnaba, F. Balsano, and R. Cortesini. 1991. Enhanced production of interferon-gamma by T lymphocytes cloned from rejected kidney grafts. *Transplantation 51:887.*

Berke, G., W. Ax, H. Ginsburg, and M. Feldman. 1969a. Graft reaction in tissue culture. II. Quantification of the lytic action on mouse fibroblasts by rat lymphocytes sensitized on mouse embryo monolayers. *Immunology 16:643.*

Berke, G., H. Ginsburg, and M. Feldman. 1969b. Graft reaction in tissue culture. 3. Effect of phytohaemagglutinin. *Immunology 16:659.*

Berke, G., G. Yagil, H. Ginsburg, and M. Feldman. 1969c. Kinetic analysis of a graft reaction induced in cell culture. *Immunology 17:723.*

Berke, G., and R. H. Levey. 1972. Cellular immunoabsorbents in transplantation immunity. Specific in vitro deletion and recovery of mouse lymphoid cells sensitized against allogeneic tumors. *J Exp Med 135:972.*

Berke, G., K. A. Sullivan, and B. Amos. 1972a. Rejection of ascites tumor allografts. II. A pathway for cell-mediated tumor destruction in vitro by peritoneal exudate lymphoid cells. *J Exp Med 136:1594.*

Berke, G., K. A. Sullivan, and B. Amos. 1972b. Rejection of ascites tumor allografts. I. Isolation, characterization, and in vitro reactivity of peritoneal lymphoid effector cells from BALB-c mice immune to EL4 leukosis. *J Exp Med 135:1334.*

Berke, G., K. A. Sullivan, and D. B. Amos. 1972c. Tumor immunity in vitro: destruction of a mouse ascites tumor through a cycling pathway. *Science 177:433.*

Berke, G., and D. B. Amos. 1973. Cytotoxic lymphocytes in the absence of detectable antibody. *Nat New Biol 242:237.*

Berke, G., and D. Gabison. 1975. Energy requirements of the binding and lytic steps of T lymphocyte-mediated cytolysis of leukemic cells in vitro. *Eur J Immunol 5:671.*

Berke, G., D. Gabison, and M. Feldman. 1975. The frequency of effector cells in populations containing cytotoxic T lymphocytes. *Eur J Immunol 5:813.*

Berke, G., and Z. Fishelson. 1977. T-lymphocyte-mediated cytolysis: contribution of intracellular components to target cell susceptibility. *Transplant Proc 9:671.*

Berke, G. 1980. Interaction of cytotoxic T lymphocytes and target cells. *Prog Allergy 27:69.*

Berke, G., V. Hu, E. McVey, and W. R. Clark. 1981a. T lymphocyte-mediated cytolysis. I. A common mechanism for target recognition in specific and lectin-dependent cytolysis. *J Immunol 127:776.*

Berke, G., E. McVey, V. Hu, and W. R. Clark. 1981b. T lymphocyte-mediated cytolysis. II. Role of target cell histocompatibility antigens in recognition and lysis. *J Immunol 127:782.*

Berke, G., Clark, W. 1982. How cytotoxic lymphocytes kill target cells: a theory. *Mechanisms of Lymphocyte Activation. Elsevier Biomedical Press.:439.*

Berke, G., and W. R. Clark. 1982. T lymphocyte-mediated cytolysis - a comprehensive theory. I. The mechanism of CTL-mediated cytolysis. *Adv Exp Med Biol 146:57.*

Berke, G., V. Hu, E. McVey, and W. Clark. 1982. A common role for target cell histocompatibility antigens in both nonspecific and specific T lymphocyte-mediated cytolysis. *J Cell Biochem 18:337.*

Berke, G., D. Rosen, and M. Moscovitch. 1983. T lymphocyte-mediated cytolysis. III. Delineation of mechanisms whereby mitogenic and non-mitogenic lectins mediate lymphocyte-target interaction. *Immunology 49:585.*

Berke, G. 1985. Enumeration of lymphocyte-target cell conjugates by cytofluorometry. *Eur J Immunol 15:337.*

Berke, G., and D. Rosen. 1987. Are lytic granules and perforin 1 involved in lysis induced by in vivo-primed peritoneal exudate cytolytic T lymphocytes? *Transplant Proc 19:412.*

Berke, G., and D. Rosen. 1988. Highly lytic in vivo primed cytolytic T lymphocytes devoid of lytic granules and BLT-esterase activity acquire these constituents in the presence of T cell growth factors upon blast transformation in vitro. *J Immunol 141:1429.*

Berke, G. 1991. Debate: the mechanism of lymphocyte-mediated killing. Lymphocyte-triggered internal target disintegration. *Immunol Today 12:396.*

Berke, G. 1991. T-cell-mediated cytotoxicity. *Curr Opin Immunol 3:320.*

Berke, G. S., D. C. Green, M. E. Smith, D. P. Arnstein, V. Honrubia, M. Natividad, and W. A. Conrad. 1991. Experimental evidence in the in vivo canine for the collapsible tube model of phonation. *J Acoust Soc Am 89:1358.*

Berke, G., D. Rosen, and D. Ronen. 1993. Mechanism of lymphocyte-mediated cytolysis: functional cytolytic T cells lacking perforin and granzymes. *Immunology 78:105.*

Berke, G. 1995. The CTL's kiss of death. *Cell 81:9.*

Bernard, A., Lamy, and I. Alberti. 2002. The two-signal model of T-cell activation after 30 years. *Transplantation 73:S31.*

Berthou, C., J. F. Bourge, Y. Zhang, A. Soulie, D. Geromin, Y. Denizot, F. Sigaux, and M. Sasportes. 2000. Interferon-gamma-induced membrane PAF-receptor expression confers tumor cell susceptibility to NK perforin-dependent lysis. *Blood 95:2329.*

Bertotto, A., R. Gerli, F. Spinozzi, C. Muscat, F. Scalise, G. Castellucci, M. Sposito, F. Candio, and R. Vaccaro. 1993. Lymphocytes bearing the gamma delta T cell receptor in acute Brucella melitensis infection. *Eur J Immunol 23:1177.*

Beutler, B., and C. van Huffel. 1994. Unraveling function in the TNF ligand and receptor families. *Science 264:667.*

Bevan, M. J., and M. Cohn. 1975. Cytotoxic effects of antigen- and mitogen-induced T cells on various targets. *J Immunol 114:559.*

Beyan, H., L. R. Buckley, N. Yousaf, M. Londei, and R. D. Leslie. 2003. A role for innate immunity in type 1 diabetes? *Diabetes Metab Res Rev 19:89.*

Beyer, T., M. Herrmann, C. Reiser, W. Bertling, and J. Hess. 2001. Bacterial carriers and virus-like-particles as antigen delivery devices: role of dendritic cells in antigen presentation. *Curr Drug Targets Infect Disord 1:287.*

Beyer, T., M. Herrmann, C. Reiser, W. Bertling, and J. Hess. 2001. Bacterial carriers and virus-like particles as antigen delivery devices: role of dendritic cells in antigen presentation. *Curr Drug Targets Infect Disorder 1:287.*

Bhardwaj, N. 1997. Interactions of viruses with dendritic cells: a double-edged sword. *J Exp Med 186:795.*

Bialas, T., J. Kolitz, E. Levi, A. Polivka, S. Oez, G. Miller, and K. Welte. 1988. Distinction of partially purified human natural killer cytotoxic factor from recombinant human tumor necrosis factor and recombinant human lymphotoxin. *Cancer Res 48:891.*

Biedermann, B. C., and J. S. Pober. 1998. Human endothelial cells induce and regulate cytolytic T cell differentiation. *J Immunol 161:4679.*

Bienzle, D., F. M. Smaill, and K. L. Rosenthal. 1996. Cytotoxic T-lymphocytes from HIV-infected individuals recognize an activation-dependent, non-polymorphic molecule on uninfected CD4+ lymphocytes. *Aids 10:247.*

Biniaminov, M., B. Ramot, E. Rosenthal, and A. Novogrodsky. 1975. Galactose oxidase-induced blastogenesis of human lymphocytes and the effect of macrophages on the reaction. *Clin Exp Immunol 19:93.*

Binz, H., H. Wigzell, and H. Bazin. 1976. T-cell idiotypes are linked to immunoglobulin heavy chain genes. *Nature 264:639.*

Biron, C. A., K. B. Nguyen, and G. C. Pien. 2002. Innate immune responses to LCMV infections: natural killer cells and cytokines. *Curr Top Microbiol Immunol 263:7.*

Bitton, N., P. Debre, Z. Eshhar, and G. Gorochov. 2001. T-bodies as anti-viral agents. *Curr Top Microbiol Immunol 260:271.*

Bix, M., and D. Raulet. 1992. Functionally conformed free class I heavy chains exist on the surface of beta 2 microglobulin negative cells. *J Exp Med 176:829.*

Bjorkman, P. J., M. A. Saper, B. Samraoui, W. S. Bennett, J. L. Strominger, and D. C. Wiley. 1987a. Structure of the human class I histocompatibility antigen, HLA-A2. *Nature 329:506.*

Bjorkman, P. J., M. A. Saper, B. Samraoui, W. S. Bennett, J. L. Strominger, and D. C. Wiley. 1987b. The foreign antigen binding site and T cell recognition regions of class I histocompatibility antigens. *Nature 329:512.*

Blakely, A., K. Gorman, H. Ostergaard, K. Svoboda, C. C. Liu, J. D. Young, and W. R. Clark. 1987. Resistance of cloned cytotoxic T lymphocytes to cell-mediated cytotoxicity. *J Exp Med 166:1070.*

Blanchard, D. K., M. B. Michelini-Norris, H. Friedman, and J. Y. Djeu. 1989. Lysis of mycobacteria-infected monocytes by IL-2-activated killer cells: role of LFA-1. *Cell Immunol 119:402.*

Blanchard, D. K., S. Wei, C. Duan, F. Pericle, J. I. Diaz, and J. Y. Djeu. 1995. Role of extracellular adenosine triphosphate in the cytotoxic T-lymphocyte-mediated lysis of antigen presenting cells. *Blood 85:3173.*

Blauvelt, A., M. Clerici, D. R. Lucey, S. M. Steinberg, R. Yarchoan, R. Walker, G. M. Shearer, and S. I. Katz. 1995. Functional studies of epidermal Langerhans cells and blood monocytes in HIV-infected persons. *J Immunol 154:3506.*

Blink, E. J., J. A. Trapani, and D. A. Jans. 1999. Perforin-dependent nuclear targeting of granzymes: A central role in the nuclear events of granule-exocytosis-mediated apoptosis? *Immunol Cell Biol 77:206.*

Bluman, E. M., K. J. Bartynski, B. R. Avalos, and M. A. Caligiuri. 1996. Human natural killer cells produce abundant macrophage inflammatory protein-1 alpha in response to monocyte-derived cytokines. *J Clin Invest 97:2722.*

Bogner, U., H. Schleusener, and J. R. Wall. 1984. Antibody-dependent cell mediated cytotoxicity against human thyroid cells in Hashimoto's thyroiditis but not Graves' disease. *J Clin Endocrinol Metab 59:734.*

Boise, L. H., A. J. Minn, P. J. Noel, C. H. June, M. A. Accavitti, T. Lindsten, and C. B. Thompson. 1995. CD28 costimulation can promote T cell survival by enhancing the expression of Bcl-XL. *Immunity 3:87.*

Bonaparte, M. I., and E. Barker. 2003. Inability of natural killer cells to destroy autologous HIV-infected T lymphocytes. *Aids 17:487.*

Bonavida, B., and T. P. Bradley. 1976. Studies on the induction and expression of T cell-mediated immunity. V. Lectin-induced nonspecific cell-mediated cytotoxicity by alloimmune lymphocytes. *Transplantation 21:94.*

Bonavida, B., and S. C. Wright. 1986. Role of natural killer cytotoxic factors in the mechanism of target-cell killing by natural killer cells. *J Clin Immunol 6:1.*

Bongrand, P., and P. Golstein. 1983. Reproducible dissociation of cellular aggregates with a wide range of calibrated shear forces: application to cytolytic lymphocyte target cell conjugates. *J Immunol Methods 58:209.*

Boon, T., and O. Kellermann. 1977. Rejection by syngeneic mice of cell variants obtained by mutagenesis of a malignant teratocarcinoma cell line. *Proc Natl Acad Sci U S A 74:272.*

Boon, T., P. G. Coulie, and B. Van den Eynde. 1997. Tumor antigens recognized by T cells. *Immunol Today 18:267.*

Booss, J., M. M. Esiri, W. W. Tourtellotte, and D. Y. Mason. 1983. Immunohistological analysis of T lymphocyte subsets in the central nervous system in chronic progressive multiple sclerosis. *J Neurol Sci 62:219.*

Borrego, F., J. Kabat, D. K. Kim, L. Lieto, K. Maasho, J. Pena, R. Solana, and J. E. Coligan. 2002. Structure and function of major histocompatibility complex (MHC) class I specific receptors expressed on human natural killer (NK) cells. *Mol Immunol 38:637.*

Borrow, P., H. Lewicki, B. H. Hahn, G. M. Shaw, and M. B. Oldstone. 1994. Virus-specific CD8+ cytotoxic T-lymphocyte activity associated with control of viremia in primary human immunodeficiency virus type 1 infection. *J Virol 68:6103.*

Borrow, P., H. Lewicki, X. Wei, M. S. Horwitz, N. Peffer, H. Meyers, J. A. Nelson, J. E. Gairin, B. H. Hahn, M. B. Oldstone, and G. M. Shaw. 1997. Antiviral pressure exerted by HIV-1-specific cytotoxic T lymphocytes (CTLs) during primary infection demonstrated by rapid selection of CTL escape virus. *Nat Med 3:205.*

Borrow, P., D. F. Tough, D. Eto, A. Tishon, I. S. Grewal, J. Sprent, R. A. Flavell, and M. B. Oldstone. 1998. CD40 ligand-mediated interactions are involved in the generation of memory CD8(+) cytotoxic T lymphocytes (CTL) but are not required for the maintenance of CTL memory following virus infection. *J Virol 72:7440.*

Borthwick, N. J., M. Lowdell, M. Salmon, and A. N. Akbar. 2000. Loss of CD28 expression on CD8(+) T cells is induced by IL-2 receptor gamma chain signalling cytokines and type I IFN, and increases susceptibility to activation-induced apoptosis. *Int Immunol 12:1005.*

Bossi, G., and G. M. Griffiths. 1999. Degranulation plays an essential part in regulating cell surface expression of Fas ligand in T cells and natural killer cells. *Nat Med 5:90.*

Bottazzo, G. F., B. M. Dean, J. M. McNally, E. H. MacKay, P. G. Swift, and D. R. Gamble. 1985. In situ characterization of autoimmune phenomena and expression of HLA molecules in the pancreas in diabetic insulitis. *N Engl J Med 313:353.*

Boullier, S., M. Cochet, F. Poccia, and M. L. Gougeon. 1995. CDR3-independent gamma delta V delta 1+ T cell expansion in the peripheral blood of HIV-infected persons. *J Immunol 154:1418.*

Bourgault, I., A. Gomez, E. Gomard, F. Picard, J. P. Levy, and E. Gomrad. 1989. A virus-specific CD4+ cell-mediated cytolytic activity revealed by CD8+ cell elimination regularly develops in uncloned human antiviral cell lines. *J Immunol 142:252.*

Bouwer, H. G., A. Bai, J. Forman, S. H. Gregory, E. J. Wing, R. A. Barry, and D. J. Hinrichs. 1998. Listeria monocytogenes-infected hepatocytes are targets of major histocompatibility complex class Ib-restricted antilisterial cytotoxic T lymphocytes. *Infect Immun 66:2814.*

Boylston, A. W., and S. J. May. 1986. Human T lymphocytes expressing a defined T-cell antigen receptor family specifically kill the hybridoma that makes the anti-receptor monoclonal antibody. *Immunology 59:383.*

Braciale, T. J., L. A. Morrison, M. T. Sweetser, J. Sambrook, M. J. Gething, and V. L. Braciale. 1987. Antigen presentation pathways to class I and class II MHC-restricted T lymphocytes. *Immunol Rev 98:95.*

Bradley, J. A., D. W. Mason, and P. J. Morris. 1985. Evidence that rat renal allografts are rejected by cytotoxic T cells and not by nonspecific effectors. *Transplantation 39:169.*

Braud, V. M., D. S. Allan, C. A. O'Callaghan, K. Soderstrom, A. D'Andrea, G. S. Ogg, S. Lazetic, N. T. Young, J. I. Bell, J. H. Phillips, L. L. Lanier, and A. J. McMichael. 1998. HLA-E binds to natural killer cell receptors CD94/NKG2A, B and C. *Nature 391:795.*

Braud, V. M., H. Aldemir, H. Breart, and W. Ferlin. 2003. Expression of CD94-NKG2A inhibitory receptor is restricted to a subset of CD8+ T cells. *Trends Immunol 24:162.*

Braun, M. Y., B. Lowin, L. French, H. Acha-Orbea, and J. Tschopp. 1996. Cytotoxic T cells deficient in both functional fas ligand and perforin show residual cytolytic activity yet lose their capacity to induce lethal acute graft-versus-host disease. *J Exp Med 183:657.*

Bregenholt, S., P. Berche, F. Brombacher, and J. P. Di Santo. 2001. Conventional alpha beta T cells are sufficient for innate and adaptive immunity against enteric Listeria monocytogenes. *J Immunol 166:1871.*

Breitmeyer, J. B., S. O. Oppenheim, J. F. Daley, H. B. Levine, and S. F. Schlossman. 1987. Growth inhibition of human T cells by antibodies recognizing the T cell antigen receptor complex. *J Immunol 138:726.*

Brennan, J., D. Mager, W. Jefferies, and F. Takei. 1994. Expression of different members of the Ly-49 gene family defines distinct natural killer cell subsets and cell adhesion properties. *J Exp Med 180:2287.*

Brent, L., J. Brown, and P. Medwar. 1962. *Proc Roy Soc London B Biol Sci 156:187.*

Brent, L., and P. Medawar. 1966. Quantitative studies on tissue transplantation immunity. *Proc Roy Soc London B Biol Sci 165:281.*

Bretscher, P., and M. Cohn. 1970. A theory of self and non-self discrimination. *Science 69:1042.*

Brocke, S., M. Dayan, L. Steinman, J. Rothbard, and E. Mozes. 1990. Inhibition of T cell proliferation specific for acetylcholine receptor epitopes related to myasthenia gravis with antibody to T cell receptor or with competitive synthetic polymers. *Int Immunol 2:735.*

Brocker, E. B., K. M. Kuhlencordt, and W. Muller-Ruchholtz. 1977. Microcytotoxicity test in allograft immunity: specificity and kinetics of effetor T cells. *Int Arch Allergy Appl Immunol 53:234.*

Bromley, S. K., W. R. Burack, K. G. Johnson, K. Somersalo, T. N. Sims, C. Sumen, M. M. Davis, A. S. Shaw, P. M. Allen, and M. L. Dustin. 2001. The immunological synapse. *Annu Rev Immunol 19:375.*

Brondz, B. D. 1968. Complex specificity of immune lymphocytes in allogeneic cell cultures. *Folia Biol (Praha) 14:115.*

Brondz, B. D., I. K. Egorov, and G. I. Drizlikh. 1975. Private specificities of H-2K and H-2D loci as possible selective targets for effector lymphocytes in cell-mediated immunity. *J Exp Med 141:11.*

Brooks, C. G., D. L. Urdal, and C. S. Henney. 1983. Lymphokine-driven "differentiation" of cytotoxic T-cell clones into cells with NK-like specificity: correlations with display of membrane macromolecules. *Immunol Rev 72:43.*

Brooks, A. G., P. E. Posch, C. J. Scorzelli, F. Borrego, and J. E. Coligan. 1997. NKG2A complexed with CD94 defines a novel inhibitory natural killer cell receptor. *J Exp Med 185:795.*

Brossart, P., and M. J. Bevan. 1997. Presentation of exogenous protein antigens on major histocompatibility complex class I molecules by dendritic cells: pathway of presentation and regulation by cytokines. *Blood 90:1594.*

Brower, R. C., R. England, T. Takeshita, S. Kozlowski, D. H. Margulies, J. A. Berzofsky, and C. Delisi. 1994. Minimal requirements for peptide mediated activation of CD8+ CTL. *Mol Immunol 31:1285.*

Brown, C., and J. O'Connell. 1995. Myocarditis and idiopathic dilated cardiomyopathy. *Am J Med 99:309.*

Brown, M. G., A. A. Scalzo, K. Matsumoto, and W. M. Yokoyama. 1997. The natural killer gene complex: a genetic basis for understanding natural killer cell function and innate immunity. *Immunol Rev 155:53.*

Brown, M. G., A. O. Dokun, J. W. Heusel, H. R. Smith, D. L. Beckman, E. A. Blattenberger, C. E. Dubbelde, L. R. Stone, A. A. Scalzo, and W. M. Yokoyama. 2001. Vital involvement of a natural killer cell activation receptor in resistance to viral infection. *Science 292:934.*

Browne, K. A., E. Blink, V. R. Sutton, C. J. Froelich, D. A. Jans, and J. A. Trapani. 1999. Cytosolic delivery of granzyme B by bacterial toxins: evidence that endosomal disruption, in addition to transmembrane pore formation, is an important function of perforin. *Mol Cell Biol 19:8604.*

Brunet, J. F., F. Denizot, M. Suzan, W. Haas, J. M. Mencia-Huerta, G. Berke, M. F. Luciani, and P. Golstein. 1987. CTLA-1 and CTLA-3 serine esterase transcripts are detected mostly in cytotoxic T cells, but not only and not always. *J Immunol 138:4102.*

Brunner, K. T., J. Mauel, J. C. Cerottini, and B. Chapuis. 1968. Quantitative assay of the lytic action of immune lymphoid cells on 51-Cr-labelled allogeneic target cells in vitro; inhibition by isoantibody and by drugs. *Immunology 14:181.*

Brunner, K. T., J. Mauel, H. Rudolf, and B. Chapuis. 1970. Studies of allograft immunity in mice. I. Induction, development and in vitro assay of cellular immunity. *Immunology 18:501.*

Brunner, T., R. J. Mogil, D. LaFace, N. J. Yoo, A. Mahboubi, F. Echeverri, S. J. Martin, W. R. Force, D. H. Lynch, C. F. Ware, and et al. 1995. Cell-autonomous Fas (CD95)/Fas-ligand interaction mediates activation-induced apoptosis in T-cell hybridomas. *Nature 373:441.*

Brutkiewicz, R. R., and V. Sriram. 2002. Natural killer T (NKT) cells and their role in antitumor immunity. *Crit Rev Oncol Hematol 41:287.*

Bryson, J. S., and D. L. Flanagan. 2000. Role of natural killer cells in the development of graft-versus-host disease. *J Hematother Stem Cell Res 9:307.*

Bubbers, J. E., and C. S. Henney. 1975. Studies on the synthetic capacity and antigenic expression of glutaraldehyde-fixed target cells. *J Immunol 114:1126.*

Buchmeier, M. J., R. M. Welsh, F. J. Dutko, and M. B. Oldstone. 1980. The virology and immunobiology of lymphocytic choriomeningitis virus infection. *Adv Immunol 30:275.*

Buisman, H. P., T. H. Steinberg, J. Fischbarg, S. C. Silverstein, S. A. Vogelzang, C. Ince, D. L. Ypey, and P. C. Leijh. 1988. Extracellular ATP induces a large nonselective conductance in macrophage plasma membranes. *Proc Natl Acad Sci U S A 85:7988.*

Bukowski, J. F., B. A. Woda, S. Habu, K. Okumura, and R. M. Welsh. 1983. Natural killer cell depletion enhances virus synthesis and virus-induced hepatitis in vivo. *J Immunol 131:1531.*

Bukowski, J. F., B. A. Woda, and R. M. Welsh. 1984. Pathogenesis of murine cytomegalovirus infection in natural killer cell-depleted mice. *J Virol 52:119.*

Bukowski, J. F., J. F. Warner, G. Dennert, and R. M. Welsh. 1985. Adoptive transfer studies demonstrating the antiviral effect of natural killer cells in vivo. *J Exp Med 161:40.*

Bukreyev, A., I. M. Belyakov, J. A. Berzofsky, B. R. Murphy, and P. L. Collins. 2001. Granulocyte-macrophage colony-stimulating factor expressed by recombinant respiratory syncytial virus attenuates viral replication and increases the level of pulmonary antigen-presenting cells. *J Virol 75:12128.*

Bulfield, G., W. G. Siller, P. A. Wight, and K. J. Moore. 1984. X chromosome-linked muscular dystrophy (mdx) in the mouse. *Proc Natl Acad Sci U S A 81:1189.*

Buller, R. M., K. L. Holmes, A. Hugin, T. N. Frederickson, and H. C. Morse, 3rd. 1987. Induction of cytotoxic T-cell responses in vivo in the absence of CD4 helper cells. *Nature 328:77.*

Burdin, N., and M. Kronenberg. 1999. CD1-mediated immune responses to glycolipids. *Curr Opin Immunol 11:326.*

Burkhardt, J. K., S. Hester, C. K. Lapham, and Y. Argon. 1990. The lytic granules of natural killer cells are dual-function organelles combining secretory and pre-lysosomal compartments. *J Cell Biol 111:2327.*

Burnet, M. 1957a. Cancer - a biological approach. *Brit Med J 1:779.*

Burnet, M. 1957b. Cancer - a biological approach. *Brit med J 1:841.*

Burnet, F. M. 1970. The concept of immunological surveillance. *Prog Exp Tumor Res 13:1.*

Burrows, S. R., A. Fernan, V. Argaet, and A. Suhrbier. 1993. Bystander apoptosis induced by CD8+ cytotoxic T cell (CTL) clones: implications for CTL lytic mechanisms. *Int Immunol 5:1049.*

Burshtyn, D. N., W. Yang, T. Yi, and E. O. Long. 1997. A novel phosphotyrosine motif with a critical amino acid at position -2 for the SH2 domain-mediated activation of the tyrosine phosphatase SHP-1. *J Biol Chem 272:13066.*

Burshtyn, D. N., A. S. Lam, M. Weston, N. Gupta, P. A. Warmerdam, and E. O. Long. 1999. Conserved residues amino-terminal of cytoplasmic tyrosines contribute to the SHP-1-mediated inhibitory function of killer cell Ig-like receptors. *J Immunol 162:897.*

Busch, G. J., E. S. Reynolds, E. G. Galvanek, W. E. Braun, and G. J. Dammin. 1971. Human renal allografts. The role of vascular injury in early graft failure. *Medicine (Baltimore) 50:29.*

Busch, D. H., I. Pilip, and E. G. Pamer. 1998. Evolution of a complex T cell receptor repertoire during primary and recall bacterial infection. *J Exp Med 188:61.*

Butz, E. A., and M. J. Bevan. 1998. Massive expansion of antigen-specific CD8+ T cells during an acute virus infection. *Immunity 8:167.*

Bykovskaja, S. N., A. N. Rytenko, M. O. Rauschenbach, and A. F. Bykovsky. 1978a. Ultrastructural alteration of cytolytic T lymphocytes following their interaction with target cells. I. Hypertrophy and change of orientation of the Golgi apparatus. *Cell Immunol 40:164.*

Bykovskaja, S. N., A. N. Rytenko, M. O. Rauschenbach, and A. F. Bykovsky. 1978b. Ultrastructural alteration of cytolytic T lymphocytes following their interaction with target cells. II. Morphogenesis of secretory granules and intracellular vacuoles. *Cell Immunol 40:175.*

Byrne, J. A., and M. B. Oldstone. 1986. Biology of cloned cytotoxic T lymphocytes specific for lymphocytic choriomeningitis virus. VI.

Migration and activity in vivo in acute and persistent infection. *J Immunol 136:698.*
Caldwell, C. W., E. D. Everett, G. McDonald, Y. W. Yesus, and W. E. Roland. 1995. Lymphocytosis of gamma/delta T cells in human ehrlichiosis. *Am J Clin Pathol 103:761.*
Caligiuri, M. A., C. Murray, M. J. Robertson, E. Wang, K. Cochran, C. Cameron, P. Schow, M. E. Ross, T. R. Klumpp, R. J. Soiffer, and et al. 1993. Selective modulation of human natural killer cells in vivo after prolonged infusion of low dose recombinant interleukin 2. *J Clin Invest 91:123.*
Campbell, A. K. 1987. Intracellular calcium: friend or foe? *Clin Sci (Lond) 72:1.*
Campbell, J. J., S. Qin, D. Unutmaz, D. Soler, K. E. Murphy, M. R. Hodge, L. Wu, and E. C. Butcher. 2001. Unique subpopulations of CD56+ NK and NK-T peripheral blood lymphocytes identified by chemokine receptor expression repertoire. *J Immunol 166:6477.*
Canaday, D. H., R. J. Wilkinson, Q. Li, C. V. Harding, R. F. Silver, and W. H. Boom. 2001. CD4(+) and CD8(+) T cells kill intracellular Mycobacterium tuberculosis by a perforin and Fas/Fas ligand-independent mechanism. *J Immunol 167:2734.*
Cardillo, F., J. C. Voltarclli, S. G. Rccd, and J. S. Silva. 1996. Regulation of Trypanosoma cruzi infection in mice by gamma interferon and interleukin 10: role of NK cells. *Infect Immun 64:128.*
Carding, S. R., and P. J. Egan. 2002. Gammadelta T cells: functional plasticity and heterogeneity. *Nature Rev Immunol 2:336.*
Carlyle, J. R., A. Martin, A. Mehra, L. Attisano, F. W. Tsui, and J. C. Zuniga-Pflucker. 1999. Mouse NKR-P1B, a novel NK1.1 antigen with inhibitory function. *J Immunol 162:5917.*
Carnaud, C., D. Lee, O. Donnars, S. H. Park, A. Beavis, Y. Koezuka, and A. Bendelac. 1999. Cutting edge: Cross-talk between cells of the innate immune system: NKT cells rapidly activate NK cells. *J Immunol 163:4647.*
Carpen, O., I. Virtanen, and E. Saksela. 1981. The cytotoxic activity of human natural killer cells requires an intact secretory apparatus. *Cell Immunol 58:97.*
Carpen, O., I. Virtanen, and E. Saksela. 1982. Ultrastructure of human natural killer cells: nature of the cytolytic contacts in relation to cellular secretion. *J Immunol 128:2691.*
Carson, W. E., J. G. Giri, M. J. Lindemann, M. L. Linett, M. Ahdieh, R. Paxton, D. Anderson, J. Eisenmann, K. Grabstein, and M. A. Caligiuri. 1994. Interleukin (IL) 15 is a novel cytokine that activates human natural killer cells via components of the IL-2 receptor. *J Exp Med 180:1395.*
Catalfamo, M., C. Roura-Mir, M. Sospedra, P. Aparicio, S. Costagliola, M. Ludgate, R. Pujol-Borrell, and D. Jaraquemada. 1996. Self-reactive cytotoxic gamma delta T lymphocytes in Graves' disease specifically recognize thyroid epithelial cells. *J Immunol 156:804.*
Cerottini, J. C., and K. T. Brunner. 1974. Cell-mediated cytotoxicity, allograft rejection, and tumor immunity. *Adv Immunol 18:67.*
Cerottini, J. C., and H. R. MacDonald. 1989. The cellular basis of T-cell memory. *Annu Rev Immunol 7:77.*

Chakravarty, A. K., and W. R. Clark. 1977. Lectin-driven maturation of cytotoxic effector cells: the nature of effector memory. *J Exp Med 146:230.*

Chambers, B. J., M. Salcedo, and H. G. Ljunggren. 1996. Triggering of natural killer cells by the costimulatory molecule CD80 (B7-1). *Immunity 5:311.*

Chambers, C. A., and J. P. Allison. 1999. Costimulatory regulation of T cell function. *Curr Opin Cell Biol 11:203.*

Chang, T. W., and S. P. Gingras. 1981. OKT3 monoclonal antibody inhibits cytotoxic T lymphocyte mediated cell lysis. *Int J Immunopharmacol 3:183.*

Chase, M. 1945. The cellular transfer of cutaneous hypersensitivity to tuberculin. *Proc Soc Exp Biol Med 59:134.*

Chen, B. P., M. Malkovsky, J. A. Hank, and P. M. Sondel. 1987. Nonrestricted cytotoxicity mediated by interleukin 2-expanded leukocytes is inhibited by anti-LFA-1 monoclonal antibodies (MoAb) but potentiated by anti-CD3 MoAb. *Cell Immunol 110:282.*

Chen, Y., N. Chiu, M. Mandal, N. Wang, and C. Wang. 1997. Impaired NK1.1+ T cells in a TH2 response and in immunoglobulin E production in CD1-deficient mice. *Immunity 6:459.*

Chen, W., W. Jin, and S. M. Wahl. 1998. Engagement of cytotoxic T lymphocyte-associated antigen 4 (CTLA-4) induces transforming growth factor beta (TGF-beta) production by murine CD4(+) T cells. *J Exp Med 188:1849.*

Chervonsky, A. V., Y. Wang, F. S. Wong, I. Visintin, R. A. Flavell, C. A. Janeway, Jr., and L. A. Matis. 1997. The role of Fas in autoimmune diabetes. *Cell 89:17.*

Chiplunkar, S., G. De Libero, and S. H. Kaufmann. 1986. Mycobacterium leprae-specific Lyt-2+ T lymphocytes with cytolytic activity. *Infect Immun 54:793.*

Chirmule, N., A. D. Moscioni, Y. Qian, R. Qian, Y. Chen, and J. M. Wilson. 1999. Fas-Fas ligand interactions play a major role in effector functions of cytotoxic T lymphocytes after adenovirus vector-mediated gene transfer. *Hum Gene Ther 10:259.*

Christensen, J. P., C. Ropke, and A. R. Thomsen. 1996. Virus-induced polyclonal T cell activation is followed by apoptosis: partitioning of CD8+ T cells based on alpha 4 integrin expression. *Int Immunol 8:707.*

Chun, T. W., J. S. Justement, S. Moir, C. W. Hallahan, L. A. Ehler, S. Liu, M. McLaughlin, M. Dybul, J. M. Mican, and A. S. Fauci. 2001. Suppression of HIV replication in the resting CD4+ T cell reservoir by autologous CD8+ T cells: implications for the development of therapeutic strategies. *Proc Natl Acad Sci U S A 98:253.*

Ciurea, A., P. Klenerman, L. Hunziker, E. Horvath, B. Odermatt, A. F. Ochsenbein, H. Hengartner, and R. M. Zinkernagel. 1999. Persistence of lymphocytic choriomeningitis virus at very low levels in immune mice. *Proc Natl Acad Sci U S A 96:11964.*

Clark, W. R. 1975. An antigen-specific component of lectin-mediated cytotoxicity. *Cell Immunol 17:505.*

Clark, W. R., and G. Berke. 1982. T lymphocyte-mediated cytolysis - a comprehensive theory. II. Lytic vs. nonlytic interactions of T lymphocytes. *Adv Exp Med Biol 146:69.*

Clark, W. R., C. M. Walsh, A. A. Glass, F. Hayashi, M. Matloubian, and R. Ahmed. 1995a. Molecular pathways of CTL-mediated cytotoxicity. *Immunol Rev 146:33.*

Clark, W. R., C. M. Walsh, A. A. Glass, M. T. Huang, R. Ahmed, and M. Matloubian. 1995b. Cell-mediated cytotoxicity in perforin-less mice. *Int Rev Immunol 13:1.*

Clark, R. B., and E. G. Lingenheld. 1998. Adoptively transferred EAE in gamma delta T cell-knockout mice. *J Autoimmun 11:105.*

Clarke, S. R. 2000. The critical role of CD40/CD40L in the CD4-dependent generation of CD8+ T cell immunity. *J Leukoc Biol 67:607.*

Clement, M. V., S. Legros-Maida, D. Israel-Biet, F. Carnot, A. Soulie, P. Reynaud, J. Guillet, I. Gandjbakch, and M. Sasportes. 1994. Perforin and granzyme B expression is associated with severe acute rejection. Evidence for in situ localization in alveolar lymphocytes of lung-transplanted patients. *Transplantation 57:322.*

Clerici, M., D. R. Lucey, R. A. Zajac, R. N. Boswell, H. M. Gebel, H. Takahashi, J. A. Berzofsky, and G. M. Shearer. 1991. Detection of cytotoxic T lymphocytes specific for synthetic peptides of gp160 in HIV-seropositive individuals. *J Immunol 146:2214.*

Cockcroft, S., and B. D. Gomperts. 1979. ATP induces nucleotide permeability in rat mast cells. *Nature 279:541.*

Cohen C. J., G. Denkberg, Y. S. Schiffenbauer, D. Segal, E. Trubniykov, G. Berke and Y. Reiter 2003. Simultaneous monitoring of binding to and activation of tumor-specific T lymphocytes by peptide-MHC. J Immunol Methods. 277:39-52.

Cohen, J. J., R. C. Duke, R. Chervenak, K. S. Sellins, and L. K. Olson. 1985. DNA fragmentation in targets of CTL: an example of programmed cell death in the immune system. *Adv Exp Med Biol 184:493.*

Cohen, G. B., R. T. Gandhi, D. M. Davis, O. Mandelboim, B. K. Chen, J. L. Strominger, and D. Baltimore. 1999. The selective downregulation of class I major histocompatibility complex proteins by HIV-1 protects HIV-infected cells from NK cells. *Immunity 10:661.*

Coles, M. C., C. W. McMahon, H. Takizawa, and D. H. Raulet. 2000. Memory CD8 T lymphocytes express inhibitory MHC-specific Ly49 receptors. *Eur J Immunol 30:236.*

Coles, M. C., and D. H. Raulet. 2000. NK1.1+ T cells in the liver arise in the thymus and are selected by interactions with class I molecules on CD4+CD8+ cells. *J Immunol 164:2412.*

Collins, T., A. M. Krensky, C. Clayberger, W. Fiers, M. A. Gimbrone, Jr., S. J. Burakoff, and J. S. Pober. 1984. Human cytolytic T lymphocyte interactions with vascular endothelium and fibroblasts: role of effector and target cell molecules. *J Immunol 133:1878.*

Collins, K. L., and D. Baltimore. 1999. HIV's evasion of the cellular immune response. *Immunol Rev 168:65.*

Colonna, M., F. Navarro, T. Bellon, M. Llano, P. Garcia, J. Samaridis, L. Angman, M. Cella, and M. Lopez-Botet. 1997. A common inhibitory receptor for major histocompatibility complex class I molecules on human lymphoid and myelomonocytic cells. *J Exp Med 186:1809.*

Comoli, P., M. Labirio, S. Basso, F. Baldanti, P. Grossi, M. Furione, M. Vigano, R. Fiocchi, G. Rossi, F. Ginevri, B. Gridelli, A. Moretta, D. Montagna, F. Locatelli, G. Gerna, and R. Maccario. 2002. Infusion of autologous Epstein-Barr virus (EBV)-specific cytotoxic T cells for prevention of EBV-related lymphoproliferative disorder in solid organ transplant recipients with evidence of active virus replication. *Blood 99:2592.*

Conceicao-Silva, F., M. Hahne, M. Schroter, J. Louis, and J. Tschopp. 1998. The resolution of lesions induced by Leishmania major in mice requires a functional Fas (APO-1, CD95) pathway of cytotoxicity. *Eur J Immunol 28:237.*

Constant, P., F. Davodeau, M. A. Peyrat, Y. Poquet, G. Puzo, M. Bonneville, and J. J. Fournie. 1994. Stimulation of human gamma delta T cells by nonpeptidic mycobacterial ligands. *Science 264:267.*

Cooper, M. D., D. Y. Perey, A. E. Gabrielsen, D. E. Sutherland, M. F. McKneally, and R. A. Good. 1968. Production of an antibody deficiency syndrome in rabbits by neonatal removal of organized intestinal lymphoid tissues. *Int Arch Allergy Appl Immunol 33:65.*

Cooper, A. M., C. D'Souza, A. A. Frank, and I. M. Orme. 1997. The course of Mycobacterium tuberculosis infection in the lungs of mice lacking expression of either perforin- or granzyme-mediated cytolytic mechanisms. *Infect Immun 65:1317.*

Cooper, M. A., T. A. Fehniger, and M. A. Caligiuri. 2001. The biology of human natural killer-cell subsets. *Trends Immunol 22:633.*

Copeland, K. F. 2002. The role of CD8+ T cell soluble factors in human immunodeficiency virus infection. *Curr Med Chem 9:1781.*

Coscoy, L., and D. Ganem. 2001. A viral protein that selectively downregulates ICAM-1 and B7-2 and modulates T cell costimulation. *J Clin Invest 107:1599.*

Cosman, D., N. Fanger, L. Borges, M. Kubin, W. Chin, L. Peterson, and M. L. Hsu. 1997. A novel immunoglobulin superfamily receptor for cellular and viral MHC class I molecules. *Immunity 7:273.*

Costello, R., S. Sivori, F. Mallet, D. Sainty, C. Arnoulet, D. Reviron, J. Gastaut, A. Moretta, and D. Olive. 2002. A novel mechanism of antitumor response involving the expansion of CD3/CD56 large granular lymphocytes triggered by a tumor-expressed activating ligand. *Leukemia 16:855.*

Cox, F. E., and F. Y. Liew. 1992. T-cell subsets and cytokines in parasitic infections. *Immunol Today 13:445.*

Cretney, E., K. Takeda, H. Yagita, M. Glaccum, J. J. Peschon, and M. J. Smyth. 2002. Increased susceptibility to tumor initiation and metastasis in TNF-related apoptosis-inducing ligand-deficient mice. *J Immunol 168:1356.*

Croft, M., L. Carter, S. L. Swain, and R. W. Dutton. 1994. Generation of polarized antigen-specific CD8 effector populations: reciprocal action of interleukin (IL)-4 and IL-12 in promoting type 2 versus type 1 cytokine profiles. *J Exp Med 180:1715.*

Crowley, M. P., A. M. Fahrer, N. Baumgarth, J. Hampl, I. Gutgemann, L. Teyton, and Y. Chien. 2000. A population of murine gammadelta T cells that recognize an inducible MHC class Ib molecule. *Science 287:314.*

Cui, J., T. Shin, T. Kawano, H. Sato, E. Kondo, I. Toura, Y. Kaneko, H. Koseki, M. Kanno, and M. Taniguchi. 1997. Requirement for Valpha14 NKT cells in IL-12-mediated rejection of tumors. *Science 278:1623.*

Cui, Y., E. Kelleher, E. Straley, E. Fuchs, K. Gorski, H. Levitsky, I. Borrello, C. I. Civin, S. P. Schoenberger, L. Cheng, D. M. Pardoll, and K. A. Whartenby. 2003. Immunotherapy of established tumors using bone marrow transplantation with antigen gene--modified hematopoietic stem cells. *Nat Med 9:952.*

Cuturi, M. C., I. Anegon, F. Sherman, R. Loudon, S. C. Clark, B. Perussia, and G. Trinchieri. 1989. Production of hematopoietic colony-stimulating factors by human natural killer cells. *J Exp Med 169:569.*

Czuprynski, C. J., and J. F. Brown. 1990. Effects of purified anti-Lyt-2 mAb treatment on murine listeriosis: comparative roles of Lyt-2+ and L3T4+ cells in resistance to primary and secondary infection, delayed-type hypersensitivity and adoptive transfer of resistance. *Immunology 71:107.*

da Conceicao-Silva, F., B. L. Perlaza, J. A. Louis, and P. Romero. 1994. Leishmania major infection in mice primes for specific major histocompatibility complex class I-restricted CD8+ cytotoxic T cell responses. *Eur J Immunol 24:2813.*

Dallal, R. M., and M. T. Lotze. 2000. The dendritic cell and human cancer vaccines. *Curr Opin Immunol 12:583.*

Dandekar, A. A., and S. Perlman. 2002. Virus-induced demyelination in nude mice is mediated by gamma delta T cells. *Am J Pathol 161:1255.*

Dannemann, B. R., V. A. Morris, F. G. Araujo, and J. S. Remington. 1989. Assessment of human natural killer and lymphokine-activated killer cell cytotoxicity against Toxoplasma gondii trophozoites and brain cysts. *J Immunol 143:2684.*

Darmon, A. J., D. W. Nicholson, and R. C. Bleackley. 1995. Activation of the apoptotic protease CPP32 by cytotoxic T-cell-derived granzyme B. *Nature 377:446.*

Darmon, A. J., T. J. Ley, D. W. Nicholson, and R. C. Bleackley. 1996. Cleavage of CPP32 by granzyme B represents a critical role for granzyme B in the induction of target cell DNA fragmentation. *J Biol Chem 271:21709.*

Davidson, W. F. 1977. Cellular requirements for the induction of cytotoxic T cells in vitro. *Immunol Rev 35:261.*

Davidson, E. J., M. D. Brown, D. J. Burt, J. L. Parish, K. Gaston, H. C. Kitchener, S. N. Stacey, and P. L. Stern. 2001. Human T cell responses to HPV 16 E2 generated with monocyte-derived dendritic cells. *Int J Cancer 94:807.*

Davis, D. M., I. Chiu, M. Fassett, G. B. Cohen, O. Mandelboim, and J. L. Strominger. 1999. The human natural killer cell immune synapse. *Proc Natl Acad Sci U S A 96:15062.*

Davis, J. E., M. J. Smyth, and J. A. Trapani. 2001. Granzyme A and B-deficient killer lymphocytes are defective in eliciting DNA fragmentation but retain potent in vivo anti-tumor capacity. *Eur J Immunol 31:39.*

Davis, S. J., and P. A. van der Merwe. 2001. The immunological synapse: required for T cell receptor signalling or directing T cell effector function? *Curr Biol 11:R289.*

Dayan, C. M., M. Londei, A. E. Corcoran, B. Grubeck-Loebenstein, R. F. James, B. Rapoport, and M. Feldmann. 1991. Autoantigen recognition by thyroid-infiltrating T cells in Graves disease. *Proc Natl Acad Sci U S A 88:7415.*

Dayan, C. M., and G. H. Daniels. 1996. Chronic autoimmune thyroiditis. *N Engl J Med 335:99.*

de Jong, S., T. Timmer, F. J. Heijenbrok, and E. G. de Vries. 2001. Death receptor ligands, in particular TRAIL, to overcome drug resistance. *Cancer Metastasis Rev 20:51.*

De Libero, G., and S. H. Kaufmann. 1986. Antigen-specific Lyt-2+ cytolytic T lymphocytes from mice infected with the intracellular bacterium Listeria monocytogenes. *J Immunol 137:2688.*

De Maria, R., and R. Testi. 1998. Fas-FasL interactions: a common pathogenetic mechanism in organ-specific autoimmunity. *Immunol Today 19:121.*

Dechanet, J., P. Merville, V. Pitard, X. Lafarge, and J. F. Moreau. 1999. Human gammadelta T cells and viruses. *Microbes Infect 1:213.*

Del Prete, G. F., D. Vercelli, A. Tiri, E. Maggi, S. Mariotti, A. Pinchera, M. Ricci, and S. Romagnani. 1986. In vivo activated cytotoxic T cells in the thyroid infiltrate of patients with Hashimoto's thyroiditis. *Clin Exp Immunol 65:140.*

Dellabona, P., E. Padovan, G. Casorati, M. Brockhaus, and A. Lanzavecchia. 1994. An invariant V alpha 24-J alpha Q/V beta 11 T cell receptor is expressed in all individuals by clonally expanded CD4-8- T cells. *J Exp Med 180:1171.*

DeLuca, D., G. W. Warr, and J. J. Marchalonis. 1979. The immunoglobulin-like T-cell receptor. II. Codistribution of Fab determinants and antigen on the surface of antigen-binding lymphocytes of mouse thymus. *J Immunogenet 6:359.*

Demkowicz, W. E., Jr., R. A. Littaua, J. Wang, and F. A. Ennis. 1996. Human cytotoxic T-cell memory: long-lived responses to vaccinia virus. *J Virol 70:2627.*

Dengler, T. J., and J. S. Pober. 2000. Human vascular endothelial cells stimulate memory but not naive CD8+ T cells to differentiate into CTL retaining an early activation phenotype. *J Immunol 164:5146.*

Denizot, F., A. Wilson, F. Battye, G. Berke, and K. Shortman. 1986. Clonal expansion of T cells: a cytotoxic T-cell response in vivo that involves precursor cell proliferation. *Proc Natl Acad Sci U S A 83:6089.*

Denkers, E. Y., R. T. Gazzinelli, D. Martin, and A. Sher. 1993. Emergence of NK1.1+ cells as effectors of IFN-gamma dependent immunity to Toxoplasma gondii in MHC class I-deficient mice. *J Exp Med 178:1465.*

Denkers, E. Y., T. Scharton-Kersten, S. Barbieri, P. Caspar, and A. Sher. 1996. A role for CD4+ NK1.1+ T lymphocytes as major histocompatibility complex class II independent helper cells in the generation of CD8+ effector function against intracellular infection. *J Exp Med 184:131.*

Denkers, E. Y., G. Yap, T. Scharton-Kersten, H. Charest, B. A. Butcher, P. Caspar, S. Heiny, and A. Sher. 1997. Perforin-mediated cytolysis plays a limited role in host resistance to Toxoplasma gondii. *J Immunol 159:1903.*

Dennert, G., G. Yogeeswaran, and S. Yamagata. 1981. Cloned cell lines with natural killer activity. Specificity, function, and cell surface markers. *J Exp Med 153:545.*

Dennert, G., and E. R. Podack. 1983. Cytolysis by H-2-specific T killer cells. Assembly of tubular complexes on target membranes. *J Exp Med 157:1483.*

Dennert, G. 2002. Elimination of virus-specific cytotoxic T cells in the liver. *Crit Rev Immunol 22:1.*

Dermime, S., A. Armstrong, R. E. Hawkins, and P. L. Stern. 2002. Cancer vaccines and immunotherapy. *Br Med Bull 62:149.*

Devendra, D., and G. S. Eisenbarth. 2003. 17. Immunologic endocrine disorders. *J Allergy Clin Immunol 111:S624.*

Dhein, J., H. Walczak, C. Baumler, K. M. Debatin, and P. H. Krammer. 1995. Autocrine T-cell suicide mediated by APO-1/(Fas/CD95). *Nature 373:438.*

Di Virgilio, F., V. Bronte, D. Collavo, and P. Zanovello. 1989. Responses of mouse lymphocytes to extracellular adenosine 5'-triphosphate (ATP). Lymphocytes with cytotoxic activity are resistant to the permeabilizing effects of ATP. *J Immunol 143:1955.*

Di Virgilio, F., P. Pizzo, P. Zanovello, V. Bronte, and D. Collavo. 1990. Extracellular ATP as a possible mediator of cell-mediated cytotoxicity. *Immunol Today 11:274.*

Diefenbach, A., E. R. Jensen, A. M. Jamieson, and D. H. Raulet. 2001. Rae1 and H60 ligands of the NKG2D receptor stimulate tumour immunity. *Nature 413:165.*

Dieli, F., M. Troye-Blomberg, J. Ivanyi, J. J. Fournie, M. Bonneville, M. A. Peyrat, G. Sireci, and A. Salerno. 2000. Vgamma9/Vdelta2 T lymphocytes reduce the viability of intracellular Mycobacterium tuberculosis. *Eur J Immunol 30:1512.*

Disis, M. L., K. H. Grabstein, P. R. Sleath, and M. A. Cheever. 1999. Generation of immunity to the HER-2/neu oncogenic protein in patients with breast and ovarian cancer using a peptide-based vaccine. *Clin Cancer Res 5:1289.*

Djeu, J. Y., E. Lanza, S. Pastore, and A. J. Hapel. 1983. Selective growth of natural cytotoxic but not natural killer effector cells in interleukin-3. *Nature 306:788.*

Doherty, P. C., and R. M. Zinkernagel. 1974. T-cell-mediated immunopathology in viral infections. *Transplant Rev 19:89.*

Doherty, P. C., J. E. Allan, and R. Ceredig. 1988. Contributions of host and donor T cells to the inflammatory process in murine lymphocytic choriomeningitis. *Cell Immunol 116:475.*

Dombrowski, K. E., Y. Ke, L. F. Thompson, and J. A. Kapp. 1995. Antigen recognition by CTL is dependent upon ectoATPase activity. *J Immunol 154:6227.*

Don, M. M., G. Ablett, C. J. Bishop, P. G. Bundesen, K. J. Donald, J. Searle, and J. F. Kerr. 1977. Death of cells by apoptosis following attachment of specifically allergized lymphocytes in vitro. *Aust J Exp Biol Med Sci 55:407.*

Dorothee, G., I. Vergnon, J. Menez, H. Echchakir, D. Grunenwald, M. Kubin, S. Chouaib, and F. Mami-Chouaib. 2002. Tumor-Infiltrating CD4(+) T Lymphocytes Express APO2 Ligand (APO2L)/TRAIL upon Specific Stimulation with Autologous Lung Carcinoma Cells: Role of IFN-alpha on APO2L/TRAIL Expression and -Mediated Cytotoxicity. *J Immunol 169:809.*

Dourmashkin, R. R., P. Deteix, C. B. Simone, and P. Henkart. 1980. Electron microscopic demonstration of lesions in target cell membranes associated with antibody-dependent cellular cytotoxicity. *Clin Exp Immunol 42:554.*

Dudley, M. E., J. R. Wunderlich, P. F. Robbins, J. C. Yang, P. Hwu, D. J. Schwartzentruber, S. L. Topalian, R. Sherry, N. P. Restifo, A. M. Hubicki, M. R. Robinson, M. Raffeld, P. Duray, C. A. Seipp, L. Rogers-Freezer, K. E. Morton, S. A. Mavroukakis, D. E. White, and S. A. Rosenberg. 2002. Cancer regression and autoimmunity in patients after clonal repopulation with antitumor lymphocytes. *Science 298:850.*

Dudley, M. E., and S. A. Rosenberg. 2003. Adoptive-cell-transfer therapy for the treatment of patients with cancer. *Nat Rev Cancer 3:666.*

Duke, R. C., R. Chervenak, and J. J. Cohen. 1983. Endogenous endonuclease-induced DNA fragmentation: an early event in cell-mediated cytolysis. *Proc Natl Acad Sci U S A 80:6361.*

Duke, R. C., J. J. Cohen, and R. Chervenak. 1986. Differences in target cell DNA fragmentation induced by mouse cytotoxic T lymphocytes and natural killer cells. *J Immunol 137:1442.*

Duke, R. C., P. M. Persechini, S. Chang, C. C. Liu, J. J. Cohen, and J. D. Young. 1989. Purified perforin induces target cell lysis but not DNA fragmentation. *J Exp Med 170:1451.*

Dunne, J., S. Lynch, C. O'Farrelly, S. Todryk, J. E. Hegarty, C. Feighery, and D. G. Doherty. 2001. Selective expansion and partial activation of human NK cells and NK receptor-positive T cells by IL-2 and IL-15. *J Immunol 167:3129.*

Dustin, M. L., and T. A. Springer. 1989. T-cell receptor cross-linking transiently stimulates adhesiveness through LFA-1. *Nature 341:619.*

Dvorak, A. M., M. C. Mihm, Jr., and H. F. Dvorak. 1976. Morphology of delayed-type hypersensitivity reactions in man. II. Ultrastructural alterations affecting the microvasculature and the tissue mast cells. *Lab Invest 34:179.*

Dvorak, H. F., M. C. Mihm, Jr., A. M. Dvorak, B. A. Barnes, E. J. Manseau, and S. J. Galli. 1979. Rejection of first-set skin allografts in man. the microvasculature is the critical target of the immune response. *J Exp Med 150:322.*

Earnshaw, W. C., L. M. Martins, and S. H. Kaufmann. 1999. Mammalian caspases: structure, activation, substrates, and functions during apoptosis. *Annu Rev Biochem 68:383.*

Eberl, G., and H. R. MacDonald. 2000. Selective induction of NK cell proliferation and cytotoxicity by activated NKT cells. *Eur J Immunol 30:985.*

Ebnet, K., M. Hausmann, F. Lehmann-Grube, A. Mullbacher, M. Kopf, M. Lamers, and M. M. Simon. 1995. Granzyme A-deficient mice retain potent cell-mediated cytotoxicity. *Embo J 14:4230.*

Eck, S. C., and L. A. Turka. 1999. Generation of protective immunity against an immunogenic carcinoma requires CD40/CD40L and B7/CD28 interactions but not CD4(+) T cells. *Cancer Immunol Immunother 48:336.*

Edwards, B. S., H. A. Nolla, and R. R. Hoffman. 1989. Relationship between target cell recognition and temporal fluctuations in intracellular Ca2+ of human NK cells. *J Immunol 143:1058.*

Edwards, B. H., A. Bansal, S. Sabbaj, J. Bakari, M. J. Mulligan, and P. A. Goepfert. 2002. Magnitude of functional CD8+ T-cell responses to the gag protein of human immunodeficiency virus type 1 correlates inversely with viral load in plasma. *J Virol 76:2298.*

Eerola, A. K., Y. Soini, and P. Paakko. 2000. A high number of tumor-infiltrating lymphocytes are associated with a small tumor size, low tumor stage, and a favorable prognosis in operated small cell lung carcinoma. *Clin Cancer Res 6:1875.*

Ehl, S., U. Hoffmann-Rohrer, S. Nagata, H. Hengartner, and R. Zinkernagel. 1996. Different susceptibility of cytotoxic T cells to CD95 (Fas/Apo-1) ligand-mediated cell death after activation in vitro versus in vivo. *J Immunol 156:2357.*

Einsele, H., G. Rauser, U. Grigoleit, H. Hebart, C. Sinzger, S. Riegler, and G. Jahn. 2002. Induction of CMV-specific T-cell lines using Ag-presenting cells pulsed with CMV protein or peptide. *Cytotherapy 4:49.*

Eisen, H. N., C. R. Verret, A. A. Firmenich, and D. M. Kranz. 1987. Resistance of cytolytic T lymphocytes to the lytic components they release. *Ann Inst Pasteur Immunol 138:328.*

Eisenbarth, G. S. 1986. Type I diabetes mellitus. A chronic autoimmune disease. *N Engl J Med 314:1360.*

Elliott, J. I., D. C. Douek, and D. M. Altmann. 1996. Mice lacking alpha beta + T cells are resistant to the induction of experimental autoimmune encephalomyelitis. *J Neuroimmunol 70:139.*

Enders, P. J., C. Yin, F. Martini, P. S. Evans, N. Propp, F. Poccia, and C. D. Pauza. 2003. HIV-mediated gammadelta T cell depletion is specific for Vgamma2+ cells expressing the Jgamma1.2 segment. *AIDS Res Hum Retroviruses 19:21.*

Engel, A. M., I. M. Svane, J. Rygaard, and O. Werdelin. 1997. MCA sarcomas induced in scid mice are more immunogenic than MCA sarcomas induced in congenic, immunocompetent mice. *Scand J Immunol 45:463.*

Engelhard, V. H., J. R. Gnarra, J. Sullivan, G. L. Mandell, and L. S. Gray. 1988. Early events in target-cell lysis by cytotoxic T cells. *Ann N Y Acad Sci 532:303.*

Erard, F., M. T. Wild, J. A. Garcia-Sanz, and G. LeGros. 1993. Switch of CD8 T cells to noncytolytic CD8-CD4 cells that make TH2 cytokines and help B cells. *Science 260:1802.*

Ettinghausen, S. E., R. K. Puri, and S. A. Rosenberg. 1988. Increased vascular permeability in organs mediated by the systemic administration of lymphokine-activated killer cells and recombinant interleukin-2 in mice. *J Natl Cancer Inst 80:177.*

Evans, D. T., D. H. O'Connor, P. Jing, J. L. Dzuris, J. Sidney, J. da Silva, T. M. Allen, H. Horton, J. E. Venham, R. A. Rudersdorf, T. Vogel, C. D. Pauza, R. E. Bontrop, R. DeMars, A. Sette, A. L. Hughes, and D. I. Watkins. 1999. Virus-specific cytotoxic T-lymphocyte responses select for amino-acid variation in simian immunodeficiency virus Env and Nef. *Nat Med 5:1270.*

Exley, M., S. Porcelli, M. Furman, J. Garcia, and S. Balk. 1998. CD161 (NKR-P1A) costimulation of CD1d-dependent activation of human T cells expressing invariant V alpha 24J alpha Q T cell receptor alpha chains. *J Exp Med 188:867.*

Exley, M. A., N. J. Bigley, O. Cheng, S. M. Tahir, S. T. Smiley, Q. L. Carter, H. F. Stills, M. J. Grusby, Y. Koezuka, M. Taniguchi, and S. P. Balk. 2001. CD1d-reactive T-cell activation leads to amelioration of disease caused by diabetogenic encephalomyocarditis virus. *J Leukoc Biol 69:713.*

Falcone, M., B. Yeung, L. Tucker, E. Rodriguez, and N. Sarvetnick. 1999. A defect in interleukin 12-induced activation and interferon gamma secretion of peripheral natural killer T cells in nonobese diabetic mice suggests new pathogenic mechanisms for insulin-dependent diabetes mellitus. *J Exp Med 190:963.*

Faltynek, C. R., G. L. Princler, and J. R. Ortaldo. 1986. Expression of IFN-alpha and IFN-gamma receptors on normal human small resting T lymphocytes and large granular lymphocytes. *J Immunol 136:4134.*

Farrell, J. P., I. Muller, and J. A. Louis. 1989. A role for Lyt-2+ T cells in resistance to cutaneous leishmaniasis in immunized mice. *J Immunol 142:2052.*

Farrell, H. E., H. Vally, D. M. Lynch, P. Fleming, G. R. Shellam, A. A. Scalzo, and N. J. Davis-Poynter. 1997. Inhibition of natural killer cells by a cytomegalovirus MHC class I homologue in vivo. *Nature 386:510.*

Faustman, D. L., R. M. Steinman, H. M. Gebel, V. Hauptfeld, J. M. Davie, and P. E. Lacy. 1984. Prevention of rejection of murine islet allografts by pretreatment with anti-dendritic cell antibody. *Proc Natl Acad Sci U S A 81:3864.*

Fenichel, G. M., J. M. Florence, A. Pestronk, J. R. Mendell, R. T. Moxley, 3rd, R. C. Griggs, M. H. Brooke, J. P. Miller, J. Robison, W. King, and et al. 1991. Long-term benefit from prednisone therapy in Duchenne muscular dystrophy. *Neurology 41:1874.*

Fenton, R. G., P. Marrack, J. W. Kappler, O. Kanagawa, and J. G. Seidman. 1988. Isotypic exclusion of gamma delta T cell receptors in transgenic mice bearing a rearranged beta-chain gene. *Science 241:1089.*

Ferlazzo, G., M. L. Tsang, L. Moretta, G. Melioli, R. M. Steinman, and C. Munz. 2002. Human dendritic cells activate resting natural killer (NK) cells and are recognized via the NKp30 receptor by activated NK cells. *J Exp Med 195:343.*

Ferluga, J., and A. C. Allison. 1974. Observations on the mechanism by which T-lymphocytes exert cytotoxic effects. *Nature 250:673.*

Fernandez, N. C., A. Lozier, C. Flament, P. Ricciardi-Castagnoli, D. Bellet, M. Suter, M. Perricaudet, T. Tursz, E. Maraskovsky, and L. Zitvogel. 1999. Dendritic cells directly trigger NK cell functions: cross-talk relevant in innate anti-tumor immune responses in vivo. *Nat Med 5:405.*

Ferrarini, M., E. Ferrero, L. Dagna, A. Poggi, and M. R. Zocchi. 2002. Human gammadelta T cells: a nonredundant system in the immune-surveillance against cancer. *Trends Immunol 23:14.*

Feuerer, M., P. Beckhove, L. Bai, E. F. Solomayer, G. Bastert, I. J. Diel, C. Pedain, M. Oberniedermayr, V. Schirrmacher, and V. Umansky. 2001. Therapy of human tumors in NOD/SCID mice with patient-derived reactivated memory T cells from bone marrow. *Nature Medicine 7:452*

Feuerer, M., M. Rocha, L. Bai, V. Umansky, E. F. Solomayer, G. Bastert, I. J. Diel, and V. Schirrmacher. 2001. Enrichment of memory T cells and other profound immunological changes in the bone marrow from untreated breast cancer patients. *Int'l. J. Cancer 92: 96.*

Feuerer, M., P. Beckhove, N. Garbi, and Y. Mahnke. 2003. Bone marrow as a priming site for T-cell responses to blood-borne antigen. *Nat Med 9:1151.*

Fiedler, P., C. Schaetzlein, and H. Eibel. 1998. (commentary). *Science 279:2015a.*

Filippini, A., R. E. Taffs, T. Agui, and M. V. Sitkovsky. 1990. Ecto-ATPase activity in cytolytic T-lymphocytes. Protection from the cytolytic effects of extracellular ATP. *J Biol Chem 265:334.*

Finberg, R., and B. Benacerraf. 1981. Induction, control and consequences of virus specific cytotoxic T cells. *Immunol Rev 58:157.*

Fink, T. M., M. Zimmer, S. Weitz, J. Tschopp, D. E. Jenne, and P. Lichter. 1992. Human perforin (PRF1) maps to 10q22, a region that is syntenic with mouse chromosome 10. *Genomics 13:1300.*

Finn, O. J. 2003. Cancer vaccines: between idea and reality. *Nat Rev Immunol 3:630.*

Fisch, P., A. Moris, H. G. Rammensee, and R. Handgretinger. 2000. Inhibitory MHC class I receptors on gammadelta T cells in tumour immunity and autoimmunity. *Immunol Today 21:187.*

Fishelson, Z., and G. Berke. 1978. T lymphocyte-mediated cytolysis: dissociation of the binding and lytic mechanisms of the effector cell. *J Immunol 120:1121.*

Fishelson, Z., and G. Berke. 1981. Tumor cell destruction by cytotoxic T lymphocytes: the basis of reduced antitumor cell activity in syngeneic hosts. *J Immunol 126:2048.*

Flax, M., and B. A. Barnes. 1966. The role of vascular injury in pulmonary allograft rejection. *Transplantation 4:66.*

Fleischer, B. 1986. Lysis of bystander target cells after triggering of human cytotoxic T lymphocytes. *Eur J Immunol 16:1021.*

Fleuridor, R., B. Wilson, R. Hou, A. Landay, H. Kessler, and L. Al-Harthi. 2003. CD1d-restricted natural killer T cells are potent targets for human immunodeficiency virus infection. *Immunology 108:3.*

Flynn, J. N., and M. Sileghem. 1994. Involvement of gamma delta T cells in immunity to trypanosomiasis. *Immunology 83:86.*

Foley, E. 1953. Antigenic properties of methylcholanthrene-induced sarcomas in mice of the strain of origin. *Cancer Research 13:835.*

Fong, T. A., and T. R. Mosmann. 1990. Alloreactive murine CD8+ T cell clones secrete the Th1 pattern of cytokines. *J Immunol 144:1744.*

Fong, L., D. Brockstedt, C. Benike, L. Wu, and E. G. Engleman. 2001. Dendritic cells injected via different routes induce immunity in cancer patients. *J Immunol 166:4254.*

Fong, L., Y. Hou, A. Rivas, C. Benike, A. Yuen, G. A. Fisher, M. M. Davis, and E. G. Engleman. 2001. Altered peptide ligand vaccination with Flt3 ligand expanded dendritic cells for tumor immunotherapy. *Proc Natl Acad Sci U S A 98:8809.*

Fonteneau, J. F., M. Larsson, and N. Bhardwaj. 2002. Interactions between dead cells and dendritic cells in the induction of antiviral CTL responses. *Curr Opin Immunol 14:471.*

Forbes, R. D., R. D. Guttmann, M. Gomersall, and J. Hibberd. 1983. A controlled serial ultrastructural tracer study of first-set cardiac allograft rejection in the rat. Evidence that the microvascular endothelium is the primary target of graft destruction. *Am J Pathol 111:184.*

Forman, J., and G. Moller. 1973. Generation of cytotoxic lymphocytes in mixed lymphocyte reactions. I. Specificity of the effector cells. *J Exp Med 138:672.*

Foulds, K. E., L. A. Zenewicz, D. J. Shedlock, J. Jiang, A. E. Troy, and H. Shen. 2002. Cutting edge: CD4 and CD8 T cells are intrinsically different in their proliferative responses. *J Immunol 168:1528.*

Fox, B. A., P. J. Spiess, A. Kasid, R. Puri, J. J. Mule, J. S. Weber, and S. A. Rosenberg. 1990. In vitro and in vivo antitumor properties of a T-cell clone generated from murine tumor-infiltrating lymphocytes. *J Biol Response Mod 9:499.*

Francke, S., C. G. Orosz, J. Hsu, and L. E. Mathes. 2002. Immunomodulatory effect of zidovudine (ZDV) on cytotoxic T lymphocytes previously exposed to ZDV. *Antimicrob Agents Chemother 46:2865.*

Franco, M. A., C. Tin, L. S. Rott, J. L. VanCott, J. R. McGhee, and H. B. Greenberg. 1997. Evidence for CD8+ T-cell immunity to murine rotavirus in the absence of perforin, fas, and gamma interferon. *J Virol 71:479.*

Fraser, S. A., R. Karimi, M. Michalak, and D. Hudig. 2000. Perforin lytic activity is controlled by calreticulin. *J Immunol 164:4150.*

Freiberg, B., H. Kupfer, W. Maslanik, J. Delli, J. Kappler, D. Zaller, and A. Kupfer. 2002. Staging and resetting T cell activation in SMACs. *Nat Immunol 3:911.*

Froelich, C. J., K. Orth, J. Turbov, P. Seth, R. Gottlieb, B. Babior, G. M. Shah, R. C. Bleackley, V. M. Dixit, and W. Hanna. 1996. New paradigm for lymphocyte granule-mediated cytotoxicity. Target cells bind and internalize granzyme B, but an endosomolytic agent is necessary for cytosolic delivery and subsequent apoptosis. *J Biol Chem 271:29073.*

Frye, L. D., and G. J. Friou. 1975. Inhibition of mammalian cytotoxic cells by phosphatidylcholine and its analogue. *Nature 258:333.*

Fuse, Y., H. Nishimura, K. Maeda, and Y. Yoshikai. 1997. CD95 (Fas) may control the expansion of activated T cells after elimination of bacteria in murine listeriosis. *Infect Immun 65:1883.*

Gadola, S. D., N. R. Zaccai, K. Harlos, D. Shepherd, J. C. Castro-Palomino, G. Ritter, R. R. Schmidt, E. Y. Jones, and V. Cerundolo. 2002. Structure of human CD1b with bound ligands at 2.3 A, a maze for alkyl chains. *Nat Immunol 3:721.*

Galandrini, R., R. De Maria, M. Piccoli, L. Frati, and A. Santoni. 1994. CD44 triggering enhances human NK cell cytotoxic functions. *J Immunol 153:4399.*
Gallo, O., M. De Carli, E. Gallina, G. Toccafondi, S. Romagnani, and G. Del Prete. 1991. T lymphocytes from tonsil and peripheral blood show different cytolytic and helper activities. *Acta Otolaryngol 111:974.*
Garcia, K. C., C. A. Scott, A. Brunmark, F. R. Carbone, P. A. Peterson, I. A. Wilson, and L. Teyton. 1996. CD8 enhances formation of stable T-cell receptor/MHC class I molecule complexes. *Nature 384:577.*
Garg, N., M. P. Nunes, and R. L. Tarleton. 1997. Delivery by Trypanosoma cruzi of proteins into the MHC class I antigen processing and presentation pathway. *J Immunol 158:3293.*
Garrett, T. P., M. A. Saper, P. J. Bjorkman, J. L. Strominger, and D. C. Wiley. 1989. Specificity pockets for the side chains of peptide antigens in HLA-Aw68. *Nature 342:692.*
Geiger, B., D. Rosen, and G. Berke. 1982. Spatial relationships of microtubule-organizing centers and the contact area of cytotoxic T lymphocytes and target cells. *J Cell Biol 95:137.*
Gell, P., and R. Coombs. 1968. *Clinical Aspects of Immunology.* Blackwell, Oxford.
Germain, R. 2002. T-cell development and the CD4-CD8 lineage decision. *Nature Reviews Immunology 2:309.*
Gerosa, F., B. Baldani-Guerra, C. Nisii, V. Marchesini, G. Carra, and G. Trinchieri. 2002. Reciprocal activating interaction between natural killer cells and dendritic cells. *J Exp Med 195:327.*
Gherardi, M. M., J. C. Ramirez, and M. Esteban. 2001. Towards a new generation of vaccines: the cytokine IL-12 as an adjuvant to enhance cellular immune responses to pathogens during prime-booster vaccination regimens. *Histol Histopathol 16:655.*
Ghiasi, H., S. Cai, G. Perng, A. B. Nesburn, and S. L. Wechsler. 1999. Perforin pathway is essential for protection of mice against lethal ocular HSV-1 challenge but not corneal scarring. *Virus Res 65:97.*
Giacomelli, R., M. Matucci-Cerinic, P. Cipriani, I. Ghersetich, R. Lattanzio, A. Pavan, A. Pignone, M. L. Cagnoni, T. Lotti, and G. Tonietti. 1998. Circulating Vdelta1+ T cells are activated and accumulate in the skin of systemic sclerosis patients. *Arthritis Rheum 41:327.*
Giacomelli, R., P. Cipriani, A. Fulminis, G. Barattelli, M. Matucci-Cerinic, S. D'Alo, G. Cifone, and G. Tonietti. 2001. Circulating gamma/delta T lymphocytes from systemic sclerosis (SSc) patients display a T helper (Th) 1 polarization. *Clin Exp Immunol 125:310.*
Gillis, S., and K. A. Smith. 1977. Long term culture of tumour-specific cytotoxic T cells. *Nature 268:154.*
Ginsburg, H., and L. Sachs. 1965. Destruction of mouse and rat embryo cells in tissue culture by lymph node cells from unsensitized rats. *J Cell Physiol 66:199.*
Ginsburg, H. 1968. Graft versus host reaction in tissue culture. I. Lysis of monolayers of embryo mouse cells from strains differing in the H-2 histocompatibility locus by rat lymphocytes sensitized in vitro. *Immunology 14:621.*

Ginsburg, H., W. Ax, and G. Berke. 1969. Graft reaction in tissue culture by normal rat lymphocytes. *Transplant Proc 1:551.*

Giordano, C., G. Stassi, R. De Maria, M. Todaro, P. Richiusa, G. Papoff, G. Ruberti, M. Bagnasco, R. Testi, and A. Galluzzo. 1997. Potential involvement of Fas and its ligand in the pathogenesis of Hashimoto's thyroiditis. *Science 275:960.*

Giordano, C., P. Richiusa, M. Bagnasco, G. Pizzolanti, F. Di Blasi, M. S. Sbriglia, A. Mattina, G. Pesce, P. Montagna, F. Capone, G. Misiano, A. Scorsone, A. Pugliese, and A. Galluzzo. 2001. Differential regulation of Fas-mediated apoptosis in both thyrocyte and lymphocyte cellular compartments correlates with opposite phenotypic manifestations of autoimmune thyroid disease. *Thyroid 11:233.*

Giorno, R., M. T. Barden, P. F. Kohler, and S. P. Ringel. 1984. Immunohistochemical characterization of the mononuclear cells infiltrating muscle of patients with inflammatory and noninflammatory myopathies. *Clin Immunol Immunopathol 30:405.*

Girardi, M., D. E. Oppenheim, C. R. Steele, J. M. Lewis, E. Glusac, R. Filler, P. Hobby, B. Sutton, R. E. Tigelaar, and A. C. Hayday. 2001. Regulation of cutaneous malignancy by gammadelta T cells. *Science 294:605.*

Gjertsen, M. K., T. Buanes, A. R. Rosseland, A. Bakka, I. Gladhaug, O. Soreide, J. A. Eriksen, M. Moller, I. Baksaas, R. A. Lothe, I. Saeterdal, and G. Gaudernack. 2001. Intradermal ras peptide vaccination with granulocyte-macrophage colony-stimulating factor as adjuvant: Clinical and immunological responses in patients with pancreatic adenocarcinoma. *Int J Cancer 92:441.*

Glass, A., C. M. Walsh, D. H. Lynch, and W. R. Clark. 1996. Regulation of the Fas lytic pathway in cloned CTL. *J Immunol 156:3638.*

Glimcher, L., F. W. Shen, and H. Cantor. 1977. Identification of a cell-surface antigen selectively expressed on the natural killer cell. *J Exp Med 145:1.*

Godfrey, D. I., K. J. Hammond, L. D. Poulton, M. J. Smyth, and A. G. Baxter. 2000. NKT cells: facts, functions and fallacies. *Immunol Today 21:573.*

Golding, H., and A. Singer. 1985. Specificity, phenotype, and precursor frequency of primary cytolytic T lymphocytes specific for class II major histocompatibility antigens. *J Immunol 135:1610.*

Golstein, P., M. D. Erik, A. J. Svedmyr, and H. Wigzell. 1971. Cells mediating specific in vitro cytotoxicity. I. Detection of receptor-bearing lymphocytes. *J Exp Med 134:1385.*

Golstein, P. 1974. Sensitivity of cytotoxic T cells to T-cell mediated cytotoxicity. *Nature 252:81.*

Golstein, P., and E. T. Smith. 1976. The lethal hit stage of mouse T and non-T cell-mediated cytolysis: differences in cation requirements and characterization of an analytical "cation pulse" method. *Eur J Immunol 6:31.*

Good, R. A. 1970. Evaluation of the evidence for immune surveillance. *Immune Surveillance. Academic Press, NY:p 439.*

Gorak-Stolinska, P., J. P. Truman, D. M. Kemeny, and A. Noble. 2001. Activation-induced cell death of human T-cell subsets is mediated by Fas and granzyme B but is independent of TNF-alpha. *J Leukoc Biol 70:756.*

Gorer, P. A. 1936. The detection of a hereditary antigenic difference in the blood of mice by means of human group A serum. *J Genetics 32:6.*

Gorman, K. C., K. P. Kane, and W. R. Clark. 1987. Target cell recognition structures in LDCC and ODCC. *J Immunol 138:1014.*

Gorman, K., C. C. Liu, A. Blakely, J. D. Young, B. E. Torbett, and W. R. Clark. 1988. Cloned cytotoxic T lymphocytes as target cells. II. Polarity of lysis revisited. *J Immunol 141:2211.*

Gotch, F. M., D. F. Nixon, N. Alp, A. J. McMichael, and L. K. Borysiewicz. 1990. High frequency of memory and effector gag specific cytotoxic T lymphocytes in HIV seropositive individuals. *Int Immunol 2:707.*

Gould, D. S., and H. Auchincloss, Jr. 1999. Direct and indirect recognition: the role of MHC antigens in graft rejection. *Immunol Today 20:77.*

Govaertz, A. 1960. Cellular antibodies in kidney homotransplantation. *J Immunol 85:516.*

Grakoui, A., S. K. Bromley, C. Sumen, M. M. Davis, A. S. Shaw, P. M. Allen, and M. L. Dustin. 1999. The immunological synapse: a molecular machine controlling T cell activation. *Science 285:221.*

Granger, G. A., and W. P. Kolb. 1968. Lymphocyte in vitro cytotoxicity: mechanisms of immune and non-immune small lymphocyte mediated target L cell destruction. *J Immunol 101:111.*

Grant, M. D., F. M. Smaill, and K. L. Rosenthal. 1993. Lysis of CD4+ lymphocytes by non-HLA-restricted cytotoxic T lymphocytes from HIV-infected individuals. *Clin Exp Immunol 93:356.*

Grant, M. D., F. M. Smail, and K. L. Rosenthal. 1994. Cytotoxic T-lymphocytes that kill autologous CD4+ lymphocytes are associated with CD4+ lymphocyte depletion in HIV-1 infection. *J Acquir Immune Defic Syndr 7:571.*

Graubert, T. A., J. H. Russell, and T. J. Ley. 1996. The role of granzyme B in murine models of acute graft-versus-host disease and graft rejection. *Blood 87:1232.*

Gravekamp, C., D. Santoli, R. Vreugdenhil, J. G. Collard, and R. L. Bolhuis. 1987. Efforts to produce human cytotoxic T-cell hybridomas by electrofusion and PEG fusion. *Hybridoma 6:121.*

Gray, L. S., J. R. Gnarra, and V. H. Engelhard. 1987. Demonstration of a calcium influx in cytolytic T lymphocytes in response to target cell binding. *J Immunol 138:63.*

Grayson, J. M., A. J. Zajac, J. D. Altman, and R. Ahmed. 2000. Cutting edge: increased expression of Bcl-2 in antigen-specific memory CD8+ T cells. *J Immunol 164:3950.*

Grayson, J. M., K. Murali-Krishna, J. D. Altman, and R. Ahmed. 2001. Gene expression in antigen-specific CD8+ T cells during viral infection. *J Immunol 166:795.*

Grazia Cifone, M., R. Giacomelli, G. Famularo, R. Paolini, C. Danese, T. Napolitano, A. Procopio, A. M. Perego, A. Santoni, and G. Tonietti. 1990. Natural killer activity and antibody-dependent cellular cytotoxicity in progressive systemic sclerosis. *Clin Exp Immunol 80:360.*

Green, H., and B. Goldberg. 1960. The action of antibody and complement on mammalian cells. *Ann N Y Acad Sci 87:352.*

Green, W. R., Z. K. Ballas, and C. S. Henney. 1978. Studies on the mechanism of lymphocyte-mediated cytolysis. XI. The role of lectin in lectin-dependent cell-mediated cytotoxicity. *J Immunol 121:1566.*

Green, D., and C. F. Ware. 1997. Fas ligand: privilege and peril. *Proc Natl Acad Sci U S A 94:5986.*

Greenberg, P. D. 1991. Adoptive T cell therapy of tumors: mechanisms operative in the recognition and elimination of tumor cells. *Adv Immunol 49:281.*

Gregory, S., and M. Kern. 1978. Adenosine and adenine nucleotides are mitogenic for mouse thymocytes. *Biochem Biophys Res Commun 83:1111.*

Grell, M., E. Douni, H. Wajant, M. Lohden, M. Clauss, B. Maxeiner, S. Georgopoulos, W. Lesslauer, G. Kollias, K. Pfizenmaier, and e. al. 1995. The transmembrane form of tumor necrosis factor is the prime activating ligand of the 80 kDa tumor necrosis factor receptor. *Cell 83:793.*

Griffith, T. S., T. Brunner, S. M. Fletcher, D. R. Green, and T. A. Ferguson. 1995. Fas ligand-induced apoptosis as a mechanism of immune privilege. *Science 270:1189.*

Griffiths, G. M., and C. Mueller. 1991. Expression of perforin and granzymes in vivo: potential diagnostic markers for activated cytotoxic cells. *Immunol Today 12:415.*

Griffiths, G. M., R. Namikawa, C. Mueller, C. C. Liu, J. D. Young, M. Billingham, and I. Weissman. 1991. Granzyme A and perforin as markers for rejection in cardiac transplantation. *Eur J Immunol 21:687.*

Griffiths, G. 1995. The cell biology of CTL killing. *Curr Opin Immunol 7:343.*

Griffiths, G. M., and Y. Argon. 1995. Structure and biogenesis of lytic granules. *Curr Top Microbiol Immunol 198:39.*

Grimm, E., Z. Price, and B. Bonavida. 1979. Studies on the induction and expression of T cell-mediated immunity. VIII. Effector-target junctions and target cell membrane disruption during cytolysis. *Cell Immunol 46:77.*

Grimm, E. A., and B. Bonavida. 1979. Studies on the induction and expression of T cell-mediated immunity. IX. Activation of alloimmune memory lymphocytes into specific secondary CTL by syngeneic NAGO-oxidized stimulator cells. *J Immunol 123:2026.*

Grimm, E. A., A. Mazumder, H. Z. Zhang, and S. A. Rosenberg. 1982. Lymphokine-activated killer cell phenomenon. Lysis of natural killer-resistant fresh solid tumor cells by interleukin 2-activated autologous human peripheral blood lymphocytes. *J Exp Med 155:1823.*

Grimm, E. A., K. M. Ramsey, A. Mazumder, D. J. Wilson, J. Y. Djeu, and S. A. Rosenberg. 1983. Lymphokine-activated killer cell phenomenon. II. Precursor phenotype is serologically distinct from peripheral T lymphocytes, memory cytotoxic thymus-derived lymphocytes, and natural killer cells. *J Exp Med 157:884.*

Groh, V., A. Steinle, S. Bauer, and T. Spies. 1998. Recognition of stress-induced MHC molecules by intestinal epithelial gammadelta T cells. *Science 279:1737.*

Groh, V., R. Rhinehart, H. Secrist, S. Bauer, K. H. Grabstein, and T. Spies. 1999. Broad tumor-associated expression and recognition by tumor-derived gamma delta T cells of MICA and MICB. *Proc Natl Acad Sci U S A 96:6879.*

Grohmann, U., R. Bianchi, E. Ayroldi, M. L. Belladonna, D. Surace, M. C. Fioretti, and P. Puccetti. 1997. A tumor-associated and self antigen peptide presented by dendritic cells may induce T cell anergy in vivo, but IL-12 can prevent or revert the anergic state. *J Immunol 158:3593.*

Gronberg, A., M. T. Ferm, J. Ng, C. W. Reynolds, and J. R. Ortaldo. 1988. IFN-gamma treatment of K562 cells inhibits natural killer cell triggering and decreases the susceptibility to lysis by cytoplasmic granules from large granular lymphocytes. *J Immunol 140:4397.*

Groscurth, P., S. Diener, R. Stahel, L. Jost, D. Kagi, and H. Hengartner. 1990. Morphologic analysis of human lymphokine-activated killer (LAK) cells. *Int J Cancer 45:694.*

Gruss, H. J. 1996. Molecular, structural, and biological characteristics of the tumor necrosis factor ligand superfamily. *Int J Clin Lab Res 26:143.*

Gryllis, C., M. A. Wainberg, M. Gornitsky, and B. Brenner. 1990. Diminution of inducible lymphokine-activated killer cell activity in individuals with AIDS-related disorders. *Aids 4:1205.*

Guerder, S., S. R. Carding, and R. A. Flavell. 1995. B7 costimulation is necessary for the activation of the lytic function in cytotoxic T lymphocyte precursors. *J Immunol 155:5167.*

Guilloux, Y., S. Lucas, V. G. Brichard, A. Van Pel, C. Viret, E. De Plaen, F. Brasseur, B. Lethe, F. Jotereau, and T. Boon. 1996. A peptide recognized by human cytolytic T lymphocytes on HLA-A2 melanomas is encoded by an intron sequence of the N-acetylglucosaminyltransferase V gene. *J Exp Med 183:1173.*

Gumperz, J. E., S. Miyake, T. Yamamura, and M. B. Brenner. 2002. Functionally distinct subsets of CD1d-restricted natural killer T cells revealed by CD1d tetramer staining. *J Exp Med 195:625.*

Gussoni, E., G. K. Pavlath, R. G. Miller, M. A. Panzara, M. Powell, H. M. Blau, and L. Steinman. 1994. Specific T cell receptor gene rearrangements at the site of muscle degeneration in Duchenne muscular dystrophy. *J Immunol 153:4798.*

Gwin, J. L., C. Gercel-Taylor, D. D. Taylor, and B. Eisenberg. 1996. Role of LFA-3, ICAM-1, and MHC class I on the sensitivity of human tumor cells to LAK cells. *J Surg Res 60:129.*

Habel, K. 1961. Resistance of polyoma virus immune animals to transplanted polyoma tumors. *Proc Soc Exp Biol Med 106:722.*

Habel, K. 1962. Immunological determinants of polyoma virus oncogenesis. *J Exp Med 115:181.*

Habel, K. 1965. Tumor viruses. *Yale J Biol Med 37:473.*

Habu, S., H. Fukui, K. Shimamura, M. Kasai, Y. Nagai, K. Okumura, and N. Tamaoki. 1981. In vivo effects of anti-asialo GM1. I. Reduction of NK activity and enhancement of transplanted tumor growth in nude mice. *J Immunol 127:34.*

Hahn, S., R. Gehri, and P. Erb. 1995. Mechanism and biological significance of CD4-mediated cytotoxicity. *Immunol Rev 146:57.*

Hahne, M., D. Rimoldi, M. Schroter, P. Romero, M. Schreier, L. E. French, P. Schneider, T. Bornand, A. Fontana, D. Lienard, J. Cerottini, and J. Tschopp. 1996. Melanoma cell expression of Fas(Apo-1/CD95) ligand: implications for tumor immune escape. *Science 274:1363.*

Haliotis, T., J. K. Ball, D. Dexter, and J. C. Roder. 1985. Spontaneous and induced primary oncogenesis in natural killer (NK)-cell-deficient beige mutant mice. *Int J Cancer 35:505.*

Hall, J. G. 1967. Studies of the cells in the afferent and efferent lymph of lymph nodes draining the site of skin homografts. *J Exp Med 125:737.*

Haller, O., and H. Wigzell. 1977. Suppression of natural killer cell activity with radioactive strontium: effector cells are marrow dependent. *J Immunol 118:1503.*

Hameed, A., K. J. Olsen, M. K. Lee, M. G. Lichtenheld, and E. R. Podack. 1989. Cytolysis by Ca-permeable transmembrane channels. Pore formation causes extensive DNA degradation and cell lysis. *J Exp Med 169:765.*

Hammond, K., W. Cain, I. van Driel, and D. Godfrey. 1998. Three day neonatal thymectomy selectively depletes NK1.1+ T cells. *Int Immunol 10:1491.*

Hammond, K. J., S. B. Pelikan, N. Y. Crowe, E. Randle-Barrett, T. Nakayama, M. Taniguchi, M. J. Smyth, I. R. van Driel, R. Scollay, A. G. Baxter, and D. I. Godfrey. 1999. NKT cells are phenotypically and functionally diverse. *Eur J Immunol 29:3768.*

Hammond, L. J., F. F. Palazzo, M. Shattock, A. W. Goode, and R. Mirakian. 2001. Thyrocyte targets and effectors of autoimmunity: a role for death receptors? *Thyroid 11:919.*

Hammond-McKibben, D. M., A. Seth, P. S. Nagarkatti, and M. Nagarkatti. 1995. Characterization of factors regulating successful immunotherapy using a tumor-specific cytotoxic T lymphocyte clone: role of interleukin-2, cycling pattern of lytic activity and adhesion molecules. *Int J Cancer 60:828.*

Hanke, T., and D. H. Raulet. 2001. Cumulative inhibition of NK cells and T cells resulting from engagement of multiple inhibitory Ly49 receptors. *J Immunol 166:3002.*

Hanninen, A., and L. C. Harrison. 2000. Gamma delta T cells as mediators of mucosal tolerance: the autoimmune diabetes model. *Immunol Rev 173:109.*

Hanon, E., J. C. Stinchcombe, M. Saito, B. E. Asquith, G. P. Taylor, Y. Tanaka, J. N. Weber, G. M. Griffiths, and C. R. Bangham. 2000. Fratricide among CD8(+) T lymphocytes naturally infected with human T cell lymphotropic virus type I. *Immunity 13:657.*

Hardy, D. A., N. R. Ling, J. Wallin, and T. Aviet. 1970. Destruction of lymphoid cells by activated human lymphocytes. *Nature 227:723.*

Harfast, B., T. Andersson, and P. Perlmann. 1975. Human lymphocyte cytotoxicity against mumps virus-infected target cells. Requirement for non-T cells. *J Immunol 114:1820.*

Harty, J. T., R. D. Schreiber, and M. J. Bevan. 1992. CD8 T cells can protect against an intracellular bacterium in an interferon gamma-independent fashion. *Proc Natl Acad Sci U S A 89:11612.*

Harty, J. T., and M. J. Bevan. 1999. Responses of CD8(+) T cells to intracellular bacteria. *Curr Opin Immunol 11:89.*

Harwell, L., B. Skidmore, P. Marrack, and J. Kappler. 1980. Concanavalin A-inducible, interleukin-2-producing T cell hybridoma. *J Exp Med 152:893.*

Haskins, K., R. Kubo, J. White, M. Pigeon, J. Kappler, and P. Marrack. 1983. The major histocompatibility complex-restricted antigen receptor on T cells. I. Isolation with a monoclonal antibody. *J Exp Med 157:1149.*

Haskins, K., and M. McDuffie. 1990. Acceleration of diabetes in young NOD mice with a CD4+ islet-specific T cell clone. *Science 249:1433.*

Hassin, D., R. Fixler, Y. Shimoni, E. Rubinstein, S. Raz, M. S. Gotsman, and Y. Hasin. 1987. Physiological changes induced in cardiac myocytes by cytotoxic T lymphocytes. *Am J Physiol 252:C10.*

Hasui, M., K. Honke, N. Takano, S. Okumura, and K. Ishikawa. 1995. [Increased gamma delta T cells in peripheral blood of patients with severe neurologic impairment]. *No To Hattatsu 27:370.*

Hauschka, R. 1952. Immunological aspects of cancer. A review. *Canc Res 12:615.*

Hauser, W. E., Jr., and V. Tsai. 1986. Acute toxoplasma infection of mice induces spleen NK cells that are cytotoxic for T. gondii in vitro. *J Immunol 136:313.*

Haverstick, D. M., V. H. Engelhard, and L. S. Gray. 1991. Three intracellular signals for cytotoxic T lymphocyte-mediated killing. Independent roles for protein kinase C, Ca2+ influx, and Ca2+ release from internal stores. *J Immunol 146:3306.*

Hayakawa, K., B. T. Lin, and R. R. Hardy. 1992. Murine thymic CD4+ T cell subsets: a subset (Thy0) that secretes diverse cytokines and overexpresses the V beta 8 T cell receptor gene family. *J Exp Med 176:269.*

Hayday, A. C. 2000. [gamma][delta] cells: a right time and a right place for a conserved third way of protection. *Annu Rev Immunol 18:975.*

Hayes, M. P., G. A. Berrebi, and P. A. Henkart. 1989. Induction of target cell DNA release by the cytotoxic T lymphocyte granule protease granzyme A. *J Exp Med 170:933.*

Haynes, B. F., M. J. Telen, L. P. Hale, and S. M. Denning. 1989. CD44 - a molecule involved in leukocyte adherence and T-cell activation. *Immunol Today 10:423.*

Hayry, P., and V. Defendi. 1970. Mixed lymphocyte cultures produce effector cells: model in vitro for allograft rejection. *Science 168:133.*

Hayry, P., and L. C. Andersson. 1974. Generation of T memory cells in one-way mixed lymphocyte culture. II. Anamnestic responses of "secondary" lymphocytes. *Scand J Immunol 3:823.*

Hedrick, S. M., D. I. Cohen, E. A. Nielsen, and M. M. Davis. 1984a. Isolation of cDNA clones encoding T cell-specific membrane-associated proteins. *Nature 308:149.*

Hedrick, S. M., E. A. Nielsen, J. Kavaler, D. I. Cohen, and M. M. Davis. 1984b. Sequence relationships between putative T-cell receptor polypeptides and immunoglobulins. *Nature 308:153.*

Heininger, D., M. Touton, A. K. Chakravarty, and W. R. Clark. 1976. Activation of cytotoxic function in T lymphocytes. *J Immunol 117:2175.*

Heintel, T., M. Sester, M. M. Rodriguez, C. Krieg, U. Sester, R. Wagner, H. W. Pees, B. Gartner, R. Maier, and A. Meyerhans. 2002. The fraction of perforin-expressing HIV-specific CD8 T cells is a marker for disease progression in HIV infection. *Aids 16:1497.*

Helgason, C. D., J. A. Prendergast, G. Berke, and R. C. Bleackley. 1992. Peritoneal exudate lymphocyte and mixed lymphocyte culture hybridomas are cytolytic in the absence of cytotoxic cell proteinases and perforin. *Eur J Immunol 22:3187.*

Hellstrom, I., and K. Hellstrom. 1968. Studies on cellular immunity and its serum-mediated inhibition in Moloney-virus-induced sarcomas. *Int J Cancer 4:587.*

Hellstrom, I., K. E. Hellstrom, G. E. Pierce, and J. P. Yang. 1968. Cellular and humoral immunity to different types of human neoplasms. *Nature 220:1352.*

Hellstrom, I., K. E. Hellstrom, A. H. Bill, G. E. Pierce, and J. P. Yang. 1970. Studies on cellular immunity to human neuroblastoma cells. *Int J Cancer 6:172.*

Hellstrom, K. E., and I. Hellstrom. 1974. Lymphocyte-mediated cytotoxicity and blocking serum activity to tumor antigens. *Adv Immunol 18:209.*

Henderson, R. A., S. C. Watkins, and J. L. Flynn. 1997. Activation of human dendritic cells following infection with Mycobacterium tuberculosis. *J Immunol 159:635.*

Hengel, H., and U. H. Koszinowski. 1997. Interference with antigen processing by viruses. *Curr Opin Immunol 9:470.*

Henkart, P., and R. Blumenthal. 1975. Interaction of lymphocytes with lipid bilayer membranes: a model for lymphocyte-mediated lysis of target cells. *Proc Natl Acad Sci U S A 72:2789.*

Henkart, M. P., and P. A. Henkart. 1982. Lymphocyte mediated cytolysis as a secretory phenomenon. *Adv Exp Med Biol 146:227.*

Henkart, P. A., P. J. Millard, C. W. Reynolds, and M. P. Henkart. 1984. Cytolytic activity of purified cytoplasmic granules from cytotoxic rat large granular lymphocyte tumors. *J Exp Med 160:75.*

Henkart, P. A. 1985. Mechanism of lymphocyte-mediated cytotoxicity. *Annu Rev Immunol 3:31.*

Henkart, P. A. 1994. Lymphocyte-mediated cytotoxicity: two pathways and multiple effector molecules. *Immunity 1:343.*

Henney, C. S. 1971. Quantitation of the cell-mediated immune response. I. The number of cytolytically active mouse lymphoid cells induced by immunization with allogeneic mastocytoma cells. *J Immunol 107:1558.*

Henney, C. S., and M. M. Mayer. 1971. Specific cytolytic activity of lymphocytes: effect of antibodies against complement components C2, C3, and C5. *Cell Immunol 2:702.*

Henney, C. S. 1973. Studies on the mechanism of lymphocyte-mediated cytolysis. II. The use of various target cell markers to study cytolytic events. *J Immunol 110:73.*

Henney, C. S. 1974. Estimation of the size of a T-cell-induced lytic lesion. *Nature 249:456.*

Henney, C. S. 1977. T-Cell-mediated cytolysis: an overview of some current issues. *Contemp Top Immunobiol 7:245.*

Henriques-Pons, A., G. M. Oliveira, M. M. Paiva, A. F. Correa, M. M. Batista, R. C. Bisaggio, C. C. Liu, V. Cotta-De-Almeida, C. M. Coutinho, P. M. Persechini, and T. C. Araujo-Jorge. 2002. Evidence for a perforin-mediated mechanism controlling cardiac inflammation in Trypanosoma cruzi infection. *Int J Exp Pathol 83:67.*

Henry, L., D. Marshall, E. Friedman, G. Dammin, and J. Merrill. 1962. The rejection of skin homografts in normal human subjects. *J Clin Invest 41:420.*

Herberman, R. B., M. E. Nunn, D. H. Lavrin, and R. Asofsky. 1973. Effect of antibody to theta antigen on cell-mediated immunity induced in syngeneic mice by murine sarcoma virus. *J Natl Cancer Inst 51:1509.*

Herberman, R. B., M. E. Nunn, and D. H. Lavrin. 1975a. Natural cytotoxic reactivity of mouse lymphoid cells against syngeneic acid allogeneic tumors. I. Distribution of reactivity and specificity. *Int J Cancer 16:216.*

Herberman, R. B., M. E. Nunn, H. T. Holden, and D. H. Lavrin. 1975b. Natural cytotoxic reactivity of mouse lymphoid cells against syngeneic and allogeneic tumors. II. Characterization of effector cells. *Int J Cancer 16:230.*

Herberman, R. B. 1981. Natural Killer (NK) cells and their possible roles in resistance to disease. *Clin Immunol Rev 1:1.*

Herberman, R. B. 1987. Activation of natural killer (NK) cells and mechanism of their cytotoxic effects. *Adv Exp Med Biol 213:275.*

Hercend, T., S. Meuer, E. L. Reinherz, S. F. Schlossman, and J. Ritz. 1982. Generation of a cloned NK cell line derived from the "null cell" fraction of human peripheral blood. *J Immunol 129:1299.*

Hercend, T., E. L. Reinherz, S. Meuer, S. F. Schlossman, and J. Ritz. 1983. Phenotypic and functional heterogeneity of human cloned natural killer cell lines. *Nature 301:158.*

Hermann, P., D. Blanchard, B. de Saint-Vis, F. Fossiez, C. Gaillard, B. Vanbervliet, F. Briere, J. Banchereau, and J. P. Galizzi. 1993. Expression of a 32-kDa ligand for the CD40 antigen on activated human T lymphocytes. *Eur J Immunol 23:961.*

Herrera, P. L., D. M. Harlan, and P. Vassalli. 2000. A mouse CD8 T cell-mediated acute autoimmune diabetes independent of the perforin and Fas cytotoxic pathways: possible role of membrane TNF. *Proc Natl Acad Sci U S A 97:279.*

Hesketh, T. R., G. A. Smith, M. D. Houslay, G. B. Warren, and J. C. Metcalfe. 1977. Is an early calcium flux necessary to stimulate lymphocytes? *Nature 267:490.*

Hesketh, T. R., G. A. Smith, J. P. Moore, M. V. Taylor, and J. C. Metcalfe. 1983. Free cytoplasmic calcium concentration and the mitogenic stimulation of lymphocytes. *J Biol Chem 258:4876.*

Heusel, J. W., R. L. Wesselschmidt, S. Shresta, J. H. Russell, and T. J. Ley. 1994. Cytotoxic lymphocytes require granzyme B for the rapid induction of DNA fragmentation and apoptosis in allogeneic target cells. *Cell 76:977.*

Hewitt, H. B., E. R. Blake, and A. S. Walder. 1976. A critique of the evidence for active host defence against cancer, based on personal studies of 27 murine tumours of spontaneous origin. *Br J Cancer 33:241.*

Hickman, C. J., J. A. Crim, H. S. Mostowski, and J. P. Siegel. 1990. Regulation of human cytotoxic T lymphocyte development by IL-7. *J Immunol 145:2415.*

Hildreth, J. E., F. M. Gotch, P. D. Hildreth, and A. J. McMichael. 1983. A human lymphocyte-associated antigen involved in cell-mediated lympholysis. *Eur J Immunol 13:202.*

Himeno, K., and H. Hisaeda. 1996. Contribution of 65-kDa heat shock protein induced by gamma and delta T cells to protection against Toxoplasma gondii infection. *Immunol Res 15:258.*

Hinz, T., D. Wesch, K. Friese, A. Reckziegel, B. Arden, and D. Kabelitz. 1994. T cell receptor gamma delta repertoire in HIV-1-infected individuals. *Eur J Immunol 24:3044.*

Hiserodt, J. C., L. J. Britvan, and S. R. Targan. 1982. Characterization of the cytolytic reaction mechanism of the human natural killer (NK) lymphocyte: resolution into binding, programming, and killer cell-independent steps. *J Immunol 129:1782.*

Ho, E. L., L. N. Carayannopoulos, J. Poursine-Laurent, J. Kinder, B. Plougastel, H. R. Smith, and W. M. Yokoyama. 2002. Costimulation of multiple NK cell activation receptors by NKG2D. *J Immunol 169:3667.*

Hodes, R. J., and E. A. Svedmyr. 1970. Specific cytotoxicity of H-2-incompatible mouse lymphocytes following mixed culture in vitro. *Transplantation 9:470.*

Hoffenbach, A., P. Langlade-Demoyen, G. Dadaglio, E. Vilmer, F. Michel, C. Mayaud, B. Autran, and F. Plata. 1989. Unusually high frequencies of HIV-specific cytotoxic T lymphocytes in humans. *J Immunol 142:452.*

Hoffman, R. W., J. A. Bluestone, O. Leo, and S. Shaw. 1985. Lysis of anti-T3-bearing murine hybridoma cells by human allospecific cytotoxic T cell clones and inhibition of that lysis by anti-T3 and anti-LFA-1 antibodies. *J Immunol 135:5.*

Hoffman, R. W., R. R. Quinones, and S. Shaw. 1985. Anti-T3 antibody both activates and inhibits the cytotoxic activity of human T cell clones. *Behring Inst Mitt:30.*

Hoglund, P., J. Mintern, C. Waltzinger, W. Heath, C. Benoist, and D. Mathis. 1999. Initiation of autoimmune diabetes by developmentally regulated presentation of islet cell antigens in the pancreatic lymph nodes. *J Exp Med 189:331.*

Holm, G., and P. Perlmann. 1967. Quantitative studies on phytohaemagglutinin-induced cytotoxicity by human lymphocytes against homologous cells in tissue culture. *Immunology 12:525.*

Holmgren, L., A. Szeles, E. Rajnavolgyi, J. Folkman, G. Klein, I. Ernberg, and K. I. Falk. 1999. Horizontal transfer of DNA by the uptake of apoptotic bodies. *Blood 93:3956.*

Homann, D., L. Teyton, and M. B. Oldstone. 2001. Differential regulation of antiviral T-cell immunity results in stable CD8+ but declining CD4+ T-cell memory. *Nat Med 7:913.*

Hou, S., L. Hyland, K. W. Ryan, A. Portner, and P. C. Doherty. 1994. Virus-specific CD8+ T-cell memory determined by clonal burst size. *Nature 369:652.*

Hou, S., X. Y. Mo, L. Hyland, and P. C. Doherty. 1995. Host response to Sendai virus in mice lacking class II major histocompatibility complex glycoproteins. *J Virol 69:1429.*

Hsu, F. J., C. Benike, F. Fagnoni, T. M. Liles, D. Czerwinski, B. Taidi, E. G. Engleman, and R. Levy. 1996. Vaccination of patients with B-cell lymphoma using autologous antigen-pulsed dendritic cells. *Nat Med 2:52.*

Huang, A. Y., P. Golumbek, M. Ahmadzadeh, E. Jaffee, D. Pardoll, and H. Levitsky. 1994. Role of bone marrow-derived cells in presenting MHC class I-restricted tumor antigens. *Science 264:961.*

Huang, B., M. Eberstadt, E. T. Olejniczak, R. P. Meadows, and S. W. Fesik. 1996. NMR structure and mutagenesis of the Fas (APO-1/CD95) death domain. *Nature 384:638.*

Huang, J. F., Y. Yang, H. Sepulveda, W. Shi, I. Hwang, P. A. Peterson, M. R. Jackson, J. Sprent, and Z. Cai. 1999. TCR-Mediated internalization of peptide-MHC complexes acquired by T cells. *Science 286:952.*

Hubbard, B. B., M. W. Glacken, J. R. Rodgers, and R. R. Rich. 1990. The role of physical forces on cytotoxic T cell-target cell conjugate stability. *J Immunol 144:4129.*

Hudrisier, D., J. Riond, H. Mazarguil, J. E. Gairin, and E. Joly. 2001. Cutting edge: CTLs rapidly capture membrane fragments from target cells in a TCR signaling-dependent manner. *J Immunol 166:3645.*

Huseby, E. S., D. Liggitt, T. Brabb, B. Schnabel, C. Ohlen, and J. Goverman. 2001. A pathogenic role for myelin-specific CD8(+) T cells in a model for multiple sclerosis. *J Exp Med 194:669.*

Hvas, J., J. R. Oksenberg, R. Fernando, L. Steinman, and C. C. Bernard. 1993. Gamma delta T cell receptor repertoire in brain lesions of patients with multiple sclerosis. *J Neuroimmunol 46:225.*

Hyoty, H., M. Hiltunen, M. Knip, M. Laakkonen, P. Vahasalo, J. Karjalainen, P. Koskela, M. Roivainen, P. Leinikki, T. Hovi, and et al. 1995. A prospective study of the role of coxsackie B and other enterovirus infections in the pathogenesis of IDDM. Childhood Diabetes in Finland (DiMe) Study Group. *Diabetes 44:652.*

Idris, A. H., H. R. Smith, L. H. Mason, J. R. Ortaldo, A. A. Scalzo, and W. M. Yokoyama. 1999. The natural killer gene complex genetic locus Chok encodes Ly-49D, a target recognition receptor that activates natural killing. *Proc Natl Acad Sci U S A 96:6330.*

Ikarashi, Y., R. Mikami, A. Bendelac, M. Terme, N. Chaput, M. Terada, T. Tursz, E. Angevin, F. A. Lemonnier, H. Wakasugi, and L. Zitvogel. 2001. Dendritic cell maturation overrules H-2D-mediated natural killer T (NKT) cell inhibition: critical role for B7 in CD1d-dependent NKT cell interferon gamma production. *J Exp Med 194:1179.*

Ikehara, S., R. N. Pahwa, G. Fernandes, C. T. Hansen, and R. A. Good. 1984. Functional T cells in athymic nude mice. *Proc Natl Acad Sci U S A 81:886.*

Illes, Z., T. Kondo, J. Newcombe, N. Oka, T. Tabira, and T. Yamamura. 2000. Differential expression of NK T cell V alpha 24J alpha Q invariant TCR chain in the lesions of multiple sclerosis and chronic inflammatory demyelinating polyneuropathy. *J Immunol 164:4375.*

Imlach, S., C. Leen, J. E. Bell, and P. Simmonds. 2003. Phenotypic analysis of peripheral blood gammadelta T lymphocytes and their targeting by human immunodeficiency virus type 1 in vivo. *Virology 305:415.*

Inaba, M., K. Kurasawa, M. Mamura, K. Kumano, Y. Saito, and I. Iwamoto. 1999. Primed T cells are more resistant to Fas-mediated activation-induced cell death than naive T cells. *J Immunol 163:1315.*

Inngjerdingen, M., B. Damaj, and A. A. Maghazachi. 2001. Expression and regulation of chemokine receptors in human natural killer cells. *Blood 97:367.*

Inverardi, L., J. C. Witson, S. A. Fuad, R. T. Winkler-Pickett, J. R. Ortaldo, and F. H. Bach. 1991. CD3 negative "small agranular lymphocytes" are natural killer cells. *J Immunol 146:4048.*

Ishikawa, H., Y. Shinkai, H. Yagita, C. C. Yue, P. A. Henkart, S. Sawada, H. A. Young, C. W. Reynolds, and K. Okumura. 1989. Molecular cloning of rat cytolysin. *J Immunol 143:3069.*

Ishiura, S., K. Matsuda, H. Koizumi, T. Tsukahara, K. Arahata, and H. Sugita. 1990. Calcium is essential for both the membrane binding and lytic activity of pore-forming protein (perforin) from cytotoxic T-lymphocyte. *Mol Immunol 27:803.*

Isobe, M., G. Russo, F. G. Haluska, and C. M. Croce. 1988. Cloning of the gene encoding the delta subunit of the human T-cell receptor reveals its physical organization within the alpha-subunit locus and its involvement in chromosome translocations in T-cell malignancy. *Proc Natl Acad Sci U S A 85:3933.*

Itoh, K., C. D. Platsoucas, and C. M. Balch. 1988. Autologous tumor-specific cytotoxic T lymphocytes in the infiltrate of human metastatic melanomas. Activation by interleukin 2 and autologous tumor cells, and involvement of the T cell receptor. *J Exp Med 168:1419.*

Itoh, N., A. Imagawa, T. Hanafusa, M. Waguri, K. Yamamoto, H. Iwahashi, M. Moriwaki, H. Nakajima, J. Miyagawa, M. Namba, S. Makino, S. Nagata, N. Kono, and Y. Matsuzawa. 1997. Requirement of Fas for the development of autoimmune diabetes in nonobese diabetic mice. *J Exp Med 186:613.*

Iwai, H., S. Kuma, M. M. Inaba, R. A. Good, T. Yamashita, T. Kumazawa, and S. Ikehara. 1989. Acceptance of murine thyroid allografts by pretreatment of anti-Ia antibody or anti-dendritic cell antibody in vitro. *Transplantation 47:45.*

Iwashiro, M., W. Jinyan, M. Toda, W. Linan, T. Kato, and K. Kuribayashi. 2002. Effective anti-tumor adoptive immunotherapy: utilization of exogenous IL-2-independent cytotoxic T lymphocyte clones. *Int Immunol 14:1459.*

Jacob, J., and D. Baltimore. 1999. Modelling T-cell memory by genetic marking of memory T cells in vivo. *Nature 399:593.*

Jacobelli J., P. G. Andres, J. Boisvert and M. F. Krummel 2004. New views of the immunological synapse: variations in assembly and function. Curr Opin Immunol 16:345-52.

Jacobs, R., G. Hintzen, A. Kemper, K. Beul, S. Kempf, G. Behrens, K. W. Sykora, and R. E. Schmidt. 2001. CD56bright cells differ in their KIR repertoire and cytotoxic features from CD56dim NK cells. *Eur J Immunol 31:3121.*

Jacobson, M., J. Burne, and M. Raff. 1994. Programmed cell death and Bcl-2 protection in the absence of a nucleus. *EMBO J 13:1899.*

Jaeckel, E., M. Manns, and M. Von Herrath. 2002. Viruses and diabetes. *Ann N Y Acad Sci 958:7.*

Jahng, A. W., I. Maricic, B. Pedersen, N. Burdin, O. Naidenko, M. Kronenberg, Y. Koezuka, and V. Kumar. 2001. Activation of natural killer T cells potentiates or prevents experimental autoimmune encephalomyelitis. *J Exp Med 194:1789.*

Jamieson, B. D., and R. Ahmed. 1989. T cell memory. Long-term persistence of virus-specific cytotoxic T cells. *J Exp Med 169:1993.*

Janeway, C. A., Jr., and R. Medzhitov. 2002. Innate immune recognition. *Annu Rev Immunol 20:197.*

Janossy, G., and M. F. Greaves. 1972. Lymphocyte activation. II. discriminating stimulation of lymphocyte subpopulations by phytomitogens and heterologous anti lymphocyte sera. *Clin Exp Immunol 10:525.*

Jans, D. A., V. R. Sutton, P. Jans, C. J. Froelich, and J. A. Trapani. 1999. BCL-2 blocks perforin-induced nuclear translocation of granzymes concomitant with protection against the nuclear events of apoptosis. *J Biol Chem 274:3953.*

Jarpe, A. J., M. R. Hickman, J. T. Anderson, W. E. Winter, and A. B. Peck. 1990. Flow cytometric enumeration of mononuclear cell populations infiltrating the islets of Langerhans in prediabetic NOD mice: development of a model of autoimmune insulitis for type I diabetes. *Reg Immunol 3:305.*

Jason, J., I. Buchanan, L. K. Archibald, O. C. Nwanyanwu, M. Bell, T. A. Green, A. Eick, A. Han, D. Razsi, P. N. Kazembe, H. Dobbie, M. Midathada, and W. R. Jarvis. 2000. Natural T, gammadelta, and NK cells in mycobacterial, Salmonella, and human immunodeficiency virus infections. *J Infect Dis 182:474.*

Jensen, E. R., A. A. Glass, W. R. Clark, E. J. Wing, J. F. Miller, and S. H. Gregory. 1998. Fas (CD95)-dependent cell-mediated immunity to Listeria monocytogenes. *Infect Immun 66:4143.*

Jewell, S. A., G. Bellomo, H. Thor, S. Orrenius, and M. Smith. 1982. Bleb formation in hepatocytes during drug metabolism is caused by disturbances in thiol and calcium ion homeostasis. *Science 217:1257.*

Jiang, S. B., P. M. Persechini, A. Zychlinsky, C. C. Liu, B. Perussia, and J. D. Young. 1988. Resistance of cytolytic lymphocytes to perforin-mediated killing. Lack of correlation with complement-associated homologous species restriction. *J Exp Med 168:2207.*

Jiang, S., P. M. Persechini, B. Perussia, and J. D. Young. 1989. Resistance of cytolytic lymphocytes to perforin-mediated killing. Murine cytotoxic T lymphocytes and human natural killer cells do not contain functional soluble homologous restriction factor or other specific soluble protective factors. *J Immunol 143:1453.*

Jiang, X., S. H. Gregory, and E. J. Wing. 1997. Immune CD8+ T lymphocytes lyse Listeria monocytogenes-infected hepatocytes by a classical MHC class I-restricted mechanism. *J Immunol 158:287.*

Jin, Y., L. Dons, K. Kristensson, and M. E. Rottenberg. 2001. Neural route of cerebral Listeria monocytogenes murine infection: role of immune response mechanisms in controlling bacterial neuroinvasion. *Infect Immun 69:1093.*

Jondal, M., and H. Pross. 1975. Surface markers on human b and t lymphocytes. VI. Cytotoxicity against cell lines as a functional marker for lymphocyte subpopulations. *Int J Cancer 15:596.*

Jongeneel, C. V., S. A. Nedospasov, G. Plaetinck, P. Naquet, and J. C. Cerottini. 1988. Expression of the tumor necrosis factor locus is not necessary for the cytolytic activity of T lymphocytes. *J Immunol 140:1916.*

Jouen-Beades, F., E. Paris, C. Dieulois, J. F. Lemeland, V. Barre-Dezelus, S. Marret, G. Humbert, J. Leroy, and F. Tron. 1997. In vivo and in vitro activation and expansion of gammadelta T cells during Listeria monocytogenes infection in humans. *Infect Immun 65:4267.*

Joyce, S., A. S. Woods, J. W. Yewdell, J. R. Bennink, A. D. De Silva, A. Boesteanu, S. P. Balk, R. J. Cotter, and R. R. Brutkiewicz. 1998. Natural ligand of mouse CD1d1: cellular glycosylphosphatidylinositol. *Science 279:1541.*

Jung, G., C. J. Honsik, R. A. Reisfeld, and H. J. Muller-Eberhard. 1986. Activation of human peripheral blood mononuclear cells by anti-T3: killing of tumor target cells coated with anti-target-anti-T3 conjugates. *Proc Natl Acad Sci U S A 83:4479.*

Jung, G., J. A. Ledbetter, and H. J. Muller-Eberhard. 1987. Induction of cytotoxicity in resting human T lymphocytes bound to tumor cells by antibody heteroconjugates. *Proc Natl Acad Sci U S A 84:4611.*

Jung, G., D. E. Martin, and H. J. Muller-Eberhard. 1987. Induction of cytotoxicity in human peripheral blood mononuclear cells by monoclonal antibody OKT3. *J Immunol 139:639.*

Jurewicz, A., W. E. Biddison, and J. P. Antel. 1998. MHC class I-restricted lysis of human oligodendrocytes by myelin basic protein peptide-specific CD8 T lymphocytes. *J Immunol 160:3056.*

Kabelitz D., W. R. Herzog, B. Zanker and H. Wagner 1985. Human cytotoxic T lymphocytes. I. Limiting-dilution analysis of alloreactive cytotoxic T-lymphocyte precursor frequencies. Scand J Immunol. 22:329-35.

Kabelitz, D., T. Pohl, and K. Pechhold. 1993. Activation-induced cell death (apoptosis) of mature peripheral T lymphocytes. *Immunol Today 14:338.*

Kabelitz, D., and O. Janssen. 1997. Antigen-induced death of T-Lymphocytes. *Front Biosci 2:d61.*

Kaech, S. M., and R. Ahmed. 2001. Memory CD8+ T cell differentiation: initial antigen encounter triggers a developmental program in naive cells. *Nat Immunol 2:415.*

Kaech, S. M., E. J. Wherry, and R. Ahmed. 2002. Effector and memory T-cell differentiation: implications for vaccine development. *Nat Rev Immunol 2:251.*

Kageyama, S., T. J. Tsomides, Y. Sykulev, and H. N. Eisen. 1995. Variations in the number of peptide-MHC class I complexes required to activate cytotoxic T cell responses. *J Immunol 154:567.*

Kagi, D., B. Ledermann, K. Burki, H. Hengartner, and R. M. Zinkernagel. 1994. CD8+ T cell-mediated protection against an intracellular bacterium by perforin-dependent cytotoxicity. *Eur J Immunol 24:3068.*

Kagi, D., B. Ledermann, K. Burki, P. Seiler, B. Odermatt, K. J. Olsen, E. R. Podack, R. M. Zinkernagel, and H. Hengartner. 1994a. Cytotoxicity mediated by T cells and natural killer cells is greatly impaired in perforin-deficient mice. *Nature 369:31.*

Kagi, D., F. Vignaux, B. Ledermann, K. Burki, V. Depraetere, S. Nagata, H. Hengartner, and P. Golstein. 1994b. Fas and perforin pathways as major mechanisms of T cell-mediated cytotoxicity. *Science 265:528.*

Kagi, D., B. Ledermann, K. Burki, R. M. Zinkernagel, and H. Hengartner. 1995a. Lymphocyte-mediated cytotoxicity in vitro and in vivo: mechanisms and significance. *Immunol Rev 146:95.*

Kagi, D., P. Seiler, J. Pavlovic, B. Ledermann, K. Burki, R. M. Zinkernagel, and H. Hengartner. 1995b. The roles of perforin- and Fas-dependent cytotoxicity in protection against cytopathic and noncytopathic viruses. *Eur J Immunol 25:3256.*

Kagi, D., B. Odermatt, P. S. Ohashi, R. M. Zinkernagel, and H. Hengartner. 1996. Development of insulitis without diabetes in transgenic mice lacking perforin-dependent cytotoxicity. *J Exp Med 183:2143.*

Kagi, D., B. Odermatt, P. Seiler, R. M. Zinkernagel, T. W. Mak, and H. Hengartner. 1997. Reduced incidence and delayed onset of diabetes in perforin-deficient nonobese diabetic mice. *J Exp Med 186:989.*

Kagi, D., B. Odermatt, and T. W. Mak. 1999. Homeostatic regulation of CD8+ T cells by perforin. *Eur J Immunol 29:3262.*

Kahaleh, M. B., P. S. Fan, and T. Otsuka. 1999. Gammadelta receptor bearing T cells in scleroderma: enhanced interaction with vascular endothelial cells in vitro. *Clin Immunol 91:188.*

Kakimi, K., L. G. Guidotti, Y. Koezuka, and F. V. Chisari. 2000. Natural killer T cell activation inhibits hepatitis B virus replication in vivo. *J Exp Med 192:921.*

Kalina, M., and G. Berke. 1976. Contact regions of cytotoxic T lymphocyte-target cell conjugates. *Cell Immunol 25:41.*

Kam, C. M., D. Hudig, and J. C. Powers. 2000. Granzymes (lymphocyte serine proteases): characterization with natural and synthetic substrates and inhibitors. *Biochim Biophys Acta 1477:307.*

Kang, Y. H., M. Carl, L. P. Watson, and L. Yaffe. 1987. Immunoultrastructural studies of human NK cells: II. Effector-target cell binding and phagocytosis. *Anat Rec 217:290.*

Kaplan, G., A. Nusrat, M. D. Witmer, I. Nath, and Z. A. Cohn. 1987. Distribution and turnover of Langerhans cells during delayed immune responses in human skin. *J Exp Med 165:763.*

Kaplan, D. H., V. Shankaran, A. S. Dighe, E. Stockert, M. Aguet, L. J. Old, and R. D. Schreiber. 1998. Demonstration of an interferon gamma-dependent tumor surveillance system in immunocompetent mice. *Proc Natl Acad Sci U S A 95:7556.*

Karlhofer, F. M., R. K. Ribaudo, and W. M. Yokoyama. 1992. MHC class I alloantigen specificity of Ly-49+ IL-2-activated natural killer cells. *Nature 358:66.*

Karre, K., H. G. Ljunggren, G. Piontek, and R. Kiessling. 1986. Selective rejection of H-2-deficient lymphoma variants suggests alternative immune defence strategy. *Nature 319:675.*

Kasai, M., M. Iwamori, Y. Nagai, K. Okumura, and T. Tada. 1980. A glycolipid on the surface of mouse natural killer cells. *Eur J Immunol 10:175.*

Kashii, Y., R. Giorda, R. B. Herberman, T. L. Whiteside, and N. L. Vujanovic. 1999. Constitutive expression and role of the TNF family ligands in apoptotic killing of tumor cells by human NK cells. *J Immunol 163:5358.*

Kaspar, A. A., S. Okada, J. Kumar, F. R. Poulain, K. A. Drouvalakis, A. Kelekar, D. A. Hanson, R. M. Kluck, Y. Hitoshi, D. E. Johnson, C. J. Froelich, C. B. Thompson, D. D. Newmeyer, A. Anel, C. Clayberger, and A. M. Krensky. 2001. A distinct pathway of cell-mediated apoptosis initiated by granulysin. *J Immunol 167:350.*

Kasper, L. H., I. A. Khan, K. H. Ely, R. Buelow, and J. C. Boothroyd. 1992. Antigen-specific (p30) mouse CD8+ T cells are cytotoxic against Toxoplasma gondii-infected peritoneal macrophages. *J Immunol 148:1493.*

Kast, W. M., L. Roux, J. Curren, H. J. Blom, A. C. Voordouw, R. H. Meloen, D. Kolakofsky, and C. J. Melief. 1991. Protection against lethal Sendai virus infection by in vivo priming of virus-specific cytotoxic T lymphocytes with a free synthetic peptide. *Proc Natl Acad Sci U S A 88:2283.*

Kastrukoff, L. F., N. G. Morgan, D. Zecchini, R. White, A. J. Petkau, J. Satoh, and D. W. Paty. 1998. A role for natural killer cells in the immunopathogenesis of multiple sclerosis. *J Neuroimmunol 86:123.*

Kataoka T., K. Takaku, J. Magae, N. Shinohara, H. Takayama, S. Kondo and Nagai K. 1994. Acidification is essential for maintaining the structure and function of lytic granules of CTL. Effect of concanamycin A, an inhibitor of vacuolar type H(+)-ATPase, on CTL-mediated cytotoxicity. J Immunol. 153:3938-47.

Kataoka, T., N. Shinohara, H. Takayama, K. Takaku, S. Kondo, S. Yonehara, and K. Nagai. 1996. Concanamycin A, a powerful tool for characterization and estimation of contribution of perforin- and Fas-based lytic pathways in cell-mediated cytotoxicity. *J Immunol 156:3678.*

Kato, Y., Y. Tanaka, F. Miyagawa, S. Yamashita, and N. Minato. 2001. Targeting of tumor cells for human gammadelta T cells by nonpeptide antigens. *J Immunol 167:5092.*

Katz, D. H., M. E. Dorf, and B. Benacerraf. 1976. Control of t-lymphocyte and B-lymphocyte activation by two complementing Ir-GLphi immune response genes. *J Exp Med 143:906.*

Katz, P., A. M. Zaytoun, and J. H. Lee, Jr. 1982. Mechanisms of human cell-mediated cytotoxicity. III. Dependence of natural killing on microtubule and microfilament integrity. *J Immunol 129:2816.*

Kaufman, D. L., M. G. Erlander, M. Clare-Salzler, M. A. Atkinson, N. K. Maclaren, and A. J. Tobin. 1992. Autoimmunity to two forms of glutamate decarboxylase in insulin-dependent diabetes mellitus. *J Clin Invest 89:283.*

Kaufman, D. L., M. Clare-Salzler, J. Tian, T. Forsthuber, G. S. Ting, P. Robinson, M. A. Atkinson, E. E. Sercarz, A. J. Tobin, and P. V. Lehmann. 1993. Spontaneous loss of T-cell tolerance to glutamic acid decarboxylase in murine insulin-dependent diabetes. *Nature 366:69.*

Kaufmann, Y., G. Berke, and Z. Eshhar. 1981. Cytotoxic T lymphocyte hybridomas that mediate specific tumor-cell lysis in vitro. *Proc Natl Acad Sci U S A 78:2502.*
Kaufmann, Y., and G. Berke. 1983. Monoclonal cytotoxic T lymphocyte hybridomas capable of specific killing activity, antigenic responsiveness, and inducible interleukin secretion. *J Immunol 131:50.*
Kaufmann, S. H. 1988. CD8+ T lymphocytes in intracellular microbial infections. *Immunol Today 9:168.*
Kaufmann, S. H., H. R. Rodewald, E. Hug, and G. De Libero. 1988. Cloned Listeria monocytogenes specific non-MHC-restricted Lyt-2+ T cells with cytolytic and protective activity. *J Immunol 140:3173.*
Kaufmann, S. H. 1993. Immunity to intracellular bacteria. *Annu Rev Immunol 11:129.*
Kawai, K., A. Shahinian, T. W. Mak, and P. S. Ohashi. 1996. Skin allograft rejection in CD28-deficient mice. *Transplantation 61:352.*
Kawakami, Y., S. Eliyahu, C. H. Delgado, P. F. Robbins, K. Sakaguchi, E. Appella, J. R. Yannelli, G. J. Adema, T. Miki, and S. A. Rosenberg. 1994a. Identification of a human melanoma antigen recognized by tumor-infiltrating lymphocytes associated with in vivo tumor rejection. *Proc Natl Acad Sci U S A 91:6458.*
Kawakami, Y., S. Eliyahu, K. Sakaguchi, P. F. Robbins, L. Rivoltini, J. R. Yannelli, E. Appella, and S. A. Rosenberg. 1994b. Identification of the immunodominant peptides of the MART-1 human melanoma antigen recognized by the majority of HLA-A2-restricted tumor infiltrating lymphocytes. *J Exp Med 180:347.*
Kawakami, Y., S. Eliyahu, C. H. Delgado, P. F. Robbins, L. Rivoltini, S. L. Topalian, T. Miki, and S. A. Rosenberg. 1994c. Cloning of the gene coding for a shared human melanoma antigen recognized by autologous T cells infiltrating into tumor. *Proc Natl Acad Sci U S A 91:3515.*
Kawano, T., J. Cui, Y. Koezuka, I. Toura, Y. Kaneko, K. Motoki, H. Ueno, R. Nakagawa, H. Sato, E. Kondo, H. Koseki, and M. Taniguchi. 1997. CD1d-restricted and TCR-mediated activation of valpha14 NKT cells by glycosylceramides. *Science 278:1626.*
Kawano, T., T. Nakayama, N. Kamada, Y. Kaneko, M. Harada, N. Ogura, Y. Akutsu, S. Motohashi, T. Iizasa, H. Endo, T. Fujisawa, H. Shinkai, and M. Taniguchi. 1999. Antitumor cytotoxicity mediated by ligand-activated human V alpha24 NKT cells. *Cancer Res 59:5102.*
Kedar, E., B. L. Ikejiri, B. Sredni, B. Bonavida, and R. B. Herberman. 1982. Propagation of mouse cytotoxic clones with characteristics of natural killer (NK) cells. *Cell Immunol 69:305.*
Kedar, E., and D. W. Weiss. 1983. The in vitro generation of effector lymphocytes and their employment in tumor immunotherapy. *Adv Cancer Res 38:171.*
Kedl, R. M., and M. F. Mescher. 1998. Qualitative differences between naive and memory T cells make a major contribution to the more rapid and efficient memory CD8+ T cell response. *J Immunol 161:674.*
Keene, J. A., and J. Forman. 1982. Helper activity is required for the in vivo generation of cytotoxic T lymphocytes. *J Exp Med 155:768.*
Kehren, J., C. Desvignes, M. Krasteva, M. T. Ducluzeau, O. Assossou, F. Horand, M. Hahne, D. Kagi, D. Kaiserlian, and J. F. Nicolas. 1999.

Cytotoxicity is mandatory for CD8(+) T cell-mediated contact hypersensitivity. *J Exp Med 189:779.*

Kelso, A., A. B. Troutt, E. Maraskovsky, N. M. Gough, L. Morris, M. H. Pech, and J. A. Thomson. 1991. Heterogeneity in lymphokine profiles of CD4+ and CD8+ T cells and clones activated in vivo and in vitro. *Immunol Rev 123: 85.*

Kennedy, M. K., M. Glaccum, S. N. Brown, E. A. Butz, J. L. Viney, M. Embers, N. Matsuki, K. Charrier, L. Sedger, C. R. Willis, K. Brasel, P. J. Morrissey, K. Stocking, J. C. Schuh, S. Joyce, and J. J. Peschon. 2000. Reversible defects in natural killer and memory CD8 T cell lineages in interleukin 15-deficient mice. *J Exp Med 191:771.*

Keren, Z., and G. Berke. 1986. Interaction of periodate-oxidized target cells and cytolytic T lymphocytes: a model system of "polyclonal MHC recognition". *Eur J Immunol 16:1049.*

Kern, P. S., M. K. Teng, A. Smolyar, J. H. Liu, J. Liu, R. E. Hussey, R. Spoerl, H. C. Chang, E. L. Reinherz, and J. H. Wang. 1998. Structural basis of CD8 coreceptor function revealed by crystallographic analysis of a murine CD8alphaalpha ectodomain fragment in complex with H-2Kb. *Immunity 9:519.*

Kerr, J. F., A. H. Wyllie, and A. R. Currie. 1972. Apoptosis: a basic biological phenomenon with wide-ranging implications in tissue kinetics. *Br J Cancer 26:239.*

Keyaki, A., K. Kuribayashi, S. Sakaguchi, T. Masuda, J. Yamashita, H. Handa, and E. Nakayama. 1985. Effector mechanisms of syngeneic anti-tumour responses in mice. II. Cytotoxic T lymphocytes mediate neutralization and rejection of radiation-induced leukaemia RL male 1 in the nude mouse system. *Immunology 56:141.*

Khan, I. A., K. A. Smith, and L. H. Kasper. 1988. Induction of antigen-specific parasiticidal cytotoxic T cell splenocytes by a major membrane protein (P30) of Toxoplasma gondii. *J Immunol 141:3600.*

Kidd, J. G. 1950. Experimental necrobiosis - a venture in cytobiology. *Proc Inst Med Chicago 18:50.*

Kidd, J. G., and H. W. Toolan. 1950. The association of lymphocytes with cancer cells undergoing distinctive necrobiosis in resistant and immune hosts. *Am J Path 26:672.*

Kiessling, R., E. Klein, and H. Wigzell. 1975a. "Natural" killer cells in the mouse. I. Cytotoxic cells with specificity for mouse Moloney leukemia cells. Specificity and distribution according to genotype. *Eur J Immunol 5:112.*

Kiessling, R., E. Klein, H. Pross, and H. Wigzell. 1975b. "Natural" killer cells in the mouse. II. Cytotoxic cells with specificity for mouse Moloney leukemia cells. Characteristics of the killer cell. *Eur J Immunol 5:117.*

Kiessling, R., and H. Wigzell. 1979. An analysis of the murine NK cell as to structure, function and biological relevance. *Immunol Rev 44:165.*

Kiessling, R., W. Wei, F. Herrmann, J. Lindencrona, and B. Seliger. 2002. Cellular immunity to the Her-2/neu protooncogene. *Adv Cancer Res 85:101.*

Kikuchi, A., M. Nieda, C. Schmidt, Y. Koezuka, S. Ishihara, Y. Ishikawa, K. Tadokoro, S. Durrant, A. Boyd, T. Juji, and A. Nicol. 2001. In vitro anti-tumour activity of alpha-galactosylceramide-stimulated human invariant Valpha24+NKT cells against melanoma. *Br J Cancer 85:741.*

Kim, J., W. Richter, H. J. Aanstoot, Y. Shi, Q. Fu, R. Rajotte, G. Warnock, and S. Baekkeskov. 1993. Differential expression of GAD65 and GAD67 in human, rat, and mouse pancreatic islets. *Diabetes 42:1799.*

Kim, J. J., M. L. Bagarazzi, N. Trivedi, Y. Hu, K. Kazahaya, D. M. Wilson, R. Ciccarelli, M. A. Chattergoon, K. Dang, S. Mahalingam, A. A. Chalian, M. G. Agadjanyan, J. D. Boyer, B. Wang, and D. B. Weiner. 1997. Engineering of in vivo immune responses to DNA immunization via codelivery of costimulatory molecule genes. *Nat Biotechnol 15:641.*

Kim, Y., S. Kim, A. Kim, H. Yagita, K. Kim, and M. K. Lee. 1999. Apoptosis of pancreatic beta cells detected in accelerated diabetes of NOD mice: no role for Fas-Fas ligand interaction in autoimmune diabetes. *Eur J Immunol 29:455.*

Kim, S., K. Iizuka, H. L. Aguila, I. L. Weissman, and W. M. Yokoyama. 2000. In vivo natural killer cell activities revealed by natural killer cell-deficient mice. *Proc Natl Acad Sci U S A 97:2731.*

Kimura, A. K., and W. R. Clark. 1974. Functional characteristics of T cell receptors during sensitization against histocompatibility antigens in vitro. *Cell Immunol 12:127.*

Kirchhoff, S., W. W. Muller, M. Li-Weber, and P. H. Krammer. 2000. Up-regulation of c-FLIPshort and reduction of activation-induced cell death in CD28-costimulated human T cells. *Eur J Immunol 30:2765.*

Kishi, A., Y. Takamori, K. Ogawa, S. Takano, S. Tomita, M. Tanigawa, M. Niman, T. Kishida, and S. Fujita. 2002. Differential expression of granulysin and perforin by NK cells in cancer patients and correlation of impaired granulysin expression with progression of cancer. *Cancer Immunol Immunother 50:604.*

Kissel, J. T., K. L. Burrow, K. W. Rammohan, and J. R. Mendell. 1991. Mononuclear cell analysis of muscle biopsies in prednisone-treated and untreated Duchenne muscular dystrophy. CIDD Study Group. *Neurology 41:667.*

Kitson, J., T. Raven, Y. P. Jiang, D. V. Goeddel, K. M. Giles, K. T. Pun, C. J. Grinham, R. Brown, and S. N. Farrow. 1996. A death-domain-containing receptor that mediates apoptosis. *Nature 384:372.*

Klagge, I. M., and S. Schneider-Schaulies. 1999. Virus interactions with dendritic cells. *J Gen Virol 80 (Pt 4):823.*

Klarnet, J. P., L. A. Matis, D. E. Kern, M. T. Mizuno, D. J. Peace, J. A. Thompson, P. D. Greenberg, and M. A. Cheever. 1987. Antigen-driven T cell clones can proliferate in vivo, eradicate disseminated leukemia, and provide specific immunologic memory. *J Immunol 138:4012.*

Klein, G., and E. Klein. 1956. Genetic studies of the relationship of tumor-host cells. *Nature 178:1389.*

Klein, E., and H. Sjogren. 1960. Humoral and cellular factors in homograft and isograft immunity against sarcoma cells. *Cancer Research 20:452.*

Klein, G., H. Sjögren, E. Klein, and K. Hellström. 1960. Demonstration of resistance against methylcholanthrene-induced sarcomas in the primary autochthonous host. *Canc Res 20:1561.*

Klein, E., and G. Klein. 1972. Specificity of homograft rejection in vivo, assessed by inoculation of artificially mixed compatible and imcompatible tumor cells. *Cell Immunol 5:201.*

Klein, J. 1975. Biology Of The Mouse Histocompatibility-2 Complex. Springer Verlag, NY.

Klenerman, P., Y. Wu, and R. Phillips. 2002. HIV: current opinion in escapology. *Curr Opin Microbiol 5:408.*

Klinman, D. 2003. Does activation of the innate immune system contribute to the development of rheumatoid arthritis? *Arthritis Rheum 48:590.*

Koch, R. 1891. Fortsetzung der mittheilungen über ein heilmittel gegen tuberculose. *Deutsche Medizinische Wochenschrift 9:101.*

Kodama, T., K. Takeda, O. Shimozato, Y. Hayakawa, M. Atsuta, K. Kobayashi, M. Ito, H. Yagita, and K. Okumura. 1999. Perforin-dependent NK cell cytotoxicity is sufficient for anti-metastatic effect of IL-12. *Eur J Immunol 29:1390.*

Koenig, M., E. P. Hoffman, C. J. Bertelson, A. P. Monaco, C. Feener, and L. M. Kunkel. 1987. Complete cloning of the Duchenne muscular dystrophy (DMD) cDNA and preliminary genomic organization of the DMD gene in normal and affected individuals. *Cell 50:509.*

Koenig, S., T. R. Fuerst, L. V. Wood, R. M. Woods, J. A. Suzich, G. M. Jones, V. F. de la Cruz, R. T. Davey, Jr., S. Venkatesan, B. Moss, and et al. 1990. Mapping the fine specificity of a cytolytic T cell response to HIV-1 nef protein. *J Immunol 145:127.*

Kohler, G., and C. Milstein. 1975. Continuous cultures of fused cells secreting antibody of predefined specificity. *Nature 256:495.*

Koike, T., Y. Itoh, T. Ishii, I. Ito, K. Takabayashi, N. Maruyama, H. Tomioka, and S. Yoshida. 1987. Preventive effect of monoclonal anti-L3T4 antibody on development of diabetes in NOD mice. *Diabetes 36:539.*

Koizumi, H., C. C. Liu, L. M. Zheng, S. V. Joag, N. K. Bayne, J. Holoshitz, and J. D. Young. 1991. Expression of perforin and serine esterases by human gamma/delta T cells. *J Exp Med 173:499.*

Kojima, H., N. Shinohara, S. Hanaoka, Y. Someya-Shirota, Y. Takagaki, H. Ohno, T. Saito, T. Katayama, H. Yagita, K. Okumura, and et al. 1994. Two distinct pathways of specific killing revealed by perforin mutant cytotoxic T lymphocytes. *Immunity 1:357.*

Kojima, H., K. Eshima, H. Takayama, and M. V. Sitkovsky. 1997. Leukocyte function-associated antigen-1-dependent lysis of Fas+ (CD95+/Apo-1+) innocent bystanders by antigen-specific CD8+ CTL. *J Immunol 159:2728.*

Kondo, T., T. Suda, H. Fukuyama, M. Adachi, and S. Nagata. 1997. Essential roles of the Fas ligand in the development of hepatitis. *Nat Med 3:409.*

Kono, K., A. Takahashi, H. Sugai, H. Fujii, and R. Kiessling. 2002. Dendritic cells pulsed with HER-2/neu-derived peptides can induce specific T-cell responses in patients with gastric cancer. *Clin Cancer Res 8:3394.*

Koo, G., A. Hatzfeld. 1980. Antigenic phenotype of mouse natural killer cells. *In, Natural Cell-Mediated Immunity Against Tumors Academic Press, NY.*

Koo, G. C., and J. R. Peppard. 1984. Establishment of monoclonal anti-Nk-1.1 antibody. *Hybridoma 3:301.*

Kopechy, S., and B. Gersh. 1987. Dilated cardiomyopathy and myocarditis: natural history, etiology, clinical manifestations and management. *Curr Probl Cardiol 12:569.*

Kornbluth, J., N. Flomenberg, and B. Dupont. 1982. Cell surface phenotype of a cloned line of human natural killer cells. *J Immunol 129:2831.*

Kos, F. J., and E. G. Engleman. 1995. Requirement for natural killer cells in the induction of cytotoxic T cells. *J Immunol 155:578.*

Kos, F. J., and E. G. Engleman. 1996. Immune regulation: a critical link between NK cells and CTLs. *Immunol Today 17:174.*

Kostense, S., G. S. Ogg, E. H. Manting, G. Gillespie, J. Joling, K. Vandenberghe, E. Z. Veenhof, D. van Baarle, S. Jurriaans, M. R. Klein, and F. Miedema. 2001. High viral burden in the presence of major HIV-specific CD8(+) T cell expansions: evidence for impaired CTL effector function. *Eur J Immunol 31:677.*

Kottilil, S., T. W. Chun, S. Moir, S. Liu, M. McLaughlin, C. W. Hallahan, F. Maldarelli, L. Corey, and A. S. Fauci. 2003. Innate immunity in human immunodeficiency virus infection: effect of viremia on natural killer cell function. *J Infect Dis 187:1038.*

Kountz, S., M. Williams, P. Williams, C. Kapros, and W. Dempster. 1963. Mechanism of rejection of homotransplanted kidneys. *Nature 199:257.*

Koup, R. A., J. T. Safrit, Y. Cao, C. A. Andrews, G. McLeod, W. Borkowsky, C. Farthing, and D. D. Ho. 1994. Temporal association of cellular immune responses with the initial control of viremia in primary human immunodeficiency virus type 1 syndrome. *J Virol 68:4650.*

Kowproski, H., and M. V. Fernandes. 1962. Autosensitization reaction in vitro. Contactual agglutination of sensitized lymph node cells in brain tissue culture accompanied by destruction of glial elements. *J Exp Med 116:467.*

Koyasu, S. 1994. CD3+CD16+NK1.1+B220+ large granular lymphocytes arise from both alpha-beta TCR+CD4-CD8- and gamma-delta TCR+CD4-CD8- cells. *J Exp Med 179:1957.*

Krammer, P. H. 1999. CD95(APO-1/Fas)-mediated apoptosis: live and let die. *Adv Immunol 71:163.*

Krammer, P. H. 2000. CD95's deadly mission in the immune system. *Nature 407:789.*

Krammer, P. H. 2000. CD95's deadly mission in the immune system. *Nature 407:789.*

Kranz, D. M., S. Tonegawa, and H. N. Eisen. 1984. Attachment of an anti-receptor antibody to non-target cells renders them susceptible to lysis by a clone of cytotoxic T lymphocytes. *Proc Natl Acad Sci U S A 81:7922.*

Kranz, D. M., and H. N. Eisen. 1987. Resistance of cytotoxic T lymphocytes to lysis by a clone of cytotoxic T lymphocytes. *Proc Natl Acad Sci U S A 84:3375.*

Krensky, A. M. 2000. Granulysin: a novel antimicrobial peptide of cytolytic T lymphocytes and natural killer cells. *Biochem Pharmacol 59:317.*

Kripke, M. L., and M. S. Fisher. 1976. Immunologic parameters of ultraviolet carcinogenesis. *J Natl Cancer Inst 57:211.*

Kronenberg, M., M. M. Davis, P. W. Early, L. E. Hood, and J. D. Watson. 1980. Helper and killer T cells do not express B cell immunoglobulin joining and constant region gene segments. *J Exp Med 152:1745.*

Krummel, M. F., and J. P. Allison. 1996. CTLA-4 engagement inhibits IL-2 accumulation and cell cycle progression upon activation of resting T cells. *J Exp Med 183:2533.*

Kubota, A., S. Kubota, S. Lohwasser, D. Mager, F. Takei. 1999. Diversity of NK cell receptor repertoire in adult and neonatal mice. *j immunol 163:212.*

Kumagai, K., K. Takeda, W. Hashimoto, S. Seki, K. Ogasawara, R. Anzai, M. Takahashi, M. Sato, and H. Rikiishi. 1997. Interleukin-12 as an inducer of cytotoxic effectors in anti-tumor immunity. *Int Rev Immunol 14:229.*

Kumar, V., E. Luevano, and M. Bennett. 1979. Hybrid resistance to EL-4 lymphoma cells. I. Characterization of natural killer cells that lyse EL-4 cells and their distinction from marrow-dependent natural killer cells. *J Exp Med 150:531.*

Kumar, S., and R. L. Tarleton. 1998. The relative contribution of antibody production and CD8+ T cell function to immune control of Trypanosoma cruzi. *Parasite Immunol 20:207.*

Kumar, J., S. Okada, C. Clayberger, and A. M. Krensky. 2001. Granulysin: a novel antimicrobial. *Expert Opin Investig Drugs 10:321.*

Kung, P., G. Goldstein, E. L. Reinherz, and S. F. Schlossman. 1979. Monoclonal antibodies defining distinctive human T cell surface antigens. *Science 206:347.*

Kunzmann, V., E. Bauer, J. Feurle, F. Weissinger, H. P. Tony, and M. Wilhelm. 2000. Stimulation of gammadelta T cells by aminobisphosphonates and induction of antiplasma cell activity in multiple myeloma. *Blood 96:384.*

Kupfer, A., G. Dennert, and S. J. Singer. 1983. Polarization of the Golgi apparatus and the microtubule-organizing center within cloned natural killer cells bound to their targets. *Proc Natl Acad Sci U S A 80:7224.*

Kupfer, A., G. Dennert, and S. J. Singer. 1985. The reorientation of the Golgi apparatus and the microtubule-organizing center in the cytotoxic effector cell is a prerequisite in the lysis of bound target cells. *J Mol Cell Immunol 2:37.*

Kupfer A. and H. Kupfer 2003. Imaging immune cell interactions and functions: SMACs and the Immunological Synapse. Semin Immunol.15:295-300.

Kupfer, A., S. J. Singer, and G. Dennert. 1986. On the mechanism of unidirectional killing in mixtures of two cytotoxic T lymphocytes. Unidirectional polarization of cytoplasmic organelles and the membrane-associated cytoskeleton in the effector cell. *J Exp Med 163:489.*

Kupfer, A., and S. J. Singer. 1989. Cell biology of cytotoxic and helper T cell functions: immunofluorescence microscopic studies of single cells and cell couples. *Annu Rev Immunol 7:309.*

Kupiec-Weglinski, J. W., J. M. Austyn, and P. J. Morris. 1988. Migration patterns of dendritic cells in the mouse. Traffic from the blood, and T cell-dependent and -independent entry to lymphoid tissues. *J Exp Med 167:632.*

Kuppers, R. C., and C. S. Henney. 1977a. Studies on the mechanism of lymphocyte-mediated cytolysis. IX. Relationships between antigen recognition and lytic expression in killer T cells. *J Immunol 118:71.*

Kuppers, R. C., and C. S. Henney. 1977b. The effects of neuraminidase and galactose oxidase on murine lymphocytes. I. Evidence for the differential delivery of signal(s) leading to cell proliferation and the differentiation of cytotoxic T cells. *J Immunol 119:2163.*

Kuribayashi, K., S. Gillis, D. E. Kern, and C. S. Henney. 1981. Murine NK cell cultures: effects of interleukin-2 and interferon on cell growth and cytotoxic reactivity. *J Immunol 126:2321.*

Kurlander, R. J., S. M. Shawar, M. L. Brown, and R. R. Rich. 1992. Specialized role for a murine class I-b MHC molecule in prokaryotic host defenses. *Science 257:678.*

Kutsch, O., T. Vey, T. Kerkau, T. Hunig, and A. Schimpl. 2002. HIV type 1 abrogates TAP-mediated transport of antigenic peptides presented by MHC class I. Transporter associated with antigen presentation. *AIDS Res Hum Retroviruses 18:1319.*

Kwon, B. S., M. Wakulchik, C. C. Liu, P. M. Persechini, J. A. Trapani, A. K. Haq, Y. Kim, and J. D. Young. 1989. The structure of the mouse lymphocyte pore-forming protein perforin. *Biochem Biophys Res Commun 158:1.*

Ladel, C. H., I. E. Flesch, J. Arnoldi, and S. H. Kaufmann. 1994. Studies with MHC-deficient knock-out mice reveal impact of both MHC I- and MHC II-dependent T cell responses on Listeria monocytogenes infection. *J Immunol 153:3116.*

Lafarge, X., P. Merville, M. C. Cazin, F. Berge, L. Potaux, J. F. Moreau, and J. Dechanet-Merville. 2001. Cytomegalovirus infection in transplant recipients resolves when circulating gammadelta T lymphocytes expand, suggesting a protective antiviral role. *J Infect Dis 184:533.*

Lafferty, K. J., A. Bootes, G. Dart, and D. W. Talmage. 1976. Effect of organ culture on the survival of thyroid allografts in mice. *Transplantation 22:138.*

Lafferty, K. J., S. J. Prowse, C. J. Simeonovic, and H. S. Warren. 1983. Immunobiology of tissue transplantation: a return to the passenger leukocyte concept. *Annu Rev Immunol 1:143.*

Lalvani, A., R. Brookes, R. J. Wilkinson, A. S. Malin, A. A. Pathan, P. Andersen, H. Dockrell, G. Pasvol, and A. V. Hill. 1998. Human cytolytic and interferon gamma-secreting CD8+ T lymphocytes specific for Mycobacterium tuberculosis. *Proc Natl Acad Sci U S A 95:270.*

Lancki, D. W., B. P. Kaper, and F. W. Fitch. 1989. The requirements for triggering of lysis by cytolytic T lymphocyte clones. II. Cyclosporin A inhibits TCR-mediated exocytosis by only selectively inhibits TCR-mediated lytic activity by cloned CTL. *J Immunol 142:416.*

Lancki, D. W., C. S. Hsieh, and F. W. Fitch. 1991. Mechanisms of lysis by cytotoxic T lymphocyte clones. Lytic activity and gene expression in cloned antigen-specific CD4+ and CD8+ T lymphocytes. *J Immunol 146:3242.*

Landsteiner, K., and M. Chase. 1942. Experiments on transfer of cutaneous sensitivity to simple compounds. *Proc Soc Exp Biol Med 49:688.*

Lane, P., A. Traunecker, S. Hubele, S. Inui, A. Lanzavecchia, and D. Gray. 1992. Activated human T cells express a ligand for the human B cell-associated antigen CD40 which participates in T cell-dependent activation of B lymphocytes. *Eur J Immunol 22:2573.*

Lanier, L. L., A. M. Le, C. I. Civin, M. R. Loken, and J. H. Phillips. 1986. The relationship of CD16 (Leu-11) and Leu-19 (NKH-1) antigen expression on human peripheral blood NK cells and cytotoxic T lymphocytes. *J Immunol 136:4480.*

Lanier, L. L., R. Testi, J. Bindl, and J. H. Phillips. 1989. Identity of Leu-19 (CD56) leukocyte differentiation antigen and neural cell adhesion molecule. *J Exp Med 169:2233.*

Lanier, L. L., R. Testi, J. Bindl, and J. H. Phillips. 1989a. Identity of Leu-19 (CD56) leukocyte differentiation antigen and neural cell adhesion molecule. *J Exp Med 169:2233.*

Lanier, L. L., G. Yu, and J. H. Phillips. 1989b. Co-association of CD3 zeta with a receptor (CD16) for IgG Fc on human natural killer cells. *Nature 342:803.*

Lanier, L. L. 1998. NK cell receptors. *Annu Rev Immunol 16:359.*

Lanier, L. L., B. Corliss, J. Wu, and J. H. Phillips. 1998. Association of DAP12 with activating CD94/NKG2C NK cell receptors. *Immunity 8:693.*

Lanier, L. L. 2000. Turning on natural killer cells. *J Exp Med 191:1259.*

Lanier, L. L., and A. B. Bakker. 2000. The ITAM-bearing transmembrane adaptor DAP12 in lymphoid and myeloid cell function. *Immunol Today 21:611.*

Lanier, L. L. 2001. On guard--activating NK cell receptors. *Nat Immunol 2:23.*

Lanzavecchia, A. 1986. Is the T-cell receptor involved in T-cell killing? *Nature 319:778.*

Lanzavecchia, A., and U. D. Staerz. 1987. Lysis of nonnucleated red blood cells by cytotoxic T lymphocytes. *Eur J Immunol 17:1073.*

Lanzavecchia, A., and F. Sallusto. 2000. From synapses to immunological memory: the role of sustained T cell stimulation. *Curr Opin Immunol 12:92.*

Lanzavecchia, A., and F. Sallusto. 2002. Progressive differentiation and selection of the fittest in the immune response. *Nat Rev Immunol 2:982.*

Laochumroonvorapong, P., J. Wang, C. C. Liu, W. Ye, A. L. Moreira, K. B. Elkon, V. H. Freedman, and G. Kaplan. 1997. Perforin, a cytotoxic molecule which mediates cell necrosis, is not required for the early control of mycobacterial infection in mice. *Infect Immun 65:127.*

Larsen, C. P., H. Barker, P. J. Morris, and J. M. Austyn. 1990a. Failure of mature dendritic cells of the host to migrate from the blood into cardiac or skin allografts. *Transplantation 50:294.*

Larsen, C. P., P. J. Morris, and J. M. Austyn. 1990b. Migration of dendritic leukocytes from cardiac allografts into host spleens. A novel pathway for initiation of rejection. *J Exp Med 171:307.*

Lau, L. L., B. D. Jamieson, T. Somasundaram, and R. Ahmed. 1994. Cytotoxic T-cell memory without antigen. *Nature 369:648.*

Lauvau, G., S. Vijh, P. Kong, T. Horng, K. Kerksiek, N. Serbina, R. A. Tuma, and E. G. Pamer. 2001. Priming of memory but not effector CD8 T cells by a killed bacterial vaccine. *Science 294:1735.*

Lazetic, S., C. Chang, J. P. Houchins, L. L. Lanier, and J. H. Phillips. 1996. Human natural killer cell receptors involved in MHC class I recognition are disulfide-linked heterodimers of CD94 and NKG2 subunits. *J Immunol 157:4741.*

Lebow, L. T., C. C. Stewart, A. S. Perelson, and B. Bonavida. 1986. Analysis of lymphocyte-target conjugates by flow cytometry. I. Discrimination between killer and non-killer lymphocytes bound to targets and sorting of conjugates containing one or multiple lymphocytes. *Nat Immun Cell Growth Regul 5:221.*

Lechler, R. I., and J. R. Batchelor. 1982. Restoration of immunogenicity to passenger cell-depleted kidney allografts by the addition of donor strain dendritic cells. *J Exp Med 155:31.*

Lee, S. C., and C. S. Raine. 1989. Multiple sclerosis: oligodendrocytes in active lesions do not express class II major histocompatibility complex molecules. *J Neuroimmunol 25:261.*

Lee, R. K., J. Spielman, D. Y. Zhao, K. J. Olsen, and E. R. Podack. 1996. Perforin, Fas ligand, and tumor necrosis factor are the major cytotoxic molecules used by lymphokine-activated killer cells. *J Immunol 157:1919.*

Lee, R. S., M. J. Grusby, T. M. Laufer, R. Colvin, L. H. Glimcher, and H. Auchincloss, Jr. 1997. CD8+ effector cells responding to residual class I antigens, with help from CD4+ cells stimulated indirectly, cause rejection of "major histocompatibility complex-deficient" skin grafts. *Transplantation 63:1123.*

Lee, N., M. Llano, M. Carretero, A. Ishitani, F. Navarro, M. Lopez-Botet, and D. E. Geraghty. 1998. HLA-E is a major ligand for the natural killer inhibitory receptor CD94/NKG2A. *Proc Natl Acad Sci U S A 95:5199.*

Lee, S. H., S. Girard, D. Macina, M. Busa, A. Zafer, A. Belouchi, P. Gros, and S. M. Vidal. 2001. Susceptibility to mouse cytomegalovirus is associated with deletion of an activating natural killer cell receptor of the C-type lectin superfamily. *Nat Genet 28:42.*

Lee, K. H., A. D. Holdorf, M. L. Dustin, A. C. Chan, P. M. Allen, and A. S. Shaw. 2002. T cell receptor signaling precedes immunological synapse formation. *Science 295:1539.*

Lee P. P., C. Yee, P. A. Savage, L. Fong, D. Brockstedt, J. S. Weber, D. Johnson, S. Swetter, J. Thompson, P. D. Greenberg, M. Roederer and M. M. Davis 1999. Characterization of circulating T cells specific for tumor-associated antigens in melanoma patients. Nat Med. 5:677-85.

Lee, P. T., K. Benlagha, L. Teyton, and A. Bendelac. 2002a. Distinct functional lineages of human V(alpha)24 natural killer T cells. *J Exp Med 195:637.*

Lee, P. T., A. Putnam, K. Benlagha, L. Teyton, P. A. Gottlieb, and A. Bendelac. 2002b. Testing the NKT cell hypothesis of human IDDM pathogenesis. *J Clin Invest 110:793.*

Leeuwenberg, J. F., H. Spits, W. J. Tax, and P. J. Capel. 1985. Induction of nonspecific cytotoxicity by monoclonal anti-T3 antibodies. *J Immunol 134:3770.*

Lefaucheur, J. P., C. Pastoret, and A. Sebille. 1995. Phenotype of dystrophinopathy in old mdx mice. *Anat Rec 242:70.*

Lefor, A. T., and S. A. Rosenberg. 1991. The specificity of lymphokine-activated killer (LAK) cells in vitro: fresh normal murine tissues are resistant to LAK-mediated lysis. *J Surg Res 50:15.*

LeFrancois, L., S. Olson, and D. Masopust. 1999. A critical role for CD40-CD40 ligand interactions in amplification of the mucosal CD8 T cell response. *J Exp Med 190:1275.*

Legendre, V., C. Boyer, S. Guerder, B. Arnold, G. Hammerling, and A. M. Schmitt-Verhulst. 1999. Selection of phenotypically distinct NK1.1+ T cells upon antigen expression in the thymus or in the liver. *Eur J Immunol 29:2330.*

Lehmann-Grube, F., D. Moskophidis, and J. Lohler. 1988. Recovery from acute virus infection. Role of cytotoxic T lymphocytes in the elimination of lymphocytic choriomeningitis virus from spleens of mice. *Ann N Y Acad Sci 532:238.*

Lenschow, D. J., T. L. Walunas, and J. A. Bluestone. 1996. CD28/B7 system of T cell costimulation. *Annu Rev Immunol 14:233.*

Leo, O., D. H. Sachs, L. E. Samelson, M. Foo, R. Quinones, R. Gress, and J. A. Bluestone. 1986. Identification of monoclonal antibodies specific for the T cell receptor complex by Fc receptor-mediated CTL lysis. *J Immunol 137:3874.*

Leo, O., M. Foo, P. A. Henkart, P. Perez, N. Shinohara, D. M. Segal, and J. A. Bluestone. 1987. Role of accessory molecules in signal transduction of cytolytic T lymphocyte by anti-T cell receptor and anti-Ly-6.2C monoclonal antibodies. *J Immunol 139:3556.*

LeRoy, E. C., C. Black, R. Fleischmajer, S. Jablonska, T. Krieg, T. A. Medsger, Jr., N. Rowell, and F. Wollheim. 1988. Scleroderma (systemic sclerosis): classification, subsets and pathogenesis. *J Rheumatol 15:202.*

Leszczynski, D., M. Laszczynska, J. Halttunen, and P. Hayry. 1987. Renal target structures in acute allograft rejection: a histochemical study. *Kidney Int 31:1311.*

Levine, M. M., J. Galen, E. Barry, F. Noriega, S. Chatfield, M. Sztein, G. Dougan, and C. Tacket. 1996. Attenuated Salmonella as live oral vaccines against typhoid fever and as live vectors. *J Biotechnol 44:193.*

Lewinsohn, D. A., R. A. Lines, and D. M. Lewinsohn. 2002. Human dendritic cells presenting adenovirally expressed antigen elicit Mycobacterium tuberculosis--specific CD8+ T cells. *Am J Respir Crit Care Med 166:843.*

Li, J. H., D. Rosen, D. Ronen, C. K. Behrens, P. H. Krammer, W. R. Clark, and G. Berke. 1998. The regulation of CD95 ligand expression and function in CTL. *J Immunol 161:3943.*

Li, J. H., D. Rosen, P. Sondel, and G. Berke. 2002. Immune privilege and FasL: two ways to inactivate effector cytotoxic T lymphocytes by FasL-expressing cells. *Immunology 105:267.*

Li, B. et al, 1998. Involvement of the Fas/Fas ligand pathway in AICD of mycobacteria-reactive human gamma delta T cells. *J Immunol 161:1558.*

Lichtenheld, M. G., and E. R. Podack. 1989. Structure of the human perforin gene. A simple gene organization with interesting potential regulatory sequences. *J Immunol 143:4267.*

Lieberman, J., J. A. Fabry, M. C. Kuo, P. Earl, B. Moss, and P. R. Skolnik. 1992. Cytotoxic T lymphocytes from HIV-1 seropositive individuals recognize immunodominant epitopes in Gp160 and reverse transcriptase. *J Immunol 148:2738.*

Lieberman, J. 2003. The ABCs of granule-mediated cytotoxicity: new weapons in the arsenal. *Nat Rev Immunol 3:361.*

Liepins, A., R. B. Faanes, J. Lifter, Y. S. Choi, and E. De Harven. 1977. Ultrastructural changes during T-lymphocyte-mediated cytolysis. *Cell Immunol 28:109.*

Lin, Y. L., and B. A. Askonas. 1981. Biological properties of an influenza A virus-specific killer T cell clone. Inhibition of virus replication in vivo and induction of delayed-type hypersensitivity reactions. *J Exp Med 154:225.*

Lin, M. T., S. A. Stohlman, and D. R. Hinton. 1997. Mouse hepatitis virus is cleared from the central nervous systems of mice lacking perforin-mediated cytolysis. *J Virol 71:383.*

Lisziewicz, J., D. I. Gabrilovich, G. Varga, J. Xu, P. D. Greenberg, S. K. Arya, M. Bosch, J. P. Behr, and F. Lori. 2001. Induction of potent human immunodeficiency virus type 1-specific T-cell-restricted immunity by genetically modified dendritic cells. *J Virol 75:7621.*

Little, C. C. 1956. *The genetics of tumor transplantation.* Dover Publications Inc., New York.

Liu, C. C., S. Jiang, P. M. Persechini, A. Zychlinsky, Y. Kaufmann, and J. D. Young. 1989. Resistance of cytolytic lymphocytes to perforin-mediated killing. Induction of resistance correlates with increase in cytotoxicity. *J Exp Med 169:2211.*

Liu L., A. Chahroudi , G. Silvestri, M. E. Wernett, W. J. Kaiser, J. T. Safrit, A. Komoriya, J. D. Altman, B. Z. Packard and M. B. Feinberg 2002 Visualization and quantification of T cell-mediated cytotoxicity using cell-permeable fluorogenic caspase substrates. Nat Med. 8:185-9.

Liu, Y., A. Mullbacher, and P. Waring. 1989. Natural killer cells and cytotoxic T cells induce DNA fragmentation in both human and murine target cells in vitro. *Scand J Immunol 30:31.*

Liu, A. Y., E. P. Miskovsky, P. E. Stanhope, and R. F. Siliciano. 1992. Production of transmembrane and secreted forms of tumor necrosis factor (TNF)-alpha by HIV-1-specific CD4+ cytolytic T lymphocyte clones. Evidence for a TNF-alpha-independent cytolytic mechanism. *J Immunol 148:3789.*

Liu, C. C., C. M. Walsh, N. Eto, W. R. Clark, and J. D. Young. 1995. Morphologic and functional characterization of perforin-deficient lymphokine-activated killer cells. *J Immunol 155:602.*

Liu, A. N., A. Z. Mohammed, W. R. Rice, D. T. Fiedeldey, J. S. Liebermann, J. A. Whitsett, T. J. Braciale, and R. I. Enelow. 1999. Perforin-independent CD8(+) T-cell-mediated cytotoxicity of alveolar epithelial cells is preferentially mediated by tumor necrosis factor-alpha: relative insensitivity to Fas ligand. *Am J Respir Cell Mol Biol 20:849.*

Liu, Z. X., S. Govindarajan, S. Okamoto, and G. Dennert. 2000. NK cells cause liver injury and facilitate the induction of T cell-mediated immunity to a viral liver infection. *J Immunol 164:6480.*

Ljunggren, H. G., and K. Karre. 1990. In search of the 'missing self': MHC molecules and NK cell recognition. *Immunol Today 11:237.*

Lo, W. F., H. Ong, E. S. Metcalf, and M. J. Soloski. 1999. T cell responses to Gram-negative intracellular bacterial pathogens: a role for CD8+ T cells in immunity to Salmonella infection and the involvement of MHC class Ib molecules. *J Immunol 162:5398.*

Lohman, B. L., E. S. Razvi, and R. M. Welsh. 1996. T-lymphocyte downregulation after acute viral infection is not dependent on CD95 (Fas) receptor-ligand interactions. *J Virol 70:8199.*

Lohwasser, S., A. Kubota, M. Salcedo, R. H. Lian, and F. Takei. 2001. The non-classical MHC class I molecule Qa-1(b) inhibits classical MHC class I-restricted cytotoxicity of cytotoxic T lymphocytes. *Int Immunol 13:321.*

Long, E. O. 1999. Regulation of immune responses through inhibitory receptors. *Annu Rev Immunol 17:875.*

Long, E. O. 2002. Versatile signaling through NKG2D. *Nat Immunol 3:1119.*

Lopes, M. F., M. P. Nunes, A. Henriques-Pons, N. Giese, H. C. Morse, 3rd, W. F. Davidson, T. C. Araujo-Jorge, and G. A. DosReis. 1999. Increased susceptibility of Fas ligand-deficient gld mice to Trypanosoma cruzi infection due to a Th2-biased host immune response. *Eur J Immunol 29:81.*

Los, M., C. Stroh, R. U. Janicke, I. H. Engels, and K. Schulze-Osthoff. 2001. Caspases: more than just killers? *Trends Immunol 22:31.*

Lotze, M. T., E. A. Grimm, A. Mazumder, J. L. Strausser, and S. A. Rosenberg. 1981. Lysis of fresh and cultured autologous tumor by human lymphocytes cultured in T-cell growth factor. *Cancer Res 41:4420.*

Lotze, M. T., L. W. Frana, S. O. Sharrow, R. J. Robb, and S. A. Rosenberg. 1985. In vivo administration of purified human interleukin 2. I. Half-life and immunologic effects of the Jurkat cell line-derived interleukin 2. *J Immunol 134:157.*

Lotze, M. T., Y. L. Matory, S. E. Ettinghausen, A. A. Rayner, S. O. Sharrow, C. A. Seipp, M. C. Custer, and S. A. Rosenberg. 1985. In vivo administration of purified human interleukin 2. II. Half life, immunologic effects, and expansion of peripheral lymphoid cells in vivo with recombinant IL 2. *J Immunol 135:2865.*

Lou, Z., D. Jevremovic, D. D. Billadeau, and P. J. Leibson. 2000. A balance between positive and negative signals in cytotoxic lymphocytes regulates the polarization of lipid rafts during the development of cell-mediated killing. *J Exp Med 191:347.*

Low, H. P., M. A. Santos, B. Wizel, and R. L. Tarleton. 1998. Amastigote surface proteins of Trypanosoma cruzi are targets for CD8+ CTL. *J Immunol 160:1817.*

Lowdell, M. W., L. Lamb, C. Hoyle, A. Velardi, and H. G. Prentice. 2001. Non-MHC-restricted cytotoxic cells: their roles in the control and treatment of leukaemias. *Br J Haematol 114:11.*

Loweth, A. C., G. T. Williams, R. F. James, J. H. Scarpello, and N. G. Morgan. 1998. Human islets of Langerhans express Fas ligand and undergo apoptosis in response to interleukin-1beta and Fas ligation. *Diabetes 47:727.*

Lowin, B., F. Beermann, A. Schmidt, and J. Tschopp. 1994. A null mutation in the perforin gene impairs cytolytic T lymphocyte- and natural killer cell-mediated cytotoxicity. *Proc Natl Acad Sci U S A 91:11571.*

Lowrey, D. M., T. Aebischer, K. Olsen, M. Lichtenheld, F. Rupp, H. Hengartner, and E. R. Podack. 1989. Cloning, analysis, and expression of murine perforin 1 cDNA, a component of cytolytic T-cell granules with homology to complement component C9. *Proc Natl Acad Sci U S A 86:247.*

Lucero, M. A., W. H. Fridman, M. A. Provost, C. Billardon, P. Pouillart, J. Dumont, and E. Falcoff. 1981. Effect of various interferons on the spontaneous cytotoxicity exerted by lymphocytes from normal and tumor-bearing patients. *Cancer Res 41:294.*

Luciani, M. F., J. F. Brunet, M. Suzan, F. Denizot, and P. Golstein. 1986. Self-sparing of long-term in vitro-cloned or uncloned cytotoxic T lymphocytes. *J Exp Med 164:962.*

Ludewig, B., S. Ehl, U. Karrer, B. Odermatt, H. Hengartner, and R. M. Zinkernagel. 1998. Dendritic cells efficiently induce protective antiviral immunity. *J Virol 72:3812.*

Lukacher, A. E., V. L. Braciale, and T. J. Braciale. 1984. In vivo effector function of influenza virus-specific cytotoxic T lymphocyte clones is highly specific. *J Exp Med 160:814.*

Lukacs, K., and R. J. Kurlander. 1989. MHC-unrestricted transfer of antilisterial immunity by freshly isolated immune CD8 spleen cells. *J Immunol 143:3731.*

Lundstedt, C. 1969. Interaction between antigenically different cells. Virus-induced cytotoxicity by immune lymphoid cells in vitro. *Acta Pathol Microbiol Scand 75:139.*

Lupetti, R., P. Pisarra, A. Verrecchia, C. Farina, G. Nicolini, A. Anichini, C. Bordignon, M. Sensi, G. Parmiani, and C. Traversari. 1998. Translation of a retained intron in tyrosinase-related protein (TRP) 2 mRNA generates a new cytotoxic T lymphocyte (CTL)-defined and shared human melanoma antigen not expressed in normal cells of the melanocytic lineage. *J Exp Med 188:1005.*

Lusso, P., A. Garzino-Demo, R. W. Crowley, and M. S. Malnati. 1995. Infection of gamma/delta T lymphocytes by human herpesvirus 6: transcriptional induction of CD4 and susceptibility to HIV infection. *J Exp Med 181:1303.*

Lust, J. A., V. Kumar, R. C. Burton, S. P. Bartlett, and M. Bennett. 1981. Heterogeneity of natural killer cells in the mouse. *J Exp Med 154:306.*

Lynch, D. H., M. L. Watson, M. R. Alderson, P. R. Baum, R. E. Miller, T. Tough, M. Gibson, T. Davis-Smith, C. A. Smith, K. Hunter, and et al. 1994. The mouse Fas-ligand gene is mutated in gld mice and is part of a TNF family gene cluster. *Immunity 1:131.*

Lyubchenko, T. A., G. A. Wurth, and A. Zweifach. 2001. Role of calcium influx in cytotoxic T lymphocyte lytic granule exocytosis during target cell killing. *Immunity 15:847.*

MacDonald, H. R., J. C. Cerottini, and K. T. Brunner. 1974. Generation of cytotoxic T lymphocytes in vitro. III. Velocity sedimentation studies of the differentiation and fate of effector cells in long-term mixed leukocyte cultures. *J Exp Med 140:1511.*

MacDonald, H. R., B. Sordat, J. C. Cerottini, and K. T. Brunner. 1975. Generation of cytotoxic T lymphocytes in vitro. IV. Functional activation of memory cells in the absence of DNA synthesis. *J Exp Med 142:622.*

MacDonald, H. R., J. C. Cerottini, J. E. Ryser, J. L. Maryanski, C. Taswell, M. B. Widmer, and K. T. Brunner. 1980. Quantitation and cloning of cytolytic T lymphocytes and their precursors. *Immunol Rev 51:93.*

MacDonald, H., A. Wilson, and R. Radtke. 2001. Notch1 and T-cell development: insights from conditional knockout mice. *Trends Immunol 22:155.*

Macino, B., A. Zambon, G. Milan, A. Cabrelle, M. Ruzzene, A. Rosato, S. Mandruzzato, L. Quintieri, P. Zanovello, and D. Collavo. 1996. CD45 regulates apoptosis induced by extracellular adenosine triphosphate and cytotoxic T lymphocytes. *Biochem Biophys Res Commun 226:769.*

Mackaness, G. B. 1969. The influence of immunologically committed lymphoid cells on macrophage activity in vivo. *J Exp Med 129:973.*

Mackenzie, C. D., P. M. Taylor, and B. A. Askonas. 1989. Rapid recovery of lung histology correlates with clearance of influenza virus by specific CD8+ cytotoxic T cells. *Immunology 67:375.*

Maggi, E., M. G. Giudizi, R. Biagiotti, F. Annunziato, R. Manetti, M. P. Piccinni, P. Parronchi, S. Sampognaro, L. Giannarini, G. Zuccati, and et al. 1994. Th2-like CD8+ T cells showing B cell helper function and reduced cytolytic activity in human immunodeficiency virus type 1 infection. *J Exp Med 180:489.*

Maghazachi, A. A., A. Al-Aoukaty, and T. J. Schall. 1996. CC chemokines induce the generation of killer cells from CD56+ cells. *Eur J Immunol 26:315.*

Malech, H. L., R. K. Root, and J. I. Gallin. 1977. Structural analysis of human neutrophil migration. Centriole, microtubule, and microfilament orientation and function during chemotaxis. *J Cell Biol 75:666.*

Maleckar, J. R., and L. A. Sherman. 1987. The composition of the T cell receptor repertoire in nude mice. *J Immunol 138:3873.*

Mamalaki, C., Y. Tanaka, P. Corbella, P. Chandler, E. Simpson, and D. Kioussis. 1993. T cell deletion follows chronic antigen specific T cell activation in vivo. *Int Immunol 5:1285.*

Mandelboim, O., P. Malik, D. M. Davis, C. H. Jo, J. E. Boyson, and J. L. Strominger. 1999. Human CD16 as a lysis receptor mediating direct natural killer cell cytotoxicity. *Proc Natl Acad Sci U S A 96:5640.*

Manjunath, N., P. Shankar, J. Wan, W. Weninger, M. A. Crowley, K. Hieshima, T. A. Springer, X. Fan, H. Shen, J. Lieberman, and U. H. von Andrian. 2001. Effector differentiation is not prerequisite for generation of memory cytotoxic T lymphocytes. *J Clin Invest 108:871.*

Manning, D. D., N. D. Reed, and C. F. Shaffer. 1973. Maintenance of skin xenografts of widely divergent phylogenetic origin of congenitally athymic (nude) mice. *J Exp Med 138:488.*

Mannoor, M. K., A. Weerasinghe, R. C. Halder, S. Reza, M. Morshed, A. Ariyasinghe, H. Watanabe, H. Sekikawa, and T. Abo. 2001. Resistance to malarial infection is achieved by the cooperation of NK1.1(+) and NK1.1(-) subsets of intermediate TCR cells which are constituents of innate immunity. *Cell Immunol 211:96.*

Mansour, I., C. Doinel, and P. Rouger. 1990. CD16+ NK cells decrease in all stages of HIV infection through a selective depletion of the CD16+CD8+CD3- subset. *AIDS Res Hum Retroviruses 6:1451.*

Maraskovsky, E., K. Brasel, M. Teepe, E. R. Roux, S. D. Lyman, K. Shortman, and H. J. McKenna. 1996. Dramatic increase in the numbers of functionally mature dendritic cells in Flt3 ligand-treated mice: multiple dendritic cell subpopulations identified. *J Exp Med 184:1953.*

Marchalonis, J. J., G. W. Warr, L. A. Santucci, A. Szenberg, R. von Fellenberg, and J. J. Burckhardt. 1980. The immunoglobulin-like T cell receptor--IV. Quantitative cellular assay and partial characterization of a heavy chain cross-reactive with the Fd fragment of serum mu chain. *Mol Immunol 17:985.*

Marincola, F. M., L. Rivoltini, M. L. Salgaller, M. Player, and S. A. Rosenberg. 1996. Differential anti-MART-1/MelanA CTL activity in peripheral blood of HLA-A2 melanoma patients in comparison to healthy donors: evidence of in vivo priming by tumor cells. *J Immunother Emphasis Tumor Immunol 19:266.*

Marrack, P., J. Kappler, and T. Mitchell. 1999. Type I interferons keep activated T cells alive. *J Exp Med 189:521.*

Martin, D., L. S. Zalman, and H. J. Muller-Eberhard. 1988. Induction of expression of cell-surface homologous restriction factor upon anti-CD3 stimulation of human peripheral lymphocytes. *Proc Natl Acad Sci U S A 85:213.*

Martin, S. J., C. P. Reutelingsperger, A. J. McGahon, J. A. Rader, R. C. van Schie, D. M. LaFace, and D. R. Green. 1995. Early redistribution of plasma membrane phosphatidylserine is a general feature of apoptosis regardless of the initiating stimulus: inhibition by overexpression of Bcl-2 and Abl. *J Exp Med 182:1545.*

Martin, S. J., G. P. Amarante-Mendes, L. Shi, T. H. Chuang, C. A. Casiano, G. A. O'Brien, P. Fitzgerald, E. M. Tan, G. M. Bokoch, A. H. Greenberg, and D. R. Green. 1996. The cytotoxic cell protease granzyme B initiates apoptosis in a cell-free system by proteolytic processing and activation of the ICE/CED-3 family protease, CPP32, via a novel two-step mechanism. *Embo J 15:2407.*

Martin, S., D. Wolf-Eichbaum, G. Duinkerken, W. A. Scherbaum, H. Kolb, J. G. Noordzij, and B. O. Roep. 2001. Development of type 1 diabetes despite severe hereditary B-lymphocyte deficiency. *N Engl J Med 345:1036.*

Martini, F., F. Poccia, D. Goletti, S. Carrara, D. Vincenti, G. D'Offizi, C. Agrati, G. Ippolito, V. Colizzi, L. P. Pucillo, and C. Montesano. 2002. Acute human immunodeficiency virus replication causes a rapid and persistent impairment of Vgamma9Vdelta2 T cells in chronically infected patients undergoing structured treatment interruption. *J Infect Dis 186:847.*

Martz, E., S. J. Burakoff, and B. Benacerraf. 1974. Interruption of the sequential release of small and large molecules from tumor cells by low temperature during cytolysis mediated by immune T-cells or complement. *Proc Natl Acad Sci U S A 71:177.*

Martz, E. 1975. Early steps in specific tumor cell lysis by sensitized mouse T lymphocytes. I. Resolution and characterization. *J Immunol 115:261.*

Martz, E., and B. Benacerraf. 1975. T-lymphocyte mediated cytolysis: temperature dependence of killer cell dependent and independent phases and lack of recovery from the lethal hit at low temperatures. *Cell Immunol 20:81.*

Martz, E. 1976a. Sizes of isotopically labeled molecules released during lysis of tumor cells labeled with 51Cr and [14C]nicotinamide. *Cell Immunol 26:313.*

Martz, E. 1976b. Early steps in specific tumor cell lysis by sensitized mouse T lymphocytes. II. Electrolyte permeability increase in the target cell membrane concomitant with programming for lysis. *J Immunol 117:1023.*

Martz, E. 1977. Mechanism of specific tumor-cell lysis by alloimmune T lymphocytes: resolution and characterization of discrete steps in the cellular interaction. *Contemp Top Immunobiol 7:301.*

Martz, E. 1980. Immune T lymphocyte to tumor cell adhesion. Magnesium sufficient, calcium insufficient. *J Cell Biol 84:584.*

Martz, E., and D. M. Howell. 1989. CTL: virus control cells first and cytolytic cells second? DNA fragmentation, apoptosis and the prelytic halt hypothesis. *Immunol Today 10:79.*

Martz, E., and S. R. Gamble. 1992. How do CTL control virus infections? Evidence for prelytic halt of herpes simplex. *Viral Immunol 5:81.*

Masson, D., P. J. Peters, H. J. Geuze, J. Borst, and J. Tschopp. 1990. Interaction of chondroitin sulfate with perforin and granzymes of cytolytic T-cells is dependent on pH. *Biochemistry 29:11229.*

Mastroeni, P., B. Villarreal-Ramos, and C. E. Hormaeche. 1993. Adoptive transfer of immunity to oral challenge with virulent salmonellae in innately susceptible BALB/c mice requires both immune serum and T cells. *Infect Immun 61:3981.*

Matloubian, M., R. J. Concepcion, and R. Ahmed. 1994. CD4+ T cells are required to sustain CD8+ cytotoxic T-cell responses during chronic viral infection. *J Virol 68:8056.*

Matloubian, M., M. Suresh, A. Glass, M. Galvan, K. Chow, J. K. Whitmire, C. M. Walsh, W. R. Clark, and R. Ahmed. 1999. A role for perforin in downregulating T-cell responses during chronic viral infection. *J Virol 73:2527.*

Matsumoto, Y., G. W. McCaughan, D. M. Painter, and G. A. Bishop. 1993. Evidence that portal tract microvascular destruction precedes bile duct loss in human liver allograft rejection. *Transplantation 56:69.*

Matsumoto, G., M. P. Nghiem, N. Nozaki, R. Schmits, and J. M. Penninger. 1998. Cooperation between CD44 and LFA-1/CD11a adhesion receptors in lymphokine-activated killer cell cytotoxicity. *J Immunol 160:5781.*

Matsumoto, G., Y. Omi, U. Lee, T. Nishimura, J. Shindo, and J. M. Penninger. 2000. Adhesion mediated by LFA-1 is required for efficient IL-12-induced NK and NKT cell cytotoxicity. *Eur J Immunol 30:3723.*

Matter, A. 1979. Microcinematographic and electron microscopic analysis of target cell lysis induced by cytotoxic T lymphocytes. *Immunology 36:179.*

Matzinger, P. 1981. A one-receptor view of T-cell behaviour. *Nature 292:497.*

Matzinger, P. 1991. The JAM test. A simple assay for DNA fragmentation and cell death. *J Immunol Methods 145:185.*

Maus, M. V., A. K. Thomas, D. G. Leonard, D. Allman, K. Addya, K. Schlienger, J. L. Riley, and C. H. June. 2002. Ex vivo expansion of polyclonal and antigen-specific cytotoxic T lymphocytes by artificial

APCs expressing ligands for the T-cell receptor, CD28 and 4-1BB. *Nat Biotechnol 20:143.*

Mayer, M. M., C. H. Hammer, D. W. Michaels, and M. L. Shin. 1978. Immunologically mediated membrane damage: the mechanism of complement action and the similarity of lymphocyte-mediated cytotoxicity. *Transplant Proc 10:707.*

Mazumder, A., M. Rosenstein, and S. A. Rosenberg. 1983. Lysis of fresh natural killer-resistant tumor cells by lectin-activated syngeneic and allogeneic murine splenocytes. *Cancer Res 43:5729.*

Mazumder, A., T. J. Eberlein, E. A. Grimm, D. J. Wilson, A. M. Keenan, R. Aamodt, and S. A. Rosenberg. 1984. Phase I study of the adoptive immunotherapy of human cancer with lectin activated autologous mononuclear cells. *Cancer 53:896.*

Mazumder, A., and S. A. Rosenberg. 1984. Successful immunotherapy of natural killer-resistant established pulmonary melanoma metastases by the intravenous adoptive transfer of syngeneic lymphocytes activated in vitro by interleukin 2. *J Exp Med 159:495.*

McAdam, A. J., B. E. Gewurz, E. A. Farkash, and A. H. Sharpe. 2000. Either B7 costimulation or IL-2 can elicit generation of primary alloreactive CTL. *J Immunol 165:3088.*

McConkey, D. J., S. C. Chow, S. Orrenius, and M. Jondal. 1990a. NK cell-induced cytotoxicity is dependent on a Ca2+ increase in the target. *Faseb J 4:2661.*

McConkey, D. J., S. Orrenius, and M. Jondal. 1990b. Cellular signalling in programmed cell death (apoptosis). *Immunol Today 11:120.*

McDevitt, H. O. 2000. Discovering the role of the major histocompatibility complex in the immune response. *Annu Rev Immunol 18:1.*

McIntyre, K. W., J. F. Bukowski, and R. M. Welsh. 1985. Exquisite specificity of adoptive immunization in arenavirus-infected mice. *Antiviral Res 5:299.*

McMahon, C. W., and D. H. Raulet. 2001. Expression and function of NK cell receptors in CD8+ T cells. *Curr Opin Immunol 13:465.*

McMichael, A. J., and S. L. Rowland-Jones. 2001. Cellular immune responses to HIV. *Nature 410:980.*

Medana, I. M., A. Gallimore, A. Oxenius, M. M. Martinic, H. Wekerle, and H. Neumann. 2000. MHC class I-restricted killing of neurons by virus-specific CD8+ T lymphocytes is effected through the Fas/FasL, but not the perforin pathway. *Eur J Immunol 30:3623.*

Medana, I., Z. Li, A. Flugel, J. Tschopp, H. Wekerle, and H. Neumann. 2001. Fas ligand (CD95L) protects neurons against perforin-mediated T lymphocyte cytotoxicity. *J Immunol 167:674.*

Medvedev, A. E., A. C. Johnsen, J. Haux, B. Steinkjer, K. Egeberg, D. H. Lynch, A. Sundan, and T. Espevik. 1997. Regulation of Fas and Fas-ligand expression in NK cells by cytokines and the involvement of Fas-ligand in NK/LAK cell-mediated cytotoxicity. *Cytokine 9:394.*

Meidenbauer N., T. K. Hoffmann and A. D. Donnenberg. 2003. Direct visualization of antigen-specific T cells using peptide-MHC-class I tetrameric complexes. Methods. 2003 31:160-71.

Mehta, I. K., J. Wang, J. Roland, D. H. Margulies, and W. M. Yokoyama. 2001. Ly49A allelic variation and MHC class I specificity. *Immunogenetics 53:572.*

Melero, I., R. G. Vile, and M. P. Colombo. 2000. Feeding dendritic cells with tumor antigens: self-service buffet or a la carte? *Gene Ther 7:1167.*

Melief, C. J. 1992. Tumor eradication by adoptive transfer of cytotoxic T lymphocytes. *Adv Cancer Res 58:143.*

Mendiratta, S., W. Martin, S. Hong, A. Boesteanu, S. Joyce, and L. Van Kaer. 1997. CD1d mutant mice are deficient in natural T cells that promptly produce IL-4. *Immunity 6:469.*

Mentzer, S. J., J. A. Barbosa, and S. J. Burakoff. 1985. T3 monoclonal antibody activation of nonspecific cytolysis: a mechanism of CTL inhibition. *J Immunol 135:34.*

Mercado, R., S. Vijh, S. E. Allen, K. Kerksiek, I. M. Pilip, and E. G. Pamer. 2000. Early programming of T cell populations responding to bacterial infection. *J Immunol 165:6833.*

Mercep, M., A. M. Weissman, S. J. Frank, R. D. Klausner, and J. D. Ashwell. 1989. Activation-driven programmed cell death and T cell receptor zeta eta expression. *Science 246:1162.*

Mescher, M. F., A. M. O'Rourke, P. Champoux, and K. P. Kane. 1991. Equilibrium binding of cytotoxic T lymphocytes to class I antigen. *J Immunol 147:36.*

Mescher, M. F. 1995. Molecular interactions in the activation of effector and precursor cytotoxic T lymphocytes. *Immunol Rev 146:177.*

Metelitsa, L. S., O. V. Naidenko, A. Kant, H. W. Wu, M. J. Loza, B. Perussia, M. Kronenberg, and R. C. Seeger. 2001. Human NKT cells mediate antitumor cytotoxicity directly by recognizing target cell CD1d with bound ligand or indirectly by producing IL-2 to activate NK cells. *J Immunol 167:3114.*

Mielke, M. E., S. Ehlers, and H. Hahn. 1988. T cell subsets in DTH, protection and granuloma formation in primary and secondary Listeria infection in mice: superior role of Lyt-2+ cells in acquired immunity. *Immunol Lett 19:211.*

Migueles, S. A., A. C. Laborico, W. L. Shupert, M. S. Sabbaghian, R. Rabin, C. W. Hallahan, D. Van Baarle, S. Kostense, F. Miedema, M. McLaughlin, L. Ehler, J. Metcalf, S. Liu, and M. Connors. 2002. HIV-specific CD8+ T cell proliferation is coupled to perforin expression and is maintained in nonprogressors. *Nat Immunol 3:1061.*

Millard, P. J., M. P. Henkart, C. W. Reynolds, and P. A. Henkart. 1984. Purification and properties of cytoplasmic granules from cytotoxic rat LGL tumors. *J Immunol 132:3197.*

Miller, J., G. Grant, and F. Roe. 1963. Effect of thymectomy on the induction of skin tumors by 3,4-benzpyrene. *Nature 199:920.*

Miller, J., G. Grant, and F. Roe. 1963. Effect of thymectomy on the induction of skin tumors by 3,4-benzopyrene. *Nature 199:920.*

Miller, J. F., and D. Osoba. 1967. Current concepts of the immunological function of the thymus. *Physiol Rev 47:437.*

Miller, S. C. 1984. Life history of cells mediating natural resistance to tumor cells and bone-marrow transplants: the respective roles of cell lineage commitment and host environment in determining strain characteristics of natural resistance to foreign marrow grafts. *Am J Anat 170:367.*

Miller, B. J., M. C. Appel, J. J. O'Neil, and L. S. Wicker. 1988. Both the Lyt-2+ and L3T4+ T cell subsets are required for the transfer of diabetes in nonobese diabetic mice. *J Immunol 140:52.*

Minato, N., L. Reid, and B. R. Bloom. 1981. On the heterogeneity of murine natural killer cells. *J Exp Med 154:750.*

Mine, H., H. Kawai, K. Yokoi, M. Akaike, and S. Saito. 1996. High frequencies of human T-lymphotropic virus type I (HTLV-I) infection and presence of HTLV-II proviral DNA in blood donors with anti-thyroid antibodies. *J Mol Med 74:471.*

Mintz, B., and W. K. Silvers. 1970. Histocompatibility antigens on melanoblasts and hair follicle cells. Cell-localized homograft rejection in allophenic skin grafts. *Transplantation 9:497.*

Mitchison, N. 1953. Pasive transfer of transplantation immunity. *Nature 171:267.*

Mitsiades, N., V. Poulaki, V. Kotoula, G. Mastorakos, S. Tseleni-Balafouta, D. A. Koutras, and M. Tsokos. 1998. Fas/Fas ligand up-regulation and Bcl-2 down-regulation may be significant in the pathogenesis of Hashimoto's thyroiditis. *J Clin Endocrinol Metab 83:2199.*

Mitsiades, N., V. Poulaki, C. S. Mitsiades, D. A. Koutras, and G. P. Chrousos. 2001. Apoptosis induced by FasL and TRAIL/Apo2L in the pathogenesis of thyroid diseases. *Trends Endocrinol Metab 12:384.*

Miyake, M., A. Horiuchi, K. Kimura, Y. Abe, S. Kimura, and Y. Hitsumoto. 1992. Correlation between killing activity towards the murine L929 cell line and expression of membrane-associated lymphotoxin-related molecule of human lymphokine-activated killer cells. *Eur J Immunol 22:2147.*

Miyamoto, K., S. Miyake, and T. Yamamura. 2001. A synthetic glycolipid prevents autoimmune encephalomyelitis by inducing TH2 bias of natural killer T cells. *Nature 413:531.*

Moffett-King, A. 2002. Natural killer cells and pregnancy. *Nat Rev Immunol 2:656.*

Mogil, R. J., L. Radvanyi, R. Gonzalez-Quintial, R. Miller, G. Mills, A. N. Theofilopoulos, and D. R. Green. 1995. Fas (CD95) participates in peripheral T cell deletion and associated apoptosis in vivo. *Int Immunol 7:1451.*

Molinero, L. L., M. B. Fuertes, G. A. Rabinovich, L. Fainboim, and N. W. Zwirner. 2002. Activation-induced expression of MICA on T lymphocytes involves engagement of CD3 and CD28. *J Leukoc Biol 71:791.*

Moller, G., O. Sjoberg, and J. Andersson. 1972. Mitogen-induced lymphocyte-mediated cytotoxicity in vitro: effect of mitogens selectively activating T or B cells. *Eur J Immunol 2:586.*

Moller, G., and E. Moller. 1976. The concept of immunological surveillance against neoplasia. *Transplant Rev 28:3.*

Moller, P., Y. Sun, T. Dorbic, S. Alijagic, A. Makki, K. Jurgovsky, M. Schroff, B. M. Henz, B. Wittig, and D. Schadendorf. 1998. Vaccination with IL-7 gene-modified autologous melanoma cells can enhance the anti-melanoma lytic activity in peripheral blood of patients with a good clinical performance status: a clinical phase I study. *Br J Cancer 77:1907.*

Mombaerts, P., J. Arnoldi, F. Russ, S. Tonegawa, and S. H. Kaufmann. 1993. Different roles of alpha beta and gamma delta T cells in immunity against an intracellular bacterial pathogen. *Nature 365:53.*

Monastra, G., A. Cabrelle, A. Zambon, A. Rosato, B. Macino, D. Collavo, and P. Zanovello. 1996. Membrane form of TNF alpha induces both cell lysis and apoptosis in susceptible target cells. *Cell Immunol 171:102.*

Monteiro, J., R. Hingorani, R. Peroglizzi, B. Apatoff, and P. K. Gregersen. 1996. Oligoclonality of CD8+ T cells in multiple sclerosis. *Autoimmunity 23:127.*

Montel, A. H., M. R. Bochan, W. S. Goebel, and Z. Brahmi. 1995a. Fas-mediated cytotoxicity remains intact in perforin and granzyme B antisense transfectants of a human NK-like cell line. *Cell Immunol 165:312.*

Montel, A. H., M. R. Bochan, J. A. Hobbs, D. H. Lynch, and Z. Brahmi. 1995b. Fas involvement in cytotoxicity mediated by human NK cells. *Cell Immunol 166:236.*

Montoya, J. G., K. E. Lowe, C. Clayberger, D. Moody, D. Do, J. S. Remington, S. Talib, and C. S. Subauste. 1996. Human CD4+ and CD8+ T lymphocytes are both cytotoxic to Toxoplasma gondii-infected cells. *Infect Immun 64:176.*

Moretta A., G. Pantaleo, L. Moretta, m.C. Mingari, and J. C. Cerottini 1983. Quantitative assessment of the pool size and subset distribution of cytolytic T lymphocytes within human resting or alloactivated peripheral blood T cell populations. J Exp Med. 158:571-85.

Moretta, A., S. Sivori, M. Vitale, D. Pende, L. Morelli, R. Augugliaro, C. Bottino, and L. Moretta. 1995. Existence of both inhibitory (p58) and activatory (p50) receptors for HLA-C molecules in human natural killer cells. *J Exp Med 182:875.*

Moretta, A., R. Biassoni, C. Bottino, M. C. Mingari, and L. Moretta. 2000. Natural cytotoxicity receptors that trigger human NK-cell-mediated cytolysis. *Immunol Today 21:228.*

Moretta, L., R. Biassoni, C. Bottino, M. C. Mingari, and A. Moretta. 2000. Human NK-cell receptors. *Immunol Today 21:420.*

Moretta, A. 2002. Natural killer cells and dendritic cells: rendezvous in abused tissues. *Nat Rev Immunol 2:957.*

Morgan, D. A., F. W. Ruscetti, and R. Gallo. 1976. Selective in vitro growth of T lymphocytes from normal human bone marrows. *Science 193:1007.*

Morse, R. H., R. Seguin, E. L. McCrea, and J. P. Antel. 2001. NK cell-mediated lysis of autologous human oligodendrocytes. *J Neuroimmunol 116:107.*

Moser, B., and P. Loetscher. 2001. Lymphocyte traffic control by chemokines. *Nat Immunol 2:123.*

Moser, J. M., A. M. Byers, and A. E. Lukacher. 2002. NK cell receptors in antiviral immunity. *Curr Opin Immunol 14:509.*

Moser, J. M., J. Gibbs, P. E. Jensen, and A. E. Lukacher. 2002. CD94-NKG2A receptors regulate antiviral CD8(+) T cell responses. *Nat Immunol 3:189.*

Mosier, D. E. 1967. A requirement for two cell types for antibody formation in vitro. *Science 158:1573.*

Moskophidis, D., F. Lechner, H. Pircher, and R. M. Zinkernagel. 1993. Virus persistence in acutely infected immunocompetent mice by exhaustion of antiviral cytotoxic effector T cells. *Nature 362:758.*

Mosmann, T. R., L. Li, and S. Sad. 1997. Functions of CD8 T-cell subsets secreting different cytokine patterns. *Semin Immunol 9:87.*

Moss, D. J., W. Scott, and J. H. Pope. 1977. An immunological basis for inhibition of transformation of human lymphocytes by EB virus. *Nature 268:735.*

Moss, P. A., S. L. Rowland-Jones, P. M. Frodsham, S. McAdam, P. Giangrande, A. J. McMichael, and J. I. Bell. 1995. Persistent high frequency of human immunodeficiency virus-specific cytotoxic T cells in peripheral blood of infected donors. *Proc Natl Acad Sci U S A 92:5773.*

Morishima C., L. Musey, M. Elizaga, K. Gaba, M. Allison, R. L. Carithers, D. R. Gretch and M. J. McElrath 2003. Hepatitis C virus-specific cytolytic T cell responses after antiviral therapy. Clin Immunol. 108:211-20.

Motyka, B., G. Korbutt, M. J. Pinkoski, J. A. Heibein, A. Caputo, M. Hobman, M. Barry, I. Shostak, T. Sawchuk, C. F. Holmes, J. Gauldie, and R. C. Bleackley. 2000. Mannose 6-phosphate/insulin-like growth factor II receptor is a death receptor for granzyme B during cytotoxic T cell-induced apoptosis. *Cell 103:491.*

Mrozek, E., P. Anderson, and M. A. Caligiuri. 1996. Role of interleukin-15 in the development of human CD56+ natural killer cells from CD34+ hematopoietic progenitor cells. *Blood 87:2632.*

Muderspach, L., S. Wilczynski, L. Roman, L. Bade, J. Felix, L. A. Small, W. M. Kast, G. Fascio, V. Marty, and J. Weber. 2000. A phase I trial of a human papillomavirus (HPV) peptide vaccine for women with high-grade cervical and vulvar intraepithelial neoplasia who are HPV 16 positive. *Clin Cancer Res 6:3406.*

Mukherji, B., A. Guha, N. G. Chakraborty, M. Sivanandham, A. L. Nashed, J. R. Sporn, and M. T. Ergin. 1989. Clonal analysis of cytotoxic and regulatory T cell responses against human melanoma. *J Exp Med 169:1961.*

Mule, J. J., S. Shu, S. L. Schwarz, and S. A. Rosenberg. 1984. Adoptive immunotherapy of established pulmonary metastases with LAK cells and recombinant interleukin-2. *Science 225:1487.*

Mullbacher, A. 1994. The long-term maintenance of cytotoxic T cell memory does not require persistence of antigen. *J Exp Med 179:317.*

Mullbacher, A., P. Waring, R. Tha Hla, T. Tran, S. Chin, T. Stehle, C. Museteanu, and M. M. Simon. 1999a. Granzymes are the essential downstream effector molecules for the control of primary virus infections by cytolytic leukocytes. *Proc Natl Acad Sci U S A 96:13950.*

Mullbacher, A., R. T. Hla, C. Museteanu, and M. M. Simon. 1999b. Perforin is essential for control of ectromelia virus but not related poxviruses in mice. *J Virol 73:1665.*

Muller, I., S. P. Cobbold, H. Waldmann, and S. H. Kaufmann. 1987. Impaired resistance to Mycobacterium tuberculosis infection after selective in vivo depletion of L3T4+ and Lyt-2+ T cells. *Infect Immun 55:2037.*

Muller, C., D. Kagi, T. Aebischer, B. Odermatt, W. Held, E. R. Podack, R. M. Zinkernagel, and H. Hengartner. 1989. Detection of perforin and granzyme A mRNA in infiltrating cells during infection of mice with lymphocytic choriomeningitis virus. *Eur J Immunol 19:1253.*

Muller, C., and J. Tschopp. 1994. Resistance of CTL to perforin-mediated lysis. Evidence for a lymphocyte membrane protein interacting with perforin. *J Immunol 153:2470.*

Muller-Eberhard, H. J. 1986. The membrane attack complex of complement. *Annu Rev Immunol 4:503.*

Murali-Krishna, K., J. D. Altman, M. Suresh, D. J. Sourdive, A. J. Zajac, J. D. Miller, J. Slansky, and R. Ahmed. 1998. Counting antigen-specific CD8 T cells: a reevaluation of bystander activation during viral infection. *Immunity 8:177.*

Murali-Krishna, K., L. L. Lau, S. Sambhara, F. Lemonnier, J. Altman, and R. Ahmed. 1999. Persistence of memory CD8 T cells in MHC class I-deficient mice. *Science 286:1377.*

Murphy, P. 1913. *J Exp Med 17:482.*

Murphy, W. J., V. Kumar, and M. Bennett. 1987. Rejection of bone marrow allografts by mice with severe combined immune deficiency (SCID). Evidence that natural killer cells can mediate the specificity of marrow graft rejection. *J Exp Med 165:1212.*

Murray, A. G., P. Petzelbauer, C. C. Hughes, J. Costa, P. Askenase, and J. S. Pober. 1994. Human T-cell-mediated destruction of allogeneic dermal microvessels in a severe combined immunodeficient mouse. *Proc Natl Acad Sci U S A 91:9146.*

Murray, P. D., D. B. McGavern, X. Lin, M. K. Njenga, J. Leibowitz, L. R. Pease, and M. Rodriguez. 1998. Perforin-dependent neurologic injury in a viral model of multiple sclerosis. *J Neurosci 18:7306.*

Nabholz, M., H. D. Engers, D. Collavo, and M. North. 1978. Cloned T-cell lines with specific cytolytic activity. *Curr Top Microbiol Immunol 81:176.*

Nabholz, M., M. Cianfriglia, O. Acuto, A. Conzelmann, W. Haas, H. von Boehmer, H. R. McDonald, H. Pohlit, and J. P. Johnson. 1980. Cytolytically active murine T-cell hybrids. *Nature 287:437.*

Nadeau, J. H. 1989. Maps of linkage and synteny homologies between mouse and man. *Trends Genet 5:82.*

Nagata, M., and J. W. Yoon. 1992. Studies on autoimmunity for T-cell-mediated beta-cell destruction. Distinct difference in beta-cell destruction between CD4+ and CD8+ T-cell clones derived from lymphocytes infiltrating the islets of NOD mice. *Diabetes 41:998.*

Nagata, S., and P. Golstein. 1995. The Fas death factor. *Science 267:1449.*

Nagata, S., and T. Suda. 1995. Fas and Fas ligand: lpr and gld mutations. *Immunol Today 16:39.*

Nagata, S. 1997. Apoptosis by death factor. *Cell 88:355.*

Nagata, S. 1998. Human autoimmune lymphoproliferative syndrome, a defect in the apoptosis-inducing Fas receptor: a lesson from the mouse model. *J Hum Genet 43:2.*

Nagler, A., L. L. Lanier, S. Cwirla, and J. H. Phillips. 1989. Comparative studies of human FcRIII-positive and negative natural killer cells. *J Immunol 143:3183.*

Nagorsen, D., U. Keilholz, L. Rivoltini, A. Schmittel, A. Letsch, A. M. Asemissen, G. Berger, H. J. Buhr, E. Thiel, and C. Scheibenbogen. 2000. Natural T-cell response against MHC class I epitopes of epithelial cell adhesion molecule, her-2/neu, and carcinoembryonic antigen in patients with colorectal cancer. *Cancer Res 60:4850.*

Nakajima, H., P. Golstein, and P. A. Henkart. 1995. The target cell nucleus is not required for cell-mediated granzyme- or Fas-based cytotoxicity. *J Exp Med 181:1905.*

Nakamoto, Y., L. G. Guidotti, V. Pasquetto, R. D. Schreiber, and F. V. Chisari. 1997. Differential target cell sensitivity to CTL-activated death pathways in hepatitis B virus transgenic mice. *J Immunol 158:5692.*

Nakamura, M., D. T. Ross, T. J. Briner, and M. L. Gefter. 1986. Cytolytic activity of antigen-specific T cells with helper phenotype. *J Immunol 136:44.*

Nakano, Y., H. Hisaeda, T. Sakai, and M. Zhang. 2001. Granule-dependent killing of Toxoplasma gondii by CD8 T cells. *Immunology 104:289.*

Nakata, M., M. J. Smyth, Y. Norihisa, A. Kawasaki, Y. Shinkai, K. Okumura, and H. Yagita. 1990. Constitutive expression of pore-forming protein in peripheral blood gamma/delta T cells: implication for their cytotoxic role in vivo. *J Exp Med 172:1877.*

Nakayama, M., M. Nagata, H. Yasuda, K. Arisawa, R. Kotani, K. Yamada, S. A. Chowdhury, S. Chakrabarty, Z. Z. Jin, H. Yagita, K. Yokono, and M. Kasuga. 2002. Fas/Fas ligand interactions play an essential role in the initiation of murine autoimmune diabetes. *Diabetes 51:1391.*

Nandi, D., J. A. Gross, and J. P. Allison. 1994. CD28-mediated costimulation is necessary for optimal proliferation of murine NK cells. *J Immunol 152:3361.*

Natarajan, K., L. F. Boyd, P. Schuck, W. M. Yokoyama, D. Eliat, and D. H. Margulies. 1999. Interaction of the NK cell inhibitory receptor Ly49A with H-2Dd: identification of a site distinct from the TCR site. *Immunity 11:591.*

Nau, G. J., R. L. Moldwin, D. W. Lancki, D. K. Kim, and F. W. Fitch. 1987. Inhibition of IL 2-driven proliferation of murine T lymphocyte clones by supraoptimal levels of immobilized anti-T cell receptor monoclonal antibody. *J Immunol 139:114.*

Naumov, Y. N., K. S. Bahjat, R. Gausling, R. Abraham, M. A. Exley, Y. Koezuka, S. B. Balk, J. L. Strominger, M. Clare-Salzer, and S. B. Wilson. 2001. Activation of CD1d-restricted T cells protects NOD mice from developing diabetes by regulating dendritic cell subsets. *Proc Natl Acad Sci U S A 98:13838.*

Nestle, F. O. 2000. Dendritic cell vaccination for cancer therapy. *Oncogene 19:6673.*

Neumanaitis, J., et al. 2004. GMCSF gene-modified autologous tumor vaccines in non-small-cell lung cancer. *JNCI 96:326.*

Neumann, H., A. Cavalie, D. E. Jenne, and H. Wekerle. 1995. Induction of MHC class I genes in neurons. *Science 269:549.*

Neumann, H., I. M. Medana, J. Bauer, and H. Lassmann. 2002. Cytotoxic T lymphocytes in autoimmune and degenerative CNS diseases. *Trends Neurosci 25:313.*

Newell, K. A., G. He, Z. Guo, O. Kim, G. L. Szot, I. Rulifson, P. Zhou, J. Hart, J. R. Thistlethwaite, and J. A. Bluestone. 1999. Cutting edge: blockade of the CD28/B7 costimulatory pathway inhibits intestinal allograft rejection mediated by CD4+ but not CD8+ T cells. *J Immunol 163:2358.*

Nguyen, K. B., L. P. Cousens, L. A. Doughty, G. C. Pien, J. E. Durbin, and C. A. Biron. 2000. Interferon alpha/beta-mediated inhibition and promotion of interferon gamma: STAT1 resolves a paradox. *Nat Immunol 1:70.*

Ni, H. T., M. J. Deeths, W. Li, D. L. Mueller, and M. F. Mescher. 1999. Signaling pathways activated by leukocyte function-associated Ag-1-dependent costimulation. *J Immunol 162:5183.*

Nickell, S. P., G. A. Stryker, and C. Arevalo. 1993. Isolation from Trypanosoma cruzi-infected mice of CD8+, MHC-restricted cytotoxic T cells that lyse parasite-infected target cells. *J Immunol 150:1446.*

Nickell, S. P., and D. Sharma. 2000. Trypanosoma cruzi: roles for perforin-dependent and perforin-independent immune mechanisms in acute resistance. *Exp Parasitol 94:207.*

Nicol, A., M. Nieda, Y. Koezuka, S. Porcelli, K. Suzuki, K. Tadokoro, S. Durrant, and T. Juji. 2000. Human invariant valpha24+ natural killer T cells activated by alpha-galactosylceramide (KRN7000) have cytotoxic anti-tumour activity through mechanisms distinct from T cells and natural killer cells. *Immunology 99:229.*

Nishimura, T., Y. Nakamura, Y. Takeuchi, Y. Tokuda, M. Iwasawa, A. Kawasaki, K. Okumura, and S. Habu. 1992. Generation propagation, and targeting of human CD4+ helper/killer T cells induced by anti-CD3 monoclonal antibody plus recombinant IL-2. An efficient strategy for adoptive tumor immunotherapy. *J Immunol 148:285.*

Nishiyama, T., M. Tachibana, Y. Horiguchi, K. Nakamura, Y. Ikeda, K. Takesako, and M. Murai. 2001. Immunotherapy of bladder cancer using autologous dendritic cells pulsed with human lymphocyte antigen-A24-specific MAGE-3 peptide. *Clin Cancer Res 7:23.*

Nitta, T., H. Yagita, K. Sato, and K. Okumura. 1989. Involvement of CD56 (NKH-1/Leu-19 antigen) as an adhesion molecule in natural killer-target cell interaction. *J Exp Med 170:1757.*

Noguchi, Y., Y. T. Chen, and L. J. Old. 1994. A mouse mutant p53 product recognized by CD4+ and CD8+ T cells. *Proc Natl Acad Sci U S A 91:3171.*

Norris, J. S., M. L. Hyer, C. Voelkel-Johnson, S. L. Lowe, S. Rubinchik, and J. Y. Dong. 2001. The use of Fas Ligand, TRAIL and Bax in gene therapy of prostate cancer. *Curr Gene Ther 1:123.*

Norris, P. J., M. Sumaroka, C. Brander, H. F. Moffett, S. L. Boswell, T. Nguyen, Y. Sykulev, B. D. Walker, and E. S. Rosenberg. 2001. Multiple effector functions mediated by human immunodeficiency virus-specific CD4(+) T-cell clones. *J Virol 75:9771.*

North, R. J. 1973. Cellular mediators of anti-Listeria immunity as an enlarged population of short lived, replicating T cells. Kinetics of their production. *J Exp Med 138:342.*

Nottage, M., and L. Siu. 2002. Rationale for Ras and raf-kinase as a target for cancer therapeutics. *Curr Pharm Des 8:2231.*

Novogrodsky, A., and E. Katchlaski. 1971. Induction of lymphocyte transformation by periodate. *FEBS Lett 12:297.*

Novogrodsky, A., and E. Katchalski. 1973. Induction of lymphocyte transformation by sequential treatment with neuraminidase and galactose oxidase. *Proc Natl Acad Sci U S A 70:1824.*

Novogrodsky, A. 1974. Selective activation of mouse T and B lymphocytes by periodate, galactose oxidase and soybean agglutinin. *Eur J Immunol 4:646.*

O'Brien, R. L., J. W. Parker, P. Paolilli, and J. Steiner. 1974. Periodate-induced lymphocyte transformation. IV. Mitogenic effect of NaIO4 treated lymphocytes upon autologous lymphocytes. *J Immunol 112:1884.*

O'Brien, B. A., B. V. Harmon, D. P. Cameron, and D. J. Allan. 1997. Apoptosis is the mode of beta-cell death responsible for the development of IDDM in the nonobese diabetic (NOD) mouse. *Diabetes 46:750.*

O'Connell, J., G. C. O'Sullivan, J. K. Collins, and F. Shanahan. 1996. The Fas counterattack: Fas-mediated T cell killing by colon cancer cells expressing Fas ligand. *J Exp Med 184:1075.*

O'Connell, J., M. W. Bennett, G. C. O'Sullivan, J. K. Collins, and F. Shanahan. 1999. The Fas counterattack: cancer as a site of immune privilege. *Immunol Today 20:46.*

O'Connell, J., A. Houston, M. W. Bennett, G. C. O'Sullivan, and F. Shanahan. 2001. Immune privilege or inflammation? Insights into the Fas ligand enigma. *Nat Med 7:271.*

O'Flynn, K., D. C. Linch, and P. E. Tatham. 1984. The effect of mitogenic lectins and monoclonal antibodies on intracellular free calcium concentration in human T-lymphocytes. *Biochem J 219:661.*

O'Rourke, A. M., J. Rogers, and M. F. Mescher. 1990. Activated CD8 binding to class I protein mediated by the T-cell receptor results in signalling. *Nature 346:187.*

O'Rourke, A. M., J. R. Apgar, K. P. Kane, E. Martz, and M. F. Mescher. 1991. Cytoskeletal function in CD8- and T cell receptor-mediated interaction of cytotoxic T lymphocytes with class I protein. *J Exp Med 173:241.*

Ochoa, M. T., S. Stenger, P. A. Sieling, S. Thoma-Uszynski, S. Sabet, S. Cho, A. M. Krensky, M. Rollinghoff, E. Nunes Sarno, A. E. Burdick, T. H. Rea, and R. L. Modlin. 2001. T-cell release of granulysin contributes to host defense in leprosy. *Nat Med 7:174.*

Oehler, J. R., L. R. Lindsay, M. E. Nunn, and R. B. Herberman. 1978. Natural cell-mediated cytotoxicity in rats. I. Tissue and strain distribution, and demonstration of a membrance receptor for the Fc portion of IgG. *Int J Cancer 21:204.*

Ogasawara, J., R. Watanabe-Fukunaga, M. Adachi, A. Matsuzawa, T. Kasugai, Y. Kitamura, N. Itoh, T. Suda, and S. Nagata. 1993. Lethal effect of the anti-Fas antibody in mice. *Nature 364:806.*

Ogg G. S. and A. J. McMichael 1998a. HLA-peptide tetrameric complexes. Curr Opin Immunol. 10:393-6.]

Ogg G. S., X. Jin, S. Bonhoeffer, P. R. Dunbar, M. A. Nowak, S. Monard, J. P. Segal, Y. Cao, S. L. Rowland-Jones, V. Cerundolo, A. Hurley, M. Markowitz, D. F. Ho DD, Nixon and A. J. McMichael 1998b

Quantitation of HIV-1-specific cytotoxic T lymphocytes and plasma load of viral RNA. Science. 279:2103-6.

Ohminami, H., M. Yasukawa, S. Kaneko, Y. Yakushijin, Y. Abe, Y. Kasahara, Y. Ishida, and S. Fujita. 1999. Fas-independent and nonapoptotic cytotoxicity mediated by a human CD4(+) T-cell clone directed against an acute myelogenous leukemia-associated DEK-CAN fusion peptide. *Blood 93:925.*

Ojcius, D. M., S. B. Jiang, P. M. Persechini, P. A. Detmers, and J. D. Young. 1991. Cytoplasts from cytotoxic T lymphocytes are resistant to perforin-mediated lysis. *Mol Immunol 28:1011.*

Old, L., E. Boyse, and D. Clarke. 1962. Antigenic properties of chemically induced tumors. *Ann N Y Acad Sci 101:80.*

Oliva, A., A. L. Kinter, M. Vaccarezza, A. Rubbert, A. Catanzaro, S. Moir, J. Monaco, L. Ehler, S. Mizell, R. Jackson, Y. Li, J. W. Romano, and A. S. Fauci. 1998. Natural killer cells from human immunodeficiency virus (HIV)-infected individuals are an important source of CC-chemokines and suppress HIV-1 entry and replication in vitro. *J Clin Invest 102:223.*

Operschall, E., J. Pavlovic, M. Nawrath, and K. Molling. 2000. Mechanism of protection against influenza A virus by DNA vaccine encoding the hemagglutinin gene. *Intervirology 43:322.*

Opferman, J. T., B. T. Ober, and P. G. Ashton-Rickardt. 1999. Linear differentiation of cytotoxic effectors into memory T lymphocytes. *Science 283:1745.*

Opferman, J. T., B. T. Ober, R. Narayanan, and P. G. Ashton-Rickardt. 2001. Suicide induced by cytolytic activity controls the differentiation of memory CD8(+) T lymphocytes. *Int Immunol 13:411.*

Orange, J. S., and C. A. Biron. 1996. Characterization of early IL-12, IFN-alphabeta, and TNF effects on antiviral state and NK cell responses during murine cytomegalovirus infection. *J Immunol 156:4746.*

Orange, J. S. 2002. Human natural killer cell deficiencies and susceptibility to infection. *Microbes Infect 4:1545.*

Orme, I. M., and F. M. Collins. 1983. Protection against Mycobacterium tuberculosis infection by adoptive immunotherapy. Requirement for T cell-deficient recipients. *J Exp Med 158:74.*

Orme, I. M., and F. M. Collins. 1984. Adoptive protection of the Mycobacterium tuberculosis-infected lung. Dissociation between cells that passively transfer protective immunity and those that transfer delayed-type hypersensitivity to tuberculin. *Cell Immunol 84:113.*

Orrenius, S., D. J. McConkey, G. Bellomo, and P. Nicotera. 1989. Role of Ca2+ in toxic cell killing. *Trends Pharmacol Sci 10:281.*

Ortaldo, J. R., S. O. Sharrow, T. Timonen, and R. B. Herberman. 1981. Determination of surface antigens on highly purified human NK cells by flow cytometry with monoclonal antibodies. *J Immunol 127:2401.*

Ortaldo, J. R., A. T. Mason, J. P. Gerard, L. E. Henderson, W. Farrar, R. F. Hopkins, 3rd, R. B. Herberman, and H. Rabin. 1984. Effects of natural and recombinant IL 2 on regulation of IFN gamma production and natural killer activity: lack of involvement of the Tac antigen for these immunoregulatory effects. *J Immunol 133:779.*

Ortaldo, J. R., A. Mason, and R. Overton. 1986a. Lymphokine-activated killer cells. Analysis of progenitors and effectors. *J Exp Med 164:1193.*

Ortaldo, J. R., J. R. Ransom, T. J. Sayers, and R. B. Herberman. 1986b. Analysis of cytostatic/cytotoxic lymphokines: relationship of natural killer cytotoxic factor to recombinant lymphotoxin, recombinant tumor necrosis factor, and leukoregulin. *J Immunol 137:2857.*

Ortaldo, J. R., L. H. Mason, B. J. Mathieson, S. M. Liang, D. A. Flick, and R. B. Herberman. 1986c. Mediation of mouse natural cytotoxic activity by tumour necrosis factor. *Nature 321:700.*

Ortaldo, J. R., R. Winkler-Pickett, A. C. Morgan, C. Woodhouse, R. Kantor, and C. W. Reynolds. 1987. Analysis of rat natural killer cytotoxic factor (NKCF) produced by rat NK cell lines and the production of a murine monoclonal antibody that neutralizes NKCF. *J Immunol 139:3159.*

Ortaldo, J. R., A. T. Mason, J. J. O'Shea, M. J. Smyth, L. A. Falk, I. C. Kennedy, D. L. Longo, and F. W. Ruscetti. 1991. Mechanistic studies of transforming growth factor-beta inhibition of IL-2-dependent activation of CD3- large granular lymphocyte functions. Regulation of IL-2R beta (p75) signal transduction. *J Immunol 146:3791.*

Ortaldo, J. R., R. Winkler-Pickett, W. Kopp, A. Kawasaki, K. Nagashima, K. Okumura, H. Yagita, and F. H. Bach. 1992a. Relationship of large and small CD3- CD56+ lymphocytes mediating NK-associated activities. *J Leukoc Biol 52:287.*

Ortaldo, J. R., R. T. Winkler-Pickett, K. Nagashima, H. Yagita, and K. Okumura. 1992b. Direct evidence for release of pore-forming protein during NK cellular lysis. *J Leukoc Biol 52:483.*

Ortaldo, J. R., T. A. Wiltrout, T. J. Sayers, H. Yagita, and R. T. Winkler-Pickett. 1995. Characterization of a non-granule associated pore-forming protein in agranular lymphocytes. *J Leukoc Biol 57:897.*

Ortiz de Landazuri, M., and R. B. Herberman. 1972. In vitro activation of cellular immune response to Gross virus-induced lymphoma. *J Exp Med 136:969.*

Osada, T., H. Nagawa, J. Kitayama, N. H. Tsuno, S. Ishihara, M. Takamizawa, and Y. Shibata. 2001. Peripheral blood dendritic cells, but not monocyte-derived dendritic cells, can augment human NK cell function. *Cell Immunol 213:14.*

Oshimi, Y., S. Oda, Y. Honda, S. Nagata, and S. Miyazaki. 1996a. Involvement of Fas ligand and Fas-mediated pathway in the cytotoxicity of human natural killer cells. *J Immunol 157:2909.*

Oshimi, Y., K. Oshimi, and S. Miyazaki. 1996b. Necrosis and apoptosis associated with distinct Ca2+ response patterns in target cells attacked by human natural killer cells. *J Physiol 495:319.*

Ostergaard, H., and W. R. Clark. 1987. The role of Ca2+ in activation of mature cytotoxic T lymphocytes for lysis. *J Immunol 139:3573.*

Ostergaard, H. L., K. P. Kane, M. F. Mescher, and W. R. Clark. 1987. Cytotoxic T lymphocyte mediated lysis without release of serine esterase. *Nature 330:71.*

Ostergaard, H., K. Gorman, and W. R. Clark. 1988. Cloned cytotoxic T lymphocyte target cells fail to induce early activation events in effector cytotoxic T lymphocytes. *Cell Immunol 114:188.*

Ostergaard, H. L., and W. R. Clark. 1989. Evidence for multiple lytic pathways used by cytotoxic T lymphocytes. *J Immunol 143:2120.*

Ostrov, D. A., W. Shi, J. C. Schwartz, S. C. Almo, and S. G. Nathenson. 2000. Structure of murine CTLA-4 and its role in modulating T cell responsiveness. *Science 290:816.*

Ou-Yang, P., L. H. Hwang, M. H. Tao, B. L. Chiang, and D. S. Chen. 2002. Co-delivery of GM-CSF gene enhances the immune responses of hepatitis C viral core protein-expressing DNA vaccine: role of dendritic cells. *J Med Virol 66:320.*

Overwijk, W. W., M. R. Theoret, S. E. Finkelstein, D. R. Surman, L. A. de Jong, F. A. Vyth-Dreese, T. A. Dellemijn, P. A. Antony, P. J. Spiess, D. C. Palmer, D. M. Heimann, C. A. Klebanoff, Z. Yu, L. N. Hwang, L. Feigenbaum, A. M. Kruisbeek, S. A. Rosenberg, and N. P. Restifo. 2003. Tumor regression and autoimmunity after reversal of a functionally tolerant state of self-reactive CD8+ T cells. *J Exp Med 198:569.*

Paige, C. J., E. F. Figarella, M. J. Cuttito, A. Cahan, and O. Stutman. 1978. Natural cytotoxic cells against solid tumors in mice. II. Some characteristics of the effector cells. *J Immunol 121:1827.*

Paiva, C. N., M. T. Castelo-Branco, J. Lannes-Vieira, and C. R. Gattass. 1999. Trypanosoma cruzi: protective response of vaccinated mice is mediated by CD8+ cells, prevents signs of polyclonal T lymphocyte activation, and allows restoration of a resting immune state after challenge. *Exp Parasitol 91:7.*

Paliard, X., R. de Waal Malefijt, H. Yssel, D. Blanchard, I. Chretien, J. Abrams, J. de Vries, and H. Spits. 1988. Simultaneous production of IL-2, IL-4, and IFN-gamma by activated human CD4+ and CD8+ T cell clones. *J Immunol 141:849.*

Palyi, I. 1980. Drug sensitivity studies on clonal cell lines isolated from heteroploid tumor cell population. II. Sensitivity of clones growing in suspension cultures. *Neoplasma 27:129.*

Pamer, E. G., C. R. Wang, L. Flaherty, K. F. Lindahl, and M. J. Bevan. 1992. H-2M3 presents a Listeria monocytogenes peptide to cytotoxic T lymphocytes. *Cell 70:215.*

Paolieri, F., C. Pronzato, M. Battifora, N. Fiorino, G. W. Canonica, and M. Bagnasco. 1995. Infiltrating gamma/delta T-cell receptor-positive lymphocytes in Hashimoto's thyroiditis, Graves' disease and papillary thyroid cancer. *J Endocrinol Invest 18:295.*

Pardo, J., S. Balkow, A. Anel, and M. M. Simon. 2002. Granzymes are essential for natural killer cell-mediated and perf-facilitated tumor control. *Eur J Immunol 32:2881.*

Parker, W. L., and E. Martz. 1980. Lectin-induced nonlethal adhesions between cytolytic T lymphocytes and antigenically unrecognizable tumor cells and nonspecific "triggering" of cytolysis. *J Immunol 124:25.*

Parmiani, G., C. Castelli, P. Dalerba, R. Mortarini, L. Rivoltini, F. M. Marincola, and A. Anichini. 2002. Cancer immunotherapy with peptide-based vaccines: what have we achieved? Where are we going? *J Natl Cancer Inst 94:805.*

Partridge, T. 1991. Animal models of muscular dystrophy--what can they teach us? *Neuropathol Appl Neurobiol 17:353.*

Pasetti, M. F., R. Salerno-Goncalves, and M. B. Sztein. 2002. Salmonella enterica serovar Typhi live vector vaccines delivered intranasally elicit regional and systemic specific CD8+ major histocompatibility class I-restricted cytotoxic T lymphocytes. *Infect Immun 70:4009.*

Pasternack, M. S., and H. N. Eisen. 1985. A novel serine esterase expressed by cytotoxic T lymphocytes. *Nature 314:743.*

Pasternack, M. S., C. R. Verret, M. A. Liu, and H. N. Eisen. 1986. Serine esterase in cytolytic T lymphocytes. *Nature 322:740.*

Paul, R. D., and D. M. Lopez. 1987. Induction of "innocent bystander" cytotoxicity in nonimmune mice by adoptive transfer of L3T4+ Lyt-1+2- mammary tumor immune T-cells. *Cancer Res 47:1105.*

Pemberton, R. M., D. C. Wraith, and B. A. Askonas. 1990. Influenza peptide-induced self-lysis and down-regulation of cloned cytotoxic T cells. *Immunology 70:223.*

Pena, S. V., and A. M. Krensky. 1997. Granulysin, a new human cytolytic granule-associated protein with possible involvement in cell-mediated cytotoxicity. *Semin Immunol 9:117.*

Pende, D., S. Sivori, L. Accame, L. Pareti, M. Falco, D. Geraghty, P. Le Bouteiller, L. Moretta, and A. Moretta. 1997. HLA-G recognition by human natural killer cells. Involvement of CD94 both as inhibitory and as activating receptor complex. *Eur J Immunol 27:1875.*

Pende, D., P. Rivera, S. Marcenaro, C. Chang, R. Biassoni, R. Conte, M. Kubin, D. Cosman, D. Ferrone, L. Moretta, and A. Moretta. 2002. MHC class I-related chain A and UL16-binding protein expression on tumor cell lines of different histotypes: analysis of tumor susceptibility to NKG2D-dependent NK cytotoxicity. *Cancer Res 62:6178.*

Peng, S. L., M. P. Madaio, A. C. Hayday, and J. Craft. 1996. Propagation and regulation of systemic autoimmunity by gammadelta T cells. *J Immunol 157:5689.*

Penn, I., and T. E. Starzl. 1972. A summary of the status of de novo cancer in transplant recipients. *Transplant Proc 4:719.*

Perez, P., J. A. Bluestone, D. A. Stephany, and D. M. Segal. 1985a. Quantitative measurements of the specificity and kinetics of conjugate formation between cloned cytotoxic T lymphocytes and splenic target cells by dual parameter flow cytometry. *J Immunol 134:478.*

Perez, P., R. W. Hoffman, S. Shaw, J. A. Bluestone, and D. M. Segal. 1985b. Specific targeting of cytotoxic T cells by anti-T3 linked to anti-target cell antibody. *Nature 316:354.*

Perez, P., J. A. Titus, M. T. Lotze, F. Cuttitta, D. L. Longo, E. S. Groves, H. Rabin, P. J. Durda, and D. M. Segal. 1986. Specific lysis of human tumor cells by T cells coated with anti-T3 cross-linked to anti-tumor antibody. *J Immunol 137:2069.*

Perez, C., I. Albert, K. DeFay, N. Zachariades, L. Gooding, and M. Kriegler. 1990. A nonsecretable cell surface mutant of tumor necrosis factor (TNF) kills by cell-to-cell contact. *Cell 63:251.*

Persechini, P. M., J. D. Young, and W. Almers. 1990. Membrane channel formation by the lymphocyte pore-forming protein: comparison between susceptible and resistant target cells. *J Cell Biol 110:2109.*

Perussia, B., D. Santoli, and G. Trinchieri. 1980. Interferon modulation of natural killer cell activity. *Ann N Y Acad Sci 350:55.*

Perussia, B., S. Starr, S. Abraham, V. Fanning, and G. Trinchieri. 1983. Human natural killer cells analyzed by B73.1, a monoclonal antibody blocking Fc receptor functions. I. Characterization of the lymphocyte subset reactive with B73.1. *J Immunol 130:2133.*

Perussia, B., G. Trinchieri, A. Jackson, N. L. Warner, J. Faust, H. Rumpold, D. Kraft, and L. L. Lanier. 1984. The Fc receptor for IgG on human natural killer cells: phenotypic, functional, and comparative studies with monoclonal antibodies. *J Immunol 133:180.*

Perussia, B. 1998. Fc receptors on natural killer cells. *Curr Top Microbiol Immunol 230:63.*

Peter, H. H., J. Pavie-Fischer, W. H. Fridman, C. Aubert, J. P. Cesarini, R. Roubin, and F. M. Kourilsky. 1975. Cell-mediate cytotoxicity in vitro of human lymphocytes against a tissue culture melanoma cell line (igr3). *J Immunol 115:539.*

Peters, P. M., J. R. Ortaldo, M. R. Shalaby, L. P. Svedersky, G. E. Nedwin, T. S. Bringman, P. E. Hass, B. B. Aggarwal, R. B. Herberman, D. V. Goeddel, and et al. 1986. Natural killer-sensitive targets stimulate production of TNF-alpha but not TNF-beta (lymphotoxin) by highly purified human peripheral blood large granular lymphocytes. *J Immunol 137:2592.*

Peters, P. J., H. J. Geuze, H. A. Van der Donk, J. W. Slot, J. M. Griffith, N. J. Stam, H. C. Clevers, and J. Borst. 1989. Molecules relevant for T cell-target cell interaction are present in cytolytic granules of human T lymphocytes. *Eur J Immunol 19:1469.*

Peters, R., H. Sauer, J. Tschopp, and G. Fritzsch. 1990. Transients of perforin pore formation observed by fluorescence microscopic single channel recording. *Embo J 9:2447.*

Peters, P. J., J. Borst, V. Oorschot, M. Fukuda, O. Krahenbuhl, J. Tschopp, J. W. Slot, and H. J. Geuze. 1991. Cytotoxic T lymphocyte granules are secretory lysosomes, containing both perforin and granzymes. *J Exp Med 173:1099.*

Phillips, J. H., and L. L. Lanier. 1986. Dissection of the lymphokine-activated killer phenomenon. Relative contribution of peripheral blood natural killer cells and T lymphocytes to cytolysis. *J Exp Med 164:814.*

Piccioli, D., S. Sbrana, E. Melandri, and N. M. Valiante. 2002. Contact-dependent stimulation and inhibition of dendritic cells by natural killer cells. *J Exp Med 195:335.*

Pinkoski, M. J., M. Hobman, J. A. Heibein, K. Tomaselli, F. Li, P. Seth, C. J. Froelich, and R. C. Bleackley. 1998. Entry and trafficking of granzyme B in target cells during granzyme B-perforin-mediated apoptosis. *Blood 92:1044.*

Plata, F., J. C. Cerottini, and K. T. Brunner. 1975. Primary and secondary in vitro generation of cytolytic T lymphocytes in the murine sarcoma virus system. *Eur J Immunol 5:227.*

Plata, F., V. Jongeneel, J. C. Cerottini, and K. T. Brunner. 1976a. Antigenic specificity of the cytolytic T lymphocyte (CTL) response to murine sarcoma virus-induced tumors. I. Preferential reactivity of in vitro generated secondary CTL with syngeneic tumor cells. *Eur J Immunol 6:823.*

Plata, R., H. R. MacDonald, and H. D. Engers. 1976b. Characterization of effector lymphocytes associated with immunity to murine sarcoma virus

(MSV) induced tumors. I. Physical properties of cytolytic T lymphocytes generated in vitro and of their immediate progenitors. *J Immunol 117:52.*

Platt, J. L. 2001. The immunological hurdles to cardiac xenotransplantation. *J Card Surg 16:439.*

Plaut, M., J. E. Bubbers, and C. S. Henney. 1976. Studies of the mechanism of lymphocyte-mediated cytolysis. VII. Two stages in the T cell-mediated lytic cycle with distinct cation requirements. *J Immunol 116:150.*

Podack, E. R., and G. Dennert. 1983. Assembly of two types of tubules with putative cytolytic function by cloned natural killer cells. *Nature 302:442.*

Podack, E. R., and P. J. Konigsberg. 1984. Cytolytic T cell granules. Isolation, structural, biochemical, and functional characterization. *J Exp Med 160:695.*

Podack, E. R. 1986. Molecular mechanisms of cytolysis by complement and by cytolytic lymphocytes. *J Cell Biochem 30:133.*

Podack, E. R., H. Hengartner, and M. G. Lichtenheld. 1991. A central role of perforin in cytolysis? *Annu Rev Immunol 9:129.*

Poenie, M., R. Y. Tsien, and A. M. Schmitt-Verhulst. 1987. Sequential activation and lethal hit measured by [Ca2+]i in individual cytolytic T cells and targets. *Embo J 6:2223.*

Porcelli, S., C. T. Morita, and M. B. Brenner. 1992. CD1b restricts the response of human CD4-8- T lymphocytes to a microbial antigen. *Nature 360:593.*

Portier, P., and C. Richet. 1902. De l'action anaphylactique de certains venins. *Comptes Rendus de la Societe de Biologie (Paris). 54:170.*

Potter, T. A., K. Grebe, B. Freiberg, and A. Kupfer. 2001. Formation of supramolecular activation clusters on fresh ex vivo CD8+ T cells after engagement of the T cell antigen receptor and CD8 by antigen-presenting cells. *Proc Natl Acad Sci U S A 98:12624.*

Prehn, R., and J. Main. 1957. Immunity to methylcholanthrene-induced sarcomas. *J Natl Canc Inst 18:769.*

Prehn, R. T. 1973. Destruction of tumor as an "innocent bystander" in an immune response specifically directed against nontumor antigens. *Isr J Med Sci 9:375.*

Purbhoo, M. A., J. M. Boulter, D. A. Price, A. L. Vuidepot, C. S. Hourigan, P. R. Dunbar, K. Olson, S. J. Dawson, R. E. Phillips, B. K. Jakobsen, J. I. Bell, and A. K. Sewell. 2001. The human CD8 coreceptor effects cytotoxic T cell activation and antigen sensitivity primarily by mediating complete phosphorylation of the T cell receptor zeta chain. *J Biol Chem 276:32786.*

Raftery, M. J., M. Schwab, S. M. Eibert, Y. Samstag, H. Walczak, and G. Schonrich. 2001. Targeting the function of mature dendritic cells by human cytomegalovirus: a multilayered viral defense strategy. *Immunity 15:997.*

Rahemtulla, A., W. P. Fung-Leung, M. W. Schilham, T. M. Kundig, S. R. Sambhara, A. Narendran, A. Arabian, A. Wakeham, C. J. Paige, R. M. Zinkernagel, and et al. 1991. Normal development and function of CD8+ cells but markedly decreased helper cell activity in mice lacking CD4. *Nature 353:180.*

Ramm, L. E., M. B. Whitlow, and M. M. Mayer. 1983. Size distribution and stability of the trans-membrane channels formed by complement complex C5b-9. *Mol Immunol 20:155.*

Ramsdell, F. J., and S. H. Golub. 1987. Generation of lymphokine-activated killer cell activity from human thymocytes. *J Immunol 139:1446.*

Ramsdell, F., M. S. Seaman, R. E. Miller, T. W. Tough, M. R. Alderson, and D. H. Lynch. 1994. gld/gld mice are unable to express a functional ligand for Fas. *Eur J Immunol 24:928.*

Ratner, A., and W. R. Clark. 1991. Lack of target cell participation in cytotoxic T lymphocyte-mediated lysis. *J Immunol 147:55.*

Ratner, A., and W. R. Clark. 1993. Role of TNF-alpha in CD8+ cytotoxic T lymphocyte-mediated lysis. *J Immunol 150:4303.*

Raulet, D. H., R. E. Vance, and C. W. McMahon. 2001. Regulation of the natural killer cell receptor repertoire. *Annu Rev Immunol 19:291.*

Ravetch, J. V., and L. L. Lanier. 2000. Immune inhibitory receptors. *Science 290:84.*

Redegeld, F. A., P. Smith, S. Apasov, and M. V. Sitkovsky. 1997. Phosphorylation of T-lymphocyte plasma membrane-associated proteins by ectoprotein kinases: implications for a possible role for ectophosphorylation in T-cell effector functions. *Biochim Biophys Acta 1328:151.*

Reich, A., H. Korner, J. D. Sedgwick, and H. Pircher. 2000. Immune down-regulation and peripheral deletion of CD8 T cells does not require TNF receptor-ligand interactions nor CD95 (Fas, APO-1). *Eur J Immunol 30:678.*

Renggli, J., M. Hahne, H. Matile, B. Betschart, J. Tschopp, and G. Corradin. 1997. Elimination of P. berghei liver stages is independent of Fas (CD95/Apo-I) or perforin-mediated cytotoxicity. *Parasite Immunol 19:145.*

Renkvist, N., C. Castelli, P. F. Robbins, and G. Parmiani. 2001. A listing of human tumor antigens recognized by T cells. *Cancer Immunol Immunother 50:3.*

Rensing-Ehl, A., U. Malipiero, M. Irmler, J. Tschopp, D. Constam, and A. Fontana. 1996. Neurons induced to express major histocompatibility complex class I antigen are killed via the perforin and not the Fas (APO-1/CD95) pathway. *Eur J Immunol 26:2271.*

Restifo, N. P. 2000. Not so Fas: Re-evaluating the mechanisms of immune privilege and tumor escape. *Nat Med 6:493.*

Restifo, N. P. 2001. Countering the 'counterattack' hypothesis. *Nat Med 7:259.*

Reynolds, C. W., M. Bonyhadi, R. B. Herberman, H. A. Young, and S. M. Hedrick. 1985. Lack of gene rearrangement and mRNA expression of the beta chain of the T cell receptor in spontaneous rat large granular lymphocyte leukemia lines. *J Exp Med 161:1249.*

Ridge, J. P., F. Di Rosa, and P. Matzinger. 1998. A conditioned dendritic cell can be a temporal bridge between a CD4+ T-helper and a T-killer cell. *Nature 393:474.*

Riera, L., M. Gariglio, G. Valente, A. Mullbacher, C. Museteanu, S. Landolfo, and M. M. Simon. 2000. Murine cytomegalovirus replication in salivary glands is controlled by both perforin and granzymes during acute infection. *Eur J Immunol 30:1350.*

Riera, L., M. Gariglio, M. Pagano, O. Gaiola, M. M. Simon, and S. Landolfo. 2001. Control of murine cytomegalovirus replication in salivary glands during acute infection is independent of the Fas ligand/Fas system. *New Microbiol 24:231.*

Ritz, J., T. J. Campen, R. E. Schmidt, H. D. Royer, T. Hercend, R. E. Hussey, and E. L. Reinherz. 1985. Analysis of T-cell receptor gene rearrangement and expression in human natural killer clones. *Science 228:1540.*

Ritz, J., R. E. Schmidt, J. Michon, T. Hercend, and S. F. Schlossman. 1988. Characterization of functional surface structures on human natural killer cells. *Adv Immunol 42:181.*

Rivero, M., J. Crespo, E. Fabrega, F. Casafont, M. Mayorga, M. Gomez-Fleitas, and F. Pons-Romero. 2002. Apoptosis mediated by the Fas system in the fulminant hepatitis by hepatitis B virus. *J Viral Hepat 9:107.*

Rivoltini, L., Y. Kawakami, K. Sakaguchi, S. Southwood, A. Sette, P. F. Robbins, F. M. Marincola, M. L. Salgaller, J. R. Yannelli, E. Appella, and et al. 1995. Induction of tumor-reactive CTL from peripheral blood and tumor-infiltrating lymphocytes of melanoma patients by in vitro stimulation with an immunodominant peptide of the human melanoma antigen MART-1. *J Immunol 154:2257.*

Rivoltini, L., M. Radrizzani, P. Accornero, P. Squarcina, C. Chiodoni, A. Mazzocchi, C. Castelli, P. Tarsini, V. Viggiano, F. Belli, M. P. Colombo, and G. Parmiani. 1998. Human melanoma-reactive CD4+ and CD8+ CTL clones resist Fas ligand-induced apoptosis and use Fas/Fas ligand-independent mechanisms for tumor killing. *J Immunol 161:1220.*

Robbins, P. F., M. El-Gamil, Y. F. Li, Y. Kawakami, D. Loftus, E. Appella, and S. A. Rosenberg. 1996. A mutated beta-catenin gene encodes a melanoma-specific antigen recognized by tumor infiltrating lymphocytes. *J Exp Med 183:1185.*

Roberts, P. J., and P. Hayry. 1976. Sponge matrix allografts. A model for analysis of killer cells infiltrating mouse allografts. *Transplantation 21:437.*

Roberts, P. J. 1977. Effector mechanisms in allograft rejection. III. Kinetics of killer cell activity inside the graft and in the immune system during primary and secondary allograft immune responses. *Scand J Immunol 6:635.*

Roberts, A. D., D. J. Ordway, and I. M. Orme. 1993. Listeria monocytogenes infection in beta 2 microglobulin-deficient mice. *Infect Immun 61:1113.*

Robertson, M. J., M. A. Caligiuri, T. J. Manley, H. Levine, and J. Ritz. 1990. Human natural killer cell adhesion molecules. Differential expression after activation and participation in cytolysis. *J Immunol 145:3194.*

Robertson, M. J. 2002. Role of chemokines in the biology of natural killer cells. *J Leukoc Biol 71:173.*

Robinson, H. L. 2002. New hope for an AIDS vaccine. *Nat Rev Immunol 2:239.*

Robles, D. T., and G. S. Eisenbarth. 2001. Type 1A diabetes induced by infection and immunization. *J Autoimmun 16:355.*

Rock, K. L., L. Rothstein, and S. Gamble. 1990. Generation of class I MHC-restricted T-T hybridomas. *J Immunol 145:804.*

Roder, J. C., and R. Kiessling. 1978. Target--effector interaction in the natural killer cell system. I. Covariance and genetic control of cytolytic and target-cell-binding subpopulations in the mouse. *Scand J Immunol 8:135.*

Roder, J. C., R. Kiessling, P. Biberfeld, and B. Andersson. 1978. Target-effector interaction in the natural killer (NK) cell system. II. The isolation of NK cells and studies on the mechanism of killing. *J Immunol 121:2509.*

Roder, J., and A. Duwe. 1979. The beige mutation in the mouse selectively impairs natural killer cell function. *Nature 278:451.*

Roder, J. C., M. L. Lohmann-Matthes, W. Domzig, and H. Wigzell. 1979. The beige mutation in the mouse. II. Selectivity of the natural killer (NK) cell defect. *J Immunol 123:2174.*

Roder, J. C., S. Argov, M. Klein, C. Petersson, R. Kiessling, K. Andersson, and M. Hansson. 1980. Target-effector cell interaction in the natural killer cell system. V. Energy requirements, membrane integrity, and the possible involvement of lysosomal enzymes. *Immunology 40:107.*

Rodrigues, M. M., A. S. Cordey, G. Arreaza, G. Corradin, P. Romero, J. L. Maryanski, R. S. Nussenzweig, and F. Zavala. 1991. CD8+ cytolytic T cell clones derived against the Plasmodium yoelii circumsporozoite protein protect against malaria. *Int Immunol 3:579.*

Rodriguez, A., A. Regnault, M. Kleijmeer, P. Ricciardi-Castagnoli, and S. Amigorena. 1999. Selective transport of internalized antigens to the cytosol for MHC class I presentation in dendritic cells. *Nat Cell Biol 1:362.*

Rola-Pleszczynski, M., and H. Lieu. 1983. Human natural cytotoxic lymphocytes: definition by a monoclonal antibody of a subset which kills an anchorage-dependent target cell line but not the K-562 cell line. *Cell Immunol 82:326.*

Roland, J., and P. A. Cazenave. 1992. Ly-49 antigen defines an alpha beta TCR population in i-IEL with an extrathymic maturation. *Int Immunol 4:699.*

Rollinghoff, M. 1974. Secondary cytotoxic tumor immune response induced in vitro. *J Immunol 112:1718.*

Romani, N., D. Reider, M. Heuer, S. Ebner, E. Kampgen, B. Eibl, D. Niederwieser, and G. Schuler. 1996. Generation of mature dendritic cells from human blood. An improved method with special regard to clinical applicability. *J Immunol Methods 196:137.*

Romero, P., J. L. Maryanski, G. Corradin, R. S. Nussenzweig, V. Nussenzweig, and F. Zavala. 1989. Cloned cytotoxic T cells recognize an epitope in the circumsporozoite protein and protect against malaria. *Nature 341:323.*

Romero, P., J. C. Cerottini, and G. A. Waanders. 1998. Novel methods to monitor antigen-specific cytotoxic T-cell responses in cancer immunotherapy. *Mol Med Today 4:305.*

Romero, P., P. R. Dunbar, D. Valmori, M. Pittet, G. S. Ogg, D. Rimoldi, J. L. Chen, D. Lienard, J. C. Cerottini, and V. Cerundolo. 1998. Ex vivo staining of metastatic lymph nodes by class I major histocompatibility complex tetramers reveals high numbers of antigen-experienced tumor-specific cytolytic T lymphocytes. *J Exp Med 188:1641.*

Romero, P., C. Ortega, A. Palma, I. J. Molina, J. Pena, and M. Santamaria. 2001. Expression of CD94 and NKG2 molecules on human CD4(+) T cells in response to CD3-mediated stimulation. *J Leukoc Biol 70:219.*

Romero P., J. C. Cerottini and D. E. Speiser 2004. Monitoring tumor antigen specific T-cell responses in cancer patients and phase I clinical trials of peptide-based vaccination. Cancer Immunol Immunother. 53:249-55.

Rook, A. H., J. H. Kehrl, L. M. Wakefield, A. B. Roberts, M. B. Sporn, D. B. Burlington, H. C. Lane, and A. S. Fauci. 1986. Effects of transforming growth factor beta on the functions of natural killer cells: depressed cytolytic activity and blunting of interferon responsiveness. *J Immunol 136:3916.*

Rooney, C. M., C. A. Smith, C. Y. Ng, S. K. Loftin, J. W. Sixbey, Y. Gan, D. K. Srivastava, L. C. Bowman, R. A. Krance, M. K. Brenner, and H. E. Heslop. 1998. Infusion of cytotoxic T cells for the prevention and treatment of Epstein-Barr virus-induced lymphoma in allogeneic transplant recipients. *Blood 92:1549.*

Rooney, C. M., C. A. Smith, C. Y. Ng, S. K. Loftin, J. W. Sixbey, Y. Gan, D. K. Srivastava, L. C. Bowman, R. A. Krance, M. K. Brenner, and H. E. Heslop. 1998. Infusion of cytotoxic T cells for the prevention and treatment of Epstein-Barr virus-induced lymphoma in allogeneic transplant recipients. *Blood 92:1549.*

Rosat, J. P., F. Conceicao-Silva, G. A. Waanders, F. Beermann, A. Wilson, M. J. Owen, A. C. Hayday, S. Huang, M. Aguet, H. R. MacDonald, and et al. 1995. Expansion of gamma delta+ T cells in BALB/c mice infected with Leishmania major is dependent upon Th2-type CD4+ T cells. *Infect Immun 63:3000.*

Rosat, J. P., E. P. Grant, E. M. Beckman, C. C. Dascher, P. A. Sieling, D. Frederique, R. L. Modlin, S. A. Porcelli, S. T. Furlong, and M. B. Brenner. 1999. CD1-restricted microbial lipid antigen-specific recognition found in the CD8+ alpha beta T cell pool. *J Immunol 162:366.*

Rosen, D., J. H. Li, S. Keidar, I. Markon, R. Orda, and G. Berke. 2000. Tumor immunity in perforin-deficient mice: a role for CD95 (Fas/APO-1). *J Immunol 164:3229.*

Rosenau, W., and H. D. Moon. 1961. Lysis of homologous cells by sensitized lymphocytes in tissue culture. *J Natl Cancer Inst 27:471.*

Rosenberg, E. B., R. B. Herberman, P. H. Levine, R. H. Halterman, J. L. McCoy, and J. R. Wunderlich. 1972. Lymphocyte cytotoxicity reactions to leukemia-associated antigens in identical twins. *Int J Cancer 9:648.*

Rosenberg, S. A., and W. D. Terry. 1977. Passive immunotherapy of cancer in animals and man. *Adv Cancer Res 25:323.*

Rosenberg, S. A. 1984. Immunotherapy of cancer by systemic administration of lymphoid cells plus interleukin-2. *J Biol Response Mod 3:501.*

Rosenberg, S. A., M. T. Lotze, L. M. Muul, S. Leitman, A. E. Chang, S. E. Ettinghausen, Y. L. Matory, J. M. Skibber, E. Shiloni, J. T. Vetto, and et al. 1985. Observations on the systemic administration of autologous lymphokine-activated killer cells and recombinant interleukin-2 to patients with metastatic cancer. *N Engl J Med 313:1485.*

Rosenberg, A. S., T. Mizuochi, and A. Singer. 1986a. Analysis of T-cell subsets in rejection of Kb mutant skin allografts differing at class I MHC. *Nature 322:829.*

Rosenberg, S. A., P. Spiess, and R. Lafreniere. 1986b. A new approach to the adoptive immunotherapy of cancer with tumor-infiltrating lymphocytes. *Science 233:1318.*

Rosenberg, S. A., M. T. Lotze, L. M. Muul, A. E. Chang, F. P. Avis, S. Leitman, W. M. Linehan, C. N. Robertson, R. E. Lee, J. T. Rubin, and et al. 1987. A progress report on the treatment of 157 patients with advanced cancer using lymphokine-activated killer cells and interleukin-2 or high-dose interleukin-2 alone. *N Engl J Med 316:889.*

Rosenberg, A. S., and A. Singer. 1988. Evidence that the effector mechanism of skin allograft rejection is antigen-specific. *Proc Natl Acad Sci U S A 85:7739.*

Rosenberg, S. A. 1988. Immunotherapy of cancer using interleukin 2: current status and future prospects. *Immunol Today 9:58.*

Rosenberg, S. A. 1997. Cancer vaccines based on the identification of genes encoding cancer regression antigens. *Immunol Today 18:175.*

Rosenberg, S. A., J. Yang, P. Scwartzentruber, P. Hwu, and F. M. Marincola. 1998a. Immunologic and therapeutic evaluation of a synthetic peptide vaccine for the treatment of patients with metastatic melanoma. *Nat Med 4:321.*

Rosenberg, S. A., Y. Zhai, J. C. Yang, D. J. Schwartzentruber, P. Hwu, F. M. Marincola, S. L. Topalian, N. P. Restifo, C. A. Seipp, J. H. Einhorn, B. Roberts, and D. E. White. 1998b. Immunizing patients with metastatic melanoma using recombinant adenoviruses encoding MART-1 or gp100 melanoma antigens. *J Natl Cancer Inst 90:1894.*

Rosenberg, S. A. 1999. A new era of cancer immunotherapy: converting theory to performance. *CA Cancer J Clin 49:70.*

Rosenberg, S. A., J. C. Yang, P. F. Robbins, J. R. Wunderlich, P. Hwu, R. M. Sherry, D. J. Schwartzentruber, S. L. Topalian, N. P. Restifo, A. Filie, R. Chang, and M. E. Dudley. 2003. Cell transfer therapy for cancer: lessons from sequential treatments of a patient with metastatic melanoma. *J Immunother 26:385.*

Rosenstein, M., S. E. Ettinghausen, and S. A. Rosenberg. 1986. Extravasation of intravascular fluid mediated by the systemic administration of recombinant interleukin 2. *J Immunol 137:1735.*

Rosenstreich, D. L., J. J. Farrar, and S. Dougherty. 1976. Absolute macrophage dependency of T lymphocyte activation by mitogens. *J Immunol 116:131.*

Rosenthal, A. S., and E. M. Shevach. 1973. Function of macrophages in antigen recognition by guinea pig T lymphocytes. I. Requirement for histocompatible macrophages and lymphocytes. *J Exp Med 138:1194.*

Rossi, C. P., A. McAllister, M. Tanguy, D. Kagi, and M. Brahic. 1998. Theiler's virus infection of perforin-deficient mice. *J Virol 72:4515.*

Rothenfusser, S., A. Buchwald, S. Kock, S. Ferrone, and P. Fisch. 2002. Missing HLA class I expression on Daudi cells unveils cytotoxic and proliferative responses of human gammadelta T lymphocytes. *Cell Immunol 215:32.*

Roths, J. B., E. D. Murphy, and E. M. Eicher. 1984. A new mutation, gld, that produces lymphoproliferation and autoimmunity in C3H/HeJ mice. *J Exp Med 159:1.*

Rouvier, E., M. F. Luciani, and P. Golstein. 1993. Fas involvement in Ca(2+)-independent T cell-mediated cytotoxicity. *J Exp Med 177:195.*

Ruddle, N. H., and B. H. Waksman. 1967. Cytotoxic effect of lymphocyte-antigen interaction in delayed hypersensitivity. *Science 157:1060.*

Russell, J. H., and C. B. Dobos. 1980. Mechanisms of immune lysis. II. CTL-induced nuclear disintegration of the target begins within minutes of cell contact. *J Immunol 125:1256.*

Russell, J. H., V. Masakowski, T. Rucinsky, and G. Phillips. 1982. Mechanisms of immune lysis. III. Characterization of the nature and kinetics of the cytotoxic T lymphocyte-induced nuclear lesion in the target. *J Immunol 128:2087.*

Russell, J. H. 1983. Internal disintegration model of cytotoxic lymphocyte-induced target damage. *Immunol Rev 72:97.*

Russell, J. H., D. E. Manning, D. E. McCulley, and P. Meleedy-Rey. 1988. Antigen as a positive and negative regulator of proliferation in cytotoxic lymphocytes. A model for the differential regulation of proliferation and lytic activity. *J Immunol 140:1796.*

Russell, J. H., C. L. White, D. Y. Loh, and P. Meleedy-Rey. 1991. Receptor-stimulated death pathway is opened by antigen in mature T cells. *Proc Natl Acad Sci U S A 88:2151.*

Russell, J. H., B. Rush, C. Weaver, and R. Wang. 1993. Mature T cells of autoimmune lpr/lpr mice have a defect in antigen-stimulated suicide. *Proc Natl Acad Sci U S A 90:4409.*

Russell, J. H., and T. J. Ley. 2002. Lymphocyte-mediated cytotoxicity. *Annu Rev Immunol 20:323.*

Russo, V., S. Tanzarella, P. Dalerba, D. Rigatti, P. Rovere, A. Villa, C. Bordignon, and C. Traversari. 2000. Dendritic cells acquire the MAGE-3 human tumor antigen from apoptotic cells and induce a class I-restricted T cell response. *Proc Natl Acad Sci U S A 97:2185.*

Ryan, J. C., J. Turck, E. C. Niemi, W. M. Yokoyama, and W. E. Seaman. 1992. Molecular cloning of the NK1.1 antigen, a member of the NKR-P1 family of natural killer cell activation molecules. *J Immunol 149:1631.*

Rygaard, J., and C. Povlsen. 1974. The mouse nude mutant does not develop spontaneous tumors. *Acta Pathol Microbiol Scand B 82:99.*

Ryser, J. E., and H. R. MacDonald. 1979. Limiting dilution analysis of alloantigen-reactive T lymphocytes. I. Comparison of precursor frequencies for proliferative and cytolytic responses. *J Immunol 122:1691.*

Ryser, J. E., E. Rungger-Brandle, C. Chaponnier, G. Gabbiani, and P. Vassalli. 1982. The area of attachment of cytotoxic T lymphocytes to their target cells shows high motility and polarization of actin, but not myosin. *J Immunol 128:1159.*

Saas, P., and P. Tiberghien. 2002. Dendritic cells: to where do they lead? *Transplantation 73:S12.*

Sad, S., R. Marcotte, and T. R. Mosmann. 1995. Cytokine-induced differentiation of precursor mouse CD8+ T cells into cytotoxic CD8+ T cells secreting Th1 or Th2 cytokines. *Immunity 2:271.*

Sad, S., and T. R. Mosmann. 1995. Interleukin (IL) 4, in the absence of antigen stimulation, induces an anergy-like state in differentiated CD8+ TC1 cells: loss of IL-2 synthesis and autonomous proliferation but retention of cytotoxicity and synthesis of other cytokines. *J Exp Med 182:1505.*

Sad, S., D. Kagi, and T. R. Mosmann. 1996. Perforin and Fas killing by CD8+ T cells limits their cytokine synthesis and proliferation. *J Exp Med 184:1543.*

Saiki, T., T. Ezaki, M. Ogawa, K. Maeda, H. Yagita, and K. Matsuno. 2001a. In vivo roles of donor and host dendritic cells in allogeneic immune response: cluster formation with host proliferating T cells. *J Leukoc Biol 69:705.*

Saiki, T., T. Ezaki, M. Ogawa, and K. Matsuno. 2001b. Trafficking of host- and donor-derived dendritic cells in rat cardiac transplantation: allosensitization in the spleen and hepatic nodes. *Transplantation 71:1806.*

Saleem, S., B. T. Konieczny, R. P. Lowry, F. K. Baddoura, and F. G. Lakkis. 1996. Acute rejection of vascularized heart allografts in the absence of IFNgamma. *Transplantation 62:1908.*

Salerno-Goncalves, R., M. F. Pasetti, and M. B. Sztein. 2002. Characterization of CD8(+) effector T cell responses in volunteers immunized with Salmonella enterica serovar Typhi strain Ty21a typhoid vaccine. *J Immunol 169:2196.*

Salio, M., M. Cella, M. Suter, and A. Lanzavecchia. 1999. Inhibition of dendritic cell maturation by herpes simplex virus. *Eur J Immunol 29:3245.*

Sallusto, F., and A. Lanzavecchia. 1994. Efficient presentation of soluble antigen by cultured human dendritic cells is maintained by granulocyte/macrophage colony-stimulating factor plus interleukin 4 and downregulated by tumor necrosis factor alpha. *J Exp Med 179:1109.*

Sallusto, F., D. Lenig, R. Forster, M. Lipp, and A. Lanzavecchia. 1999. Two subsets of memory T lymphocytes with distinct homing potentials and effector functions. *Nature 401:708.*

Salmaso, C., M. Bagnasco, G. Pesce, P. Montagna, R. Brizzolara, V. Altrinetti, P. Richiusa, A. Galluzzo, and C. Giordano. 2002. Regulation of apoptosis in endocrine autoimmunity: insights from Hashimoto's thyroiditis and Graves' disease. *Ann N Y Acad Sci 966:496.*

Sambhara, S., I. Switzer, A. Kurichh, R. Miranda, L. Urbanczyk, O. James, B. Underdown, M. Klein, and D. Burt. 1998. Enhanced antibody and cytokine responses to influenza viral antigens in perforin-deficient mice. *Cell Immunol 187:13.*

Sanchez, M. J., H. Spits, L. L. Lanier, and J. H. Phillips. 1993. Human natural killer cell committed thymocytes and their relation to the T cell lineage. *J Exp Med 178:1857.*

Sancho, D., M. Nieto, M. Llano, J. L. Rodriguez-Fernandez, R. Tejedor, S. Avraham, C. Cabanas, M. Lopez-Botet, and F. Sanchez-Madrid. 2000. The tyrosine kinase PYK-2/RAFTK regulates natural killer (NK) cell cytotoxic response, and is translocated and activated upon specific target cell recognition and killing. *J Cell Biol 149:1249.*

Sanderson, A. R. 1965. Quantitative titration, kinetic behaviour, and inhibition of cytotoxic mouse isoantisera. *Immunology 9:287.*

Sanderson, C. J., and A. M. Glauert. 1977. The mechanism of T cell mediated cytotoxicity. V. Morphological studies by electron microscopy. *Proc R Soc Lond B Biol Sci 198:315.*

Sanderson, C. J., P. J. Hall, and J. A. Thomas. 1977. The mechanism of T cell mediated cytotoxicity. IV. Studies on communicating junctions between cells in contact. *Proc R Soc Lond B Biol Sci 196:73.*

Sanderson, C. J., and A. M. Glauert. 1979. The mechanism of T-cell mediated cytotoxicity. VI. T-cell projections and their role in target cell killing. *Immunology 36:119.*

Sanderson, C. J. 1981. The mechanism of lymphocyte-mediated cytotoxicity. *Biol Rev Camb Philos Soc 56:153.*

Sano, G., J. C. Hafalla, A. Morrot, R. Abe, J. J. Lafaille, and F. Zavala. 2001. Swift development of protective effector functions in naive CD8(+) T cells against malaria liver stages. *J Exp Med 194:173.*

Santoli, D., G. Trinchieri, and F. S. Lief. 1978. Cell-mediated cytotoxicity against virus-infected target cells in humans. I. Characterization of the effector lymphocyte. *J Immunol 121:526.*

Saper, M. A., P. J. Bjorkman, and D. C. Wiley. 1991. Refined structure of the human histocompatibility antigen HLA-A2 at 2.6 A resolution. *J Mol Biol 219:277.*

Sarin, A., M. Conan-Cibotti, and P. A. Henkart. 1995. Cytotoxic effect of TNF and lymphotoxin on T lymphoblasts. *J Immunol 155:3716.*

Sarin, A., M. S. Williams, M. A. Alexander-Miller, J. A. Berzofsky, C. M. Zacharchuk, and P. A. Henkart. 1997. Target cell lysis by CTL granule exocytosis is independent of ICE/Ced-3 family proteases. *Immunity 6:209.*

Sarin, A., E. K. Haddad, and P. A. Henkart. 1998. Caspase dependence of target cell damage induced by cytotoxic lymphocytes. *J Immunol 161:2810.*

Sasaki, S., R. R. Amara, A. E. Oran, J. M. Smith, and H. L. Robinson. 2001. Apoptosis-mediated enhancement of DNA-raised immune responses by mutant caspases. *Nat Biotechnol 19:543.*

Sato, T., J. H. Laver, Y. Aiba, and M. Ogawa. 1999. NK cell colony formation from human fetal thymocytes. *Exp Hematol 27:726.*

Sauer, H., L. Pratsch, J. Tschopp, S. Bhakdi, and R. Peters. 1991. Functional size of complement and perforin pores compared by confocal laser scanning microscopy and fluorescence microphotolysis. *Biochim Biophys Acta 1063:137.*

Savill, J., V. Fadok, P. Henson, and C. Haslett. 1993. Phagocyte recognition of cells undergoing apoptosis. *Immunol Today 14:131.*

Savinov, A. Y., A. Tcherepanov, E. A. Green, R. A. Flavell, and A. V. Chervonsky. 2003. Contribution of Fas to diabetes development. *Proc Natl Acad Sci U S A 100:628.*

Savoldo, B., M. L. Cubbage, A. G. Durett, J. Goss, M. H. Huls, Z. Liu, L. Teresita, A. P. Gee, P. D. Ling, M. K. Brenner, H. E. Heslop, and C. M. Rooney. 2002. Generation of EBV-specific CD4+ cytotoxic T cells from virus naive individuals. *J Immunol 168:909.*

Saxena, R. K., Q. B. Saxena, and W. H. Adler. 1982. Defective T-cell response in beige mutant mice. *Nature 295:240.*

Schadendorf, D., and F. O. Nestle. 2001. Autologous dendritic cells for treatment of advanced cancer--an update. *Recent Results Cancer Res 158:236.*

Schaible, U., F. Winau, P. A. Sieling, K. Fischer, and S. H. Kaufmann. 2003. Apoptosis facilitates antigen presentation to T lymphocytes through MHC-I and CD1 in tuberculosis. *Nat Med 9:1039.*

Scharton-Kersten, T. M., T. A. Wynn, E. Y. Denkers, S. Bala, E. Grunvald, S. Hieny, R. T. Gazzinelli, and A. Sher. 1996. In the absence of endogenous IFN-gamma, mice develop unimpaired IL-12 responses to Toxoplasma gondii while failing to control acute infection. *J Immunol 157:4045.*

Schick, B., and G. Berke. 1978. Is the presence of serologically defined target cell antigens sufficient for binding of cytotoxic T lymphocytes? *Transplantation 26:14.*

Schick, B., and G. Berke. 1979. Competitive inhibition of cytotoxic T lymphocyte-target cell conjugation. A direct evaluation of membrane antigens involved in cell-mediated immunity. *Transplantation 27:365.*

Schick, B., and G. Berke. 1990. The lysis of cytotoxic T lymphocytes and their blasts by cytotoxic T lymphocytes. *Immunology 71:428.*

Schirmbeck, R., and J. Reimann. 2001. Revealing the potential of DNA-based vaccination: lessons learned from the hepatitis B virus surface antigen. *Biol Chem 382:543.*

Schlitt, H. J., R. Schwinzer, and K. Wonigeit. 1990. Different activation states of human lymphocytes after antibody-mediated stimulation via CD3 and the alpha/beta T-cell receptor. *Scand J Immunol 32:717.*

Schluns, K. S., and L. Lefrancois. 2003. Cytokine control of memory T-cell development and survival. *Nat Rev Immunol 3:269.*

Schmid, D., M. Powell, K. Mahoney, and N. H. Ruddle. 1985. A comparison of lysis mediated by Lyt2+ TNP-specific cytotoxic T lymphocyte lines with that mediated by rapidly internalized lymphotoxin-containing supernatant fluids: evidence for a role of soluble mediators in CTL-mediated killing. *Cell Immunol 93:68.*

Schmidt, R. E., C. Murray, J. F. Daley, S. F. Schlossman, and J. Ritz. 1986. A subset of natural killer cells in peripheral blood displays a mature T cell phenotype. *J Exp Med 164:351.*

Schmitt-Verhulst, A., and G. M. Shearer. 1976. Effects of sodium periodate modification of lymphocytes on the sensitization and lytic phases of T cell-mediated lympholysis. *J Immunol 116:947.*

Schmitz, J. E., M. J. Kuroda, S. Santra, V. G. Sasseville, M. A. Simon, M. A. Lifton, P. Racz, K. Tenner-Racz, M. Dalesandro, B. J. Scallon, J. Ghrayeb, M. A. Forman, D. C. Montefiori, E. P. Rieber, N. L. Letvin, and K. A. Reimann. 1999. Control of viremia in simian immunodeficiency virus infection by CD8+ lymphocytes. *Science 283:857.*

Schoenberger, S. P., R. E. Toes, E. I. van der Voort, R. Offringa, and C. J. Melief. 1998. T-cell help for cytotoxic T lymphocytes is mediated by CD40-CD40L interactions. *Nature 393:480.*

Schulz, M., R. M. Zinkernagel, and H. Hengartner. 1991. Peptide-induced antiviral protection by cytotoxic T cells. *Proc Natl Acad Sci U S A 88:991.*

Schulz, M., H. J. Schuurman, J. Joergensen, C. Steiner, T. Meerloo, D. Kagi, H. Hengartner, R. M. Zinkernagel, M. H. Schreier, K. Burki, and et al. 1995. Acute rejection of vascular heart allografts by perforin-deficient mice. *Eur J Immunol 25:474.*

Schulze-Osthoff, K., H. Walczak, W. Droge, and P. Krammer. 1994. Cell nucleus and DNA fragmentation are not required for apoptosis. *J Cell Biol 127:15.*

Seaman, M. S., B. Perarnau, K. F. Lindahl, F. A. Lemonnier, and J. Forman. 1999. Response to Listeria monocytogenes in mice lacking MHC class Ia molecules. *J Immunol 162:5429.*

Sedegah, M., R. Hedstrom, P. Hobart, and S. L. Hoffman. 1994. Protection against malaria by immunization with plasmid DNA encoding circumsporozoite protein. *Proc Natl Acad Sci U S A 91:9866.*

Seder, R. A., J. L. Boulay, F. Finkelman, S. Barbier, S. Z. Ben-Sasson, G. Le Gros, and W. E. Paul. 1992. CD8+ T cells can be primed in vitro to produce IL-4. *J Immunol 148:1652.*

Sedger, L. M., M. B. Glaccum, J. C. Schuh, S. T. Kanaly, E. Williamson, N. Kayagaki, T. Yun, P. Smolak, T. Le, R. Goodwin, and B. Gliniak. 2002. Characterization of the in vivo function of TNF-alpha-related apoptosis-inducing ligand, TRAIL/Apo2L, using TRAIL/Apo2L gene-deficient mice. *Eur J Immunol 32:2246.*

Seino, K., N. Kayagaki, H. Bashuda, K. Okumura, and H. Yagita. 1996. Contribution of Fas ligand to cardiac allograft rejection. *Int Immunol 8:1347.*

Seki, S., T. Abo, T. Ohteki, K. Sugiura, and K. Kumagai. 1991. Unusual alpha beta-T cells expanded in autoimmune lpr mice are probably a counterpart of normal T cells in the liver. *J Immunol 147:1214.*

Seki, N., A. Brooks, C. Carter, T. Back, E. Parsoneault, M. Smyth, R. Wiltrout, and T. Sayers. 2002. Tumor-specific CTL kill murine renal cancer cells using both perforin and Fas ligand-mediated lysis in vitro, but cause tumor regression in vivo in the absence of perforin. *J Immunol 168:3484.*

Sekiya, M., A. Ohwada, M. Katae, T. Dambara, I. Nagaoka, and Y. Fukuchi. 2002. Adenovirus vector-mediated transfer of 9 kDa granulysin induces DNA fragmentation in HuD antigen-expressing small cell lung cancer murine model cells. *Respirology 7:29.*

Selin, D., D. F. H. Wallach, and H. Fischer. 1971. Intercellular communication in cell-mediated cytotoxicity. Fluorescein transfer between H-2^d target cells and H-2^b lymphocytes in vitro. *Eur J Immunol 1:453.*

Selin, L. K., P. A. Santolucito, A. K. Pinto, E. Szomolanyi-Tsuda, and R. M. Welsh. 2001. Innate immunity to viruses: control of vaccinia virus infection by gamma delta T cells. *J Immunol 166:6784.*

Selvaggi, G., C. Ricordi, E. R. Podack, and L. Inverardi. 1996. The role of the perforin and Fas pathways of cytotoxicity in skin graft rejection. *Transplantation 62:1912.*

Sendo, F., T. Aoki, E. A. Boyse, and C. K. Buafo. 1975. Natural occurrence of lymphocytes showing cytotoxic activity to BALB/c radiation-induced leukemia RL male 1 cells. *J Natl Cancer Inst 55:603.*

Seth, A., L. Gote, M. Nagarkatti, and P. S. Nagarkatti. 1991. T-cell-receptor-independent activation of cytolytic activity of cytotoxic T lymphocytes mediated through CD44 and gp90MEL-14. *Proc Natl Acad Sci U S A 88:7877.*

Sevilla, C. L., G. Radcliff, N. H. Mahle, S. Swartz, M. D. Sevilla, J. Chores, and D. M. Callewaert. 1989. Multiple mechanisms of target cell disintegration are employed in cytotoxicity reactions mediated by human natural killer cells. *Nat Immun Cell Growth Regul 8:20.*

Sewell, A. K., D. A. Price, A. Oxenius, A. D. Kelleher, and R. E. Phillips. 2000. Cytotoxic T lymphocyte responses to human immunodeficiency virus: control and escape. *Stem Cells 18:230.*

Sewell, A. K., D. A. Price, A. Oxenius, A. D. Kelleher, and R. Phillips. 2000. Cytotoxic T lymphocyte response to human immunodeficiency virus: control and escape. *Stem Cells 18:230.*

Sgonc, R., M. S. Gruschwitz, H. Dietrich, H. Recheis, M. E. Gershwin, and G. Wick. 1996. Endothelial cell apoptosis is a primary pathogenetic event underlying skin lesions in avian and human scleroderma. *J Clin Invest 98:785.*

Sgonc, R., M. S. Gruschwitz, G. Boeck, N. Sepp, J. Gruber, and G. Wick. 2000. Endothelial cell apoptosis in systemic sclerosis is induced by antibody-dependent cell-mediated cytotoxicity via CD95. *Arthritis Rheum 43:2550.*

Shahinian, A., K. Pfeffer, K. P. Lee, T. M. Kundig, K. Kishihara, A. Wakeham, K. Kawai, P. S. Ohashi, C. B. Thompson, and T. W. Mak. 1993. Differential T cell costimulatory requirements in CD28-deficient mice. *Science 261:609.*

Shankar, P., Z. Xu, and J. Lieberman. 1999. Viral-specific cytotoxic T lymphocytes lyse human immunodeficiency virus-infected primary T lymphocytes by the granule exocytosis pathway. *Blood 94:3084.*

Shankaran, V., H. Ikeda, A. T. Bruce, J. M. White, P. E. Swanson, L. J. Old, and R. D. Schreiber. 2001. IFNgamma and lymphocytes prevent primary tumour development and shape tumour immunogenicity. *Nature 410:1107.*

Sharif, S., G. A. Arreaza, P. Zucker, Q. S. Mi, and T. L. Delovitch. 2002. Regulation of autoimmune disease by natural killer T cells. *J Mol Med 80:290.*

Sharon, N., and H. Lis. 1972. Lectins: cell-agglutinating and sugar-specific proteins. *Science 177:949.*

Shaw, S., G. E. Luce, R. Quinones, R. E. Gress, T. A. Springer, and M. E. Sanders. 1986. Two antigen-independent adhesion pathways used by human cytotoxic T-cell clones. *Nature 323:262.*

Shellam, G. R., J. E. Allan, J. M. Papadimitriou, and G. J. Bancroft. 1981. Increased susceptibility to cytomegalovirus infection in beige mutant mice. *Proc Natl Acad Sci U S A 78:5104.*

Shi, L., S. Mai, S. Israels, K. Browne, J. A. Trapani, and A. H. Greenberg. 1997. Granzyme B (GraB) autonomously crosses the cell membrane and perforin initiates apoptosis and GraB nuclear localization. *J Exp Med 185:855.*

Shibuya, K., L. L. Lanier, J. H. Phillips, H. D. Ochs, K. Shimizu, E. Nakayama, H. Nakauchi, and A. Shibuya. 1999. Physical and functional association of LFA-1 with DNAM-1 adhesion molecule. *Immunity 11:615.*

Shier, P., K. Ngo, and W. P. Fung-Leung. 1999. Defective CD8+ T cell activation and cytolytic function in the absence of LFA-1 cannot be restored by increased TCR signaling. *J Immunol 163:4826.*

Shimizu, Y., and R. DeMars. 1989. Demonstration by class I gene transfer that reduced susceptibility of human cells to natural killer cell-mediated lysis is inversely correlated with HLA class I antigen expression. *Eur J Immunol 19:447.*

Shimizu, M., A. Fontana, Y. Takeda, H. Yagita, T. Yoshimoto, and A. Matsuzawa. 1999. Induction of antitumor immunity with Fas/APO-1 ligand (CD95L)-transfected neuroblastoma neuro-2a cells. *J Immunol 162:7350.*

Shimizu, H., T. Matsuguchi, Y. Fukuda, I. Nakano, T. Hayakawa, O. Takeuchi, S. Akira, M. Umemura, T. Suda, and Y. Yoshikai. 2002. Toll-like receptor 2 contributes to liver injury by Salmonella infection through Fas ligand expression on NKT cells in mice. *Gastroenterology 123:1265.*

Shimonkevitz, R., C. Colburn, J. A. Burnham, R. S. Murray, and B. L. Kotzin. 1993. Clonal expansions of activated gamma/delta T cells in recent-onset multiple sclerosis. *Proc Natl Acad Sci U S A 90:923.*

Shinkai, Y., H. Ishikawa, M. Hattori, and K. Okumura. 1988a. Resistance of mouse cytolytic cells to pore-forming protein-mediated cytolysis. *Eur J Immunol 18:29.*

Shinkai, Y., K. Takio, and K. Okumura. 1988b. Homology of perforin to the ninth component of complement (C9). *Nature 334:525.*

Shinkai, Y., M. C. Yoshida, K. Maeda, T. Kobata, K. Maruyama, J. Yodoi, H. Yagita, and K. Okumura. 1989. Molecular cloning and chromosomal assignment of a human perforin (PFP) gene. *Immunogenetics 30:452.*

Shinkai, Y., G. Rathbun, K. P. Lam, E. M. Oltz, V. Stewart, M. Mendelsohn, J. Charron, M. Datta, F. Young, A. M. Stall, and et al. 1992. RAG-2-deficient mice lack mature lymphocytes owing to inability to initiate V(D)J rearrangement. *Cell 68:855.*

Shiver, J. W., and P. A. Henkart. 1991. A noncytotoxic mast cell tumor line exhibits potent IgE-dependent cytotoxicity after transfection with the cytolysin/perforin gene. *Cell 64:1175.*

Shiver, J. W., L. Su, and P. A. Henkart. 1992. Cytotoxicity with target DNA breakdown by rat basophilic leukemia cells expressing both cytolysin and granzyme A. *Cell 71:315.*

Shortman, K., and Y. J. Liu. 2002. Mouse and human dendritic cell subtypes. *Nat Rev Immunol 2:151.*

Sili, U., M. H. Huls, A. R. Davis, S. Gottschalk, M. K. Brenner, H. E. Heslop, and C. M. Rooney. 2003. Large-scale expansion of dendritic cell-primed polyclonal human cytotoxic T-lymphocyte lines using lymphoblastoid cell lines for adoptive immunotherapy. *J Immunother 26:241.*

Siliciano, R. F., and C. S. Henney. 1978. Studies on the mechanism of lymphocyte-mediated cytolysis. X. Enucleated cells as targets for cytotoxic attack. *J Immunol 121:186.*

Silva, C. L., M. F. Silva, R. C. Pietro, and D. B. Lowrie. 1994. Protection against tuberculosis by passive transfer with T-cell clones recognizing mycobacterial heat-shock protein 65. *Immunology 83:341.*

Simon, M. M., M. Hausmann, T. Tran, K. Ebnet, J. Tschopp, R. ThaHla, and A. Mullbacher. 1997. In vitro- and ex vivo-derived cytolytic leukocytes from granzyme A x B double knockout mice are defective in granule-mediated apoptosis but not lysis of target cells. *J Exp Med 186:1781.*

Simone, C. B., and P. Henkart. 1980. Permeability changes induced in erthrocyte ghost targets by antibody-dependent cytotoxic effector cells: evidence for membrane pores. *J Immunol 124:954.*

Sin, J. I., J. Kim, C. Pachuk, D. B. Weiner, and C. Patchuk. 2000. Interleukin 7 can enhance antigen-specific cytotoxic-T-lymphocyte and/or Th2-type immune responses in vivo. *Clin Diagn Lab Immunol. 7:751.*

Sinclair, N. R. 2000. Immunoreceptor tyrosine-based inhibitory motifs on activating molecules. *Crit Rev Immunol 20:89.*

Sindhu, S. T., R. Ahmad, R. Morisset, A. Ahmad, and J. Menezes. 2003. Peripheral blood cytotoxic gammadelta T lymphocytes from patients with human immunodeficiency virus type 1 infection and AIDS lyse uninfected CD4+ T cells, and their cytocidal potential correlates with viral load. *J Virol 77:1848.*

Singer, G. G., and A. K. Abbas. 1994. The fas antigen is involved in peripheral but not thymic deletion of T lymphocytes in T cell receptor transgenic mice. *Immunity 1:365.*

Singh, N., S. Hong, D. C. Scherer, I. Serizawa, N. Burdin, M. Kronenberg, Y. Koezuka, and L. Van Kaer. 1999. Cutting edge: activation of NK T cells by CD1d and alpha-galactosylceramide directs conventional T cells to the acquisition of a Th2 phenotype. *J Immunol 163:2373.*

Sjogren, H. 1964. Studies on specific transplant resistance to polyoma virus-induced tumors. II. Mechanism of resistance induced by polyoma virus infection. *J Natl Cancer Inst 32:375.*

Sjögren, H., I. Hellström, and G. Klein. 1961. Transplantation of polyoma virus-induced tumors in mice. *Canc Res 21:329.*

Skinner, M., and J. Marbrook. 1987. The most efficient cytotoxic T lymphocytes are the least susceptible to lysis. *J Immunol 139:985.*

Slansky, J. E., F. M. Rattis, L. F. Boyd, T. Fahmy, E. M. Jaffee, J. P. Schneck, D. H. Margulies, and D. M. Pardoll. 2000. Enhanced antigen-specific antitumor immunity with altered peptide ligands that stabilize the MHC-peptide-TCR complex. *Immunity 13:529.*

Slifka, M. K., R. R. Pagarigan, and J. L. Whitton. 2000. NK markers are expressed on a high percentage of virus-specific CD8+ and CD4+ T cells. *J Immunol 164:2009.*

Slovin, S. F., R. D. Lackman, S. Ferrone, P. E. Kiely, and M. J. Mastrangelo. 1986. Cellular immune response to human sarcomas: cytotoxic T cell clones reactive with autologous sarcomas. I. Development, phenotype, and specificity. *J Immunol 137:3042.*

Small, E. J., P. Fratesi, D. M. Reese, G. Strang, R. Laus, M. V. Peshwa, and F. H. Valone. 2000. Immunotherapy of hormone-refractory prostate cancer with antigen-loaded dendritic cells. *J Clin Oncol 18:3894.*

Smith, C. A., T. Farrah, and R. G. Goodwin. 1994. The TNF receptor superfamily of cellular and viral proteins: activation, costimulation, and death. *Cell 76:959.*

Smith, K. M., J. Wu, A. B. Bakker, J. H. Phillips, and L. L. Lanier. 1998. Ly-49D and Ly-49H associate with mouse DAP12 and form activating receptors. *J Immunol 161:7.*

Smith, H. R., H. H. Chuang, L. L. Wang, M. Salcedo, J. W. Heusel, and W. M. Yokoyama. 2000. Nonstochastic coexpression of activation receptors on murine natural killer cells. *J Exp Med 191:1341.*

Smith, H. R., J. W. Heusel, I. K. Mehta, S. Kim, B. G. Dorner, O. V. Naidenko, K. Iizuka, H. Furukawa, D. L. Beckman, J. T. Pingel, A. A. Scalzo, D. H. Fremont, and W. M. Yokoyama. 2002. Recognition of a virus-encoded ligand by a natural killer cell activation receptor. *Proc Natl Acad Sci U S A 99:8826.*

Smyth, M. J., M. D. O'Connor, and J. A. Trapani. 1996. Granzymes: a variety of serine protease specificities encoded by genetically distinct subfamilies. *J Leukoc Biol 60:555.*

Smyth, M. J., and J. D. Sedgwick. 1998. Delayed kinetics of tumor necrosis factor-mediated bystander lysis by peptide-specific CD8+ cytotoxic T lymphocytes. *Eur J Immunol 28:4162.*

Smyth, M. J., K. Y. Thia, E. Cretney, J. M. Kelly, M. B. Snook, C. A. Forbes, and A. A. Scalzo. 1999. Perforin is a major contributor to NK cell control of tumor metastasis. *J Immunol 162:6658.*

Smyth, M. J., and R. W. Johnstone. 2000a. Role of TNF in lymphocyte-mediated cytotoxicity. *Microsc Res Tech 50:196.*

Smyth, M. J., K. Y. Thia, S. E. Street, D. MacGregor, D. I. Godfrey, and J. A. Trapani. 2000b. Perforin-mediated cytotoxicity is critical for surveillance of spontaneous lymphoma. *J Exp Med 192:755.*

Smyth, M. J., K. Y. Thia, S. E. Street, E. Cretney, J. A. Trapani, M. Taniguchi, T. Kawano, S. B. Pelikan, N. Y. Crowe, and D. I. Godfrey. 2000c. Differential tumor surveillance by natural killer (NK) and NKT cells. *J Exp Med 191:661.*

Smyth, M. J., N. Y. Crowe, and D. I. Godfrey. 2001a. NK cells and NKT cells collaborate in host protection from methylcholanthrene-induced fibrosarcoma. *Int Immunol 13:459.*

Smyth, M. J., J. M. Kelly, V. R. Sutton, J. E. Davis, K. A. Browne, T. J. Sayers, and J. A. Trapani. 2001b. Unlocking the secrets of cytotoxic granule proteins. *J Leukoc Biol 70:18.*

Smyth, M. J., and J. A. Trapani. 2001c. Lymphocyte-mediated immunosurveillance of epithelial cancers? *Trends Immunol 22:409.*

Snell, G. 1957. The homograft reaction. *Ann Rev Microbiol 11:439.*

Sobel, R. A. 1989. T-lymphocyte subsets in the multiple sclerosis lesion. *Res Immunol 140:208.*

Solliday, S., and F. H. Bach. 1970. Cytotoxicity: specificity after in vitro sensitization. *Science 170:1406.*

Soloski, M. J., and E. S. Metcalf. 2001. The involvement of class Ib molecules in the host response to infection with Salmonella and its relevance to autoimmunity. *Microbes Infect 3:1249.*

Somersalo, K., O. Carpen, and E. Saksela. 1994. Stimulated natural killer cells secrete factors with chemotactic activity, including NAP-1/IL-8, which supports VLA-4- and VLA-5-mediated migration of T lymphocytes. *Eur J Immunol 24:2957.*

Sourdive, D. J., K. Murali-Krishna, J. D. Altman, A. J. Zajac, J. K. Whitmire, C. Pannetier, P. Kourilsky, B. Evavold, A. Sette, and R. Ahmed. 1998. Conserved T cell receptor repertoire in primary and memory CD8 T cell responses to an acute viral infection. *J Exp Med 188:71.*

Spada, F. M., E. P. Grant, P. J. Peters, M. Sugita, A. Melian, D. S. Leslie, H. K. Lee, E. van Donselaar, D. A. Hanson, A. M. Krensky, O. Majdic, S. A. Porcelli, C. T. Morita, and M. B. Brenner. 2000. Self-recognition of CD1 by gamma/delta T cells: implications for innate immunity. *J Exp Med 191:937.*

Spaeny-Dekking, E. H., A. M. Kamp, C. J. Froelich, and C. E. Hack. 2000. Extracellular granzyme A, complexed to proteoglycans, is protected against inactivation by protease inhibitors. *Blood 95:1465.*

Spaner, D., K. Raju, L. Radvanyi, Y. Lin, and R. G. Miller. 1998. A role for perforin in activation-induced cell death. *J Immunol 160:2655.*

Spaner, D., K. Raju, B. Rabinovich, R. Miller. 1999. A role for perforin in activation-induced T cell death in vivo: increased expansion of allogeneic perforin-deficient T cells in SCID mice. *J Immunol 162:1192.*

Spence, H. J., Y. J. Chen, and S. J. Winder. 2002. Muscular dystrophies, the cytoskeleton and cell adhesion. *Bioessays 24:542.*

Spencer, M. J., C. M. Walsh, K. A. Dorshkind, E. M. Rodriguez, and J. G. Tidball. 1997. Myonuclear apoptosis in dystrophic mdx muscle occurs by perforin-mediated cytotoxicity. *J Clin Invest 99:2745.*

Spencer, M. J., E. Montecino-Rodriguez, K. Dorshkind, and J. G. Tidball. 2001. Helper (CD4(+)) and cytotoxic (CD8(+)) T cells promote the pathology of dystrophin-deficient muscle. *Clin Immunol 98:235.*

Spencer, M. J., and J. G. Tidball. 2001. Do immune cells promote the pathology of dystrophin-deficient myopathies? *Neuromuscul Disord 11:556.*

Sperling, A. I., P. S. Linsley, T. A. Barrett, and J. A. Bluestone. 1993. CD28-mediated costimulation is necessary for the activation of T cell receptor-gamma delta+ T lymphocytes. *J Immunol 151:6043.*

Sprent, J., and J. F. Miller. 1971. Activation of thymus cells by histocompatibility antigens. *Nat New Biol 234:195.*

Sprent, J., and M. Schaefer. 1985. Properties of purified T cell subsets. I. In vitro responses to class I vs. class II H-2 alloantigens. *J Exp Med 162:2068.*

Sprent, J., and C. D. Surh. 2001. Generation and maintenance of memory T cells. *Curr Opin Immunol 13:248.*

Sprent, J., and C. D. Surh. 2002. T cell memory. *Annu Rev Immunol 20:551.*

Springer, T. A., M. L. Dustin, T. K. Kishimoto, and S. D. Marlin. 1987. The lymphocyte function-associated LFA-1, CD2, and LFA-3 molecules: cell adhesion receptors of the immune system. *Annu Rev Immunol 5:223.*

Spritz, R. A. 1998. Genetic defects in Chediak-Higashi syndrome and the beige mouse. *J Clin Immunol 18:97.*

Staerz, U. D., and M. J. Bevan. 1985. Cytotoxic T lymphocyte-mediated lysis via the Fc receptor of target cells. *Eur J Immunol 15:1172.*

Staerz, U. D., O. Kanagawa, and M. J. Bevan. 1985. Hybrid antibodies can target sites for attack by T cells. *Nature 314:628.*

Staerz, U. D., and M. J. Bevan. 1986. Activation of resting T lymphocytes by a monoclonal antibody directed against an allotypic determinant on the T cell receptor. *Eur J Immunol 16:263.*

Stalder, T., S. Hahn, and P. Erb. 1994. Fas antigen is the major target molecule for CD4+ T cell-mediated cytotoxicity. *J Immunol 152:1127.*

Stassi, G., M. Todaro, P. Richiusa, M. Giordano, A. Mattina, M. S. Sbriglia, A. Lo Monte, G. Buscemi, A. Galluzzo, and C. Giordano. 1995. Expression of apoptosis-inducing CD95 (Fas/Apo-1) on human beta-cells sorted by flow-cytometry and cultured in vitro. *Transplant Proc 27:3271.*

Stassi, G., R. De Maria, G. Trucco, W. Rudert, R. Testi, A. Galluzzo, C. Giordano, and M. Trucco. 1997. Nitric oxide primes pancreatic beta cells for Fas-mediated destruction in insulin-dependent diabetes mellitus. *J Exp Med 186:1193.*

Steinberg, T. H., A. S. Newman, J. A. Swanson, and S. C. Silverstein. 1987. ATP4- permeabilizes the plasma membrane of mouse macrophages to fluorescent dyes. *J Biol Chem 262:8884.*

Steinmuller, D. 1967. Immunization with skin isografts taken from tolerant mice. *Science 158:127.*

Stenger, S., R. J. Mazzaccaro, K. Uyemura, S. Cho, P. F. Barnes, J. P. Rosat, A. Sette, M. B. Brenner, S. A. Porcelli, B. R. Bloom, and R. L. Modlin. 1997. Differential effects of cytolytic T cell subsets on intracellular infection. *Science 276:1684.*

Stenger, S., D. A. Hanson, R. Teitelbaum, P. Dewan, K. R. Niazi, C. J. Froelich, T. Ganz, S. Thoma-Uszynski, A. Melian, C. Bogdan, S. A. Porcelli, B. R. Bloom, A. M. Krensky, and R. L. Modlin. 1998. An antimicrobial activity of cytolytic T cells mediated by granulysin. *Science 282:121.*

Stenger, S., D. A. Hanson, R. Teitelbaum, P. Dewan, K. R. Niazi, C. J. Froelich, T. Ganz, S. Thoma-Uszynski, A. Melian, C. Bogdan, S. A. Porcelli, B. R. Bloom, A. M. Krensky, and R. L. Modlin. 1998. An antimicrobial activity of cytolytic T cells mediated by granulysin. *Science 282:121.*

Stenger, S. 2001. Cytolytic T cells in the immune response to mycobacterium tuberculosis. *Scand J Infect Dis 33:483.*

Stepp, S. E., R. Dufourcq-Lagelouse, F. Le Deist, S. Bhawan, S. Certain, P. A. Mathew, J. I. Henter, M. Bennett, A. Fischer, G. de Saint Basile, and V. Kumar. 1999. Perforin gene defects in familial hemophagocytic lymphohistiocytosis. *Science 286:1957.*

Stepp, S. E., P. A. Mathew, M. Bennett, G. de Saint Basile, and V. Kumar. 2000. Perforin: more than just an effector molecule. *Immunol Today 21:254.*

Sterkers, G., J. Michon, Y. Henin, E. Gomard, C. Hannoun, and J. P. Levy. 1985. Fine specificity analysis of human influenza-specific cloned cell lines. *Cell Immunol 94:394.*

Stevens, R. L., M. M. Kamada, and W. E. Serafin. 1989. Structure and function of the family of proteoglycans that reside in the secretory granules of natural killer cells and other effector cells of the immune response. *Curr Top Microbiol Immunol 140:93.*

Stinchcombe, J. C., L. J. Page, and G. M. Griffiths. 2000. Secretory lysosome biogenesis in cytotoxic T lymphocytes from normal and Chediak Higashi syndrome patients. *Traffic 1:435.*

Stinchcombe, J. C., G. Bossi, S. Booth, and G. M. Griffiths. 2001. The immunological synapse of CTL contains a secretory domain and membrane bridges. *Immunity 15:751.*

Stinissen, P., J. Zhang, C. Vandevyver, G. Hermans, and J. Raus. 1998. Gammadelta T cell responses to activated T cells in multiple sclerosis patients induced by T cell vaccination. *J Neuroimmunol 87:94.*

Stokes, T., M. Rymaszewski, P. Arscott, S. Wang, J. Bretz, J. Bartron, and J. Baker. 1998. Constitutive expression of FasL in thyrocytes. *Science 279:2015a.*

Storkus, W. J., D. N. Howell, R. D. Salter, J. R. Dawson, and P. Cresswell. 1987. NK susceptibility varies inversely with target cell class I HLA antigen expression. *J Immunol 138:1657.*

Storkus, W. J., J. Alexander, J. A. Payne, J. R. Dawson, and P. Cresswell. 1989. Reversal of natural killing susceptibility in target cells expressing transfected class I HLA genes. *Proc Natl Acad Sci U S A 86:2361.*

Strand, S., W. J. Hofmann, H. Hug, M. Muller, G. Otto, D. Strand, S. M. Mariani, W. Stremmel, P. H. Krammer, and P. R. Galle. 1996. Lymphocyte apoptosis induced by CD95 (APO-1/Fas) ligand-expressing tumor cells--a mechanism of immune evasion? *Nat Med 2:1361.*

Street, S. E., E. Cretney, and M. J. Smyth. 2001. Perforin and interferon-gamma activities independently control tumor initiation, growth, and metastasis. *Blood 97:192.*

Street, S. E., J. A. Trapani, D. MacGregor, and M. J. Smyth. 2002. Suppression of lymphoma and epithelial malignancies effected by interferon gamma. *J Exp Med 196:129.*

Streilein, J. W. 1972. Pathologic lesions of GVH disease in hamsters: antigenic target versus 'innocent bystander'. *Prog Exp Tumor Res 16:396.*

Stremmel, C., M. Exley, S. Balk, W. Hohenberger, and V. K. Kuchroo. 2001. Characterization of the phenotype and function of CD8(+), alpha / beta(+) NKT cells from tumor-bearing mice that show a natural killer cell activity and lyse multiple tumor targets. *Eur J Immunol 31:2818.*

Strober, S., and J. Gowans. 1965. The role of lymphocytes in the sensitization of rats to renal homografts. *J Exp Med 122:347.*

Stulting, R. D., and G. Berke. 1973. Nature of lymphocyte-tumor interaction. A general method for cellular immunoabsorption. *J Exp Med 137:932.*

Stulting, R. D., R. Todd, and B. Amos. 1975. Lymphocyte-mediated cytolysis of allogeneic tumor cells in vitro. II. Binding of cytotoxic lymphocytes to formaldehyde-fixed target cells. *Cell Immunol 20:54.*

Stutman, O. 1974. Tumor development after 3-methylcholanthrene in immunologically deficient athymic-nude mice. *Science 183:534.*

Stutman, O., C. J. Paige, and E. F. Figarella. 1978. Natural cytotoxic cells against solid tumors in mice. I. Strain and age distribution and target cell susceptibility. *J Immunol 121:1819.*

Su, X., Q. Hu, J. M. Kristan, C. Costa, Y. Shen, D. Gero, L. A. Matis, and Y. Wang. 2000. Significant role for Fas in the pathogenesis of autoimmune diabetes. *J Immunol 164:2523.*

Suarez-Pinzon, W., O. Sorensen, R. C. Bleackley, J. F. Elliott, R. V. Rajotte, and A. Rabinovitch. 1999. Beta-cell destruction in NOD mice correlates with Fas (CD95) expression on beta-cells and proinflammatory cytokine expression in islets. *Diabetes 48:21.*

Suarez-Pinzon, W. L., R. F. Power, and A. Rabinovitch. 2000. Fas ligand-mediated mechanisms are involved in autoimmune destruction of islet beta cells in non-obese diabetic mice. *Diabetologia 43:1149.*

Suda, T., T. Takahashi, P. Golstein, and S. Nagata. 1993. Molecular cloning and expression of the Fas ligand, a novel member of the tumor necrosis factor family. *Cell 75:1169.*

Sugawa, S., D. Palliser, H. N. Eisen, and J. Chen. 2002. How do cultured CD8(+) murine T cell clones survive repeated ligation of the TCR? *Int Immunol 14:23.*

Sumida, T., M. Furukawa, A. Sakamoto, T. Namekawa, T. Maeda, M. Zijlstra, I. Iwamoto, T. Koike, S. Yoshida, H. Tomioka, and et al. 1994. Prevention of insulitis and diabetes in beta 2-microglobulin-deficient non-obese diabetic mice. *Int Immunol 6:1445.*

Sumida, T., A. Sakamoto, H. Murata, Y. Makino, H. Takahashi, S. Yoshida, K. Nishioka, I. Iwamoto, and M. Taniguchi. 1995. Selective reduction of T cells bearing invariant V alpha 24J alpha Q antigen receptor in patients with systemic sclerosis. *J Exp Med 182:1163.*

Sun, D., J. N. Whitaker, Z. Huang, D. Liu, C. Coleclough, H. Wekerle, and C. S. Raine. 2001. Myelin antigen-specific CD8+ T cells are encephalitogenic and produce severe disease in C57BL/6 mice. *J Immunol 166:7579.*

Sun J.C., M. A. Williams and m. J. Bevan 2004. CD4(+) T cells are required for the maintenance, not programming, of memory CD8(+) T cells after acute infection. Nat Immunol. 5:927-33.

Sung, K. L., L. A. Sung, M. Crimmins, S. J. Burakoff, and S. Chien. 1986. Determination of junction avidity of cytolytic T cell and target cell. *Science 234:1405.*

Sura, S. N., I. Y. Chernyakhovskaya, Z. G. Kadaghidze, B. B. Fuks, and G. J. Svet-Moldavsky. 1967. Cytochemical study of interaction between lymphocytes and target cells in tissue culture. *Exp Cell Res 48:656.*

Suzuki, G., Y. Kawase, S. Koyasu, I. Yahara, Y. Kobayashi, and R. H. Schwartz. 1988. Antigen-induced suppression of the proliferative response of T cell clones. *J Immunol 140:1359.*

Suzuki, Y., and J. S. Remington. 1988. Dual regulation of resistance against Toxoplasma gondii infection by Lyt-2+ and Lyt-1+, L3T4+ T cells in mice. *J Immunol 140:3943.*

Suzuki, I., and P. Fink. 2000. The dual functions of fas ligand in the regulation of peripheral CD8 and CD4 T cells. *Proc Natl Acad Sci U S A 97:1707.*

Svane, I. M., A. M. Engel, M. B. Nielsen, H. G. Ljunggren, J. Rygaard, and O. Werdelin. 1996. Chemically induced sarcomas from nude mice are more immunogenic than similar sarcomas from congenic normal mice. *Eur J Immunol 26:1844.*

Swain, S. L., G. Dennert, S. Wormsley, and R. W. Dutton. 1981. The Lyt phenotype of a long-term allospecific T cell line. Both helper and killer activities to IA are mediated by Ly-1 cells. *Eur J Immunol 11:175.*

Swain, S. L., H. Hu, and G. Huston. 1999. Class II-independent generation of CD4 memory T cells from effectors. *Science 286:1381.*

Sykes, M. 1990. Unusual T cell populations in adult murine bone marrow. Prevalence of CD3+CD4-CD8- and alpha beta TCR+NK1.1+ cells. *J Immunol 145:3209.*

Sykulev, Y., M. Joo, I. Vturina, T. J. Tsomides, and H. N. Eisen. 1996. Evidence that a single peptide-MHC complex on a target cell can elicit a cytolytic T cell response. *Immunity 4:565.*

Sytwu, H. K., R. S. Liblau, and H. O. McDevitt. 1996. The roles of Fas/APO-1 (CD95) and TNF in antigen-induced programmed cell death in T cell receptor transgenic mice. *Immunity 5:17.*

Szopa, T. M., P. A. Titchener, N. D. Portwood, and K. W. Taylor. 1993. Diabetes mellitus due to viruses--some recent developments. *Diabetologia 36:687.*

Takahashi, T., M. Tanaka, C. I. Brannan, N. A. Jenkins, N. G. Copeland, T. Suda, and S. Nagata. 1994. Generalized lymphoproliferative disease in mice, caused by a point mutation in the Fas ligand. *Cell 76:969.*

Takasugi, M., M. R. Mickey, and P. I. Terasaki. 1973. Reactivity of lymphocytes from normal persons on cultured tumor cells. *Cancer Res 33:2898.*

Takayama, H., N. Shinohara, A. Kawasaki, Y. Someya, S. Hanaoka, H. Kojima, H. Yagita, K. Okumura, and Y. Shinkai. 1991. Antigen-specific directional target cell lysis by perforin-negative T lymphocyte clones. *Int Immunol 3:1149.*

Takeda, K., Y. Hayakawa, M. J. Smyth, N. Kayagaki, N. Yamaguchi, S. Kakuta, Y. Iwakura, H. Yagita, and K. Okumura. 2001. Involvement of tumor necrosis factor-related apoptosis-inducing ligand in surveillance of tumor metastasis by liver natural killer cells. *Nat Med 7:94.*

Takeda, K., M. J. Smyth, E. Cretney, Y. Hayakawa, N. Kayagaki, H. Yagita, and K. Okumura. 2002. Critical role for tumor necrosis factor-related apoptosis-inducing ligand in immune surveillance against tumor development. *J Exp Med 195:161.*

Talanian, R. V., X. Yang, J. Turbov, P. Seth, T. Ghayur, C. A. Casiano, K. Orth, and C. J. Froelich. 1997. Granule-mediated killing: pathways for granzyme B-initiated apoptosis. *J Exp Med 186:1323.*

Talmadge, J. E., K. M. Meyers, D. J. Prieur, and J. R. Starkey. 1980. Role of NK cells in tumour growth and metastasis in beige mice. *Nature 284:622.*

Talmadge, J. E., R. H. Wiltrout, D. F. Counts, R. B. Herberman, T. McDonald, and J. R. Ortaldo. 1986. Proliferation of human peripheral blood lymphocytes induced by recombinant human interleukin 2: contribution of large granular lymphocytes and T lymphocytes. *Cell Immunol 102:261.*

Tan, J. T., E. Dudl, E. LeRoy, R. Murray, J. Sprent, K. I. Weinberg, and C. D. Surh. 2001. IL-7 is critical for homeostatic proliferation and survival of naive T cells. *Proc Natl Acad Sci U S A 98:8732.*

Tan, J. T., B. Ernst, W. C. Kieper, E. LeRoy, J. Sprent, and C. D. Surh. 2002. Interleukin (IL)-15 and IL-7 jointly regulate homeostatic proliferation of memory phenotype CD8+ cells but are not required for memory phenotype CD4+ cells. *J Exp Med 195:1523.*

Tang, Y., A. W. Hugin, N. A. Giese, L. Gabriele, S. K. Chattopadhyay, T. N. Fredrickson, D. Kagi, J. W. Hartley, and H. C. Morse, 3rd. 1997. Control of immunodeficiency and lymphoproliferation in mouse AIDS: studies of mice deficient in CD8+ T cells or perforin. *J Virol 71:1808.*

Tao, M. H., and R. Levy. 1993. Idiotype/granulocyte-macrophage colony-stimulating factor fusion protein as a vaccine for B-cell lymphoma. *Nature 362:755.*

Targan, S., E. Grimm, and B. Bonavida. 1980. A single cell marker of active NK cytotoxicity: only a fraction of target binding lymphocytes are killer cells. *J Clin Lab Immunol 4:165.*

Tarleton, R. L. 1990. Depletion of CD8+ T cells increases susceptibility and reverses vaccine-induced immunity in mice infected with Trypanosoma cruzi. *J Immunol 144:717.*

Tarleton, R. L., B. H. Koller, A. Latour, and M. Postan. 1992. Susceptibility of beta 2-microglobulin-deficient mice to Trypanosoma cruzi infection. *Nature 356:338.*

Taub, D. D., J. R. Ortaldo, S. M. Turcovski-Corrales, M. L. Key, D. L. Longo, and W. J. Murphy. 1996. Beta chemokines costimulate lymphocyte cytolysis, proliferation, and lymphokine production. *J Leukoc Biol 59:81.*

Tay, C. H., and R. M. Welsh. 1997. Distinct organ-dependent mechanisms for the control of murine cytomegalovirus infection by natural killer cells. *J Virol 71:267.*

Taylor, H. E., and C. F. A. Culling. 1963. Cytopathic effect in vitro of sensitized homologous and heterologous spleen cells on fibroblasts. *Lab Invest 12:884.*

Tew, J. G., G. J. Thorbecke, and R. M. Steinman. 1982. Dendritic cells in the immune response: characteristics and recommended nomenclature (A report from the Reticuloendothelial Society Committee on Nomenclature). *J Reticuloendothel Soc 31:371.*

Thierness, N., A. David, J. Bernard, P. Jeannesson, and D. Zagury. 1977. Active phosphatiasique acide de la cellule T cytolytique au cours du processure de cytolyse. *CR Acad Sci Paris 285:713.*

Thilenius, A. R., K. A. Sabelko-Downes, and J. H. Russell. 1999. The role of the antigen-presenting cell in Fas-mediated direct and bystander killing: potential in vivo function of Fas in experimental allergic encephalomyelitis. *J Immunol 162:643.*

Thomas, L. 1959. *In: Lawrence, H., ed. Cellular and Humoral Aspects of the Hypersensitive State.:529.*

Thomas, W. D., and P. Hersey. 1998. TNF-related apoptosis-inducing ligand (TRAIL) induces apoptosis in Fas ligand-resistant melanoma cells and mediates CD4 T cell killing of target cells. *J Immunol 161:2195.*

Thomsen, A. R., A. Nansen, J. P. Christensen, S. O. Andreasen, and O. Marker. 1998. CD40 ligand is pivotal to efficient control of virus replication in mice infected with lymphocytic choriomeningitis virus. *J Immunol 161:4583.*

Thomson, A. W., and L. Lu. 1999. Dendritic cells as regulators of immune reactivity: implications for transplantation. *Transplantation 68:1.*

Thorley-Lawson, D. A., L. Chess, and J. L. Strominger. 1977. Suppression of in vitro Epstein-Barr virus infection. A new role for adult human T lymphocytes. *J Exp Med 146:495.*

Thorn, R. M., and C. S. Henney. 1976. Studies on the mechanism of lymphocyte-mediated cytolysis. VI. A reappraisal of the requirement for protein synthesis during T cell-mediated lysis. *J Immunol 116:146.*

Thurner, B., I. Haendle, C. Roder, D. Dieckmann, P. Keikavoussi, H. Jonuleit, A. Bender, C. Maczek, D. Schreiner, P. von den Driesch, E. B. Brocker, R. M. Steinman, A. Enk, E. Kampgen, and G. Schuler. 1999. Vaccination with mage-3A1 peptide-pulsed mature, monocyte-derived dendritic cells expands specific cytotoxic T cells and induces regression of some metastases in advanced stage IV melanoma. *J Exp Med 190:1669.*

Tidball, J. G., D. E. Albrecht, B. E. Lokensgard, and M. J. Spencer. 1995. Apoptosis precedes necrosis of dystrophin-deficient muscle. *J Cell Sci 108 (Pt 6):2197.*

Tilden, A. B., R. Cauda, C. E. Grossi, C. M. Balch, A. D. Lakeman, and R. J. Whitley. 1986. Demonstration of NK cell-mediated lysis of varicella-zoster virus (VZV)-infected cells: characterization of the effector cells. *J Immunol 136:4243.*

Tilney, N. L., and J. L. Gowans. 1971. The sensitization of rats by allografts transplanted to alymphatic pedicles of skin. *J Exp Med 133:951.*

Timmerman, N., and e. al. 1996. Rapid shuttling of NF-AT in discrimination of Ca signals and immunosuppression. *Nature 383:837.*

Timonen, T., A. Ranki, E. Saksela, and P. Hayry. 1979. Human natural cell-mediated cytotoxicity against fetal fibroblasts. III. Morphological and functional characterization of the effector cells. *Cell Immunol 48:121.*

Timonen, T., J. R. Ortaldo, and R. B. Herberman. 1981. Characteristics of human large granular lymphocytes and relationship to natural killer and K cells. *J Exp Med 153:569.*

Timonen, T., C. W. Reynolds, J. R. Ortaldo, and R. B. Herberman. 1982. Isolation of human and rat natural killer cells. *J Immunol Methods 51:269.*

Ting, R., and L. Law. 1965. The role of thymus in transplant resistance induced by polyoma virus. *J Natl Canc Inst 34:521.*

Ting, C. C., G. Shiu, D. Rodrigues, and R. B. Herberman. 1974. Cell-mediated immunity to Friend virus-induced leukemia. *Cancer Res 34:1684.*

Tirosh, R., and G. Berke. 1985a. T-Lymphocyte-mediated cytolysis as an excitatory process of the target. I. Evidence that the target cell may be the site of Ca2+ action. *Cell Immunol 95:113.*

Tirosh, R., and G. Berke. 1985b. Immune cytolysis viewed as a stimulatory process of the target. *Adv Exp Med Biol 184:473.*

Tite, J. P., and C. A. Janeway, Jr. 1984. Cloned helper T cells can kill B lymphoma cells in the presence of specific antigen: Ia restriction and cognate vs. noncognate interactions in cytolysis. *Eur J Immunol 14:878.*

Titus, R. G., G. Milon, G. Marchal, P. Vassalli, J. C. Cerottini, and J. A. Louis. 1987. Involvement of specific Lyt-2+ T cells in the immunological control of experimentally induced murine cutaneous leishmaniasis. *Eur J Immunol 17:1429.*

Tivol, E. A., F. Borriello, A. N. Schweitzer, W. P. Lynch, J. A. Bluestone, and A. H. Sharpe. 1995. Loss of CTLA-4 leads to massive lymphoproliferation and fatal multiorgan tissue destruction, revealing a critical negative regulatory role of CTLA-4. *Immunity 3:541.*

Tomasec, P., V. M. Braud, C. Rickards, M. B. Powell, B. P. McSharry, S. Gadola, V. Cerundolo, L. K. Borysiewicz, A. J. McMichael, and G. W. Wilkinson. 2000. Surface expression of HLA-E, an inhibitor of natural killer cells, enhanced by human cytomegalovirus gpUL40. *Science 287:1031.*

Tomasello, E., L. Olcese, F. Vely, C. Geourgeon, M. Blery, A. Moqrich, D. Gautheret, M. Djabali, M. G. Mattei, and E. Vivier. 1998. Gene structure, expression pattern, and biological activity of mouse killer cell activating receptor-associated protein (KARAP)/DAP-12. *J Biol Chem 273:34115.*

Topalian, S. L., D. Solomon, and S. A. Rosenberg. 1989. Tumor-specific cytolysis by lymphocytes infiltrating human melanomas. *J Immunol 142:3714.*

Topham, D. J., R. A. Tripp, and P. C. Doherty. 1997. CD8+ T cells clear influenza virus by perforin or Fas-dependent processes. *J Immunol 159:5197.*

Tormo, J., K. Natarajan, D. H. Margulies, and R. A. Mariuzza. 1999. Crystal structure of a lectin-like natural killer cell receptor bound to its MHC class I ligand. *Nature 402:623.*

Tortorella, D., B. E. Gewurz, M. H. Furman, D. J. Schust, and H. L. Ploegh. 2000. Viral subversion of the immune system. *Annu Rev Immunol 18:861.*

Townsend, A. R., and J. J. Skehel. 1982. Influenza A specific cytotoxic T-cell clones that do not recognize viral glycoproteins. *Nature 300:655.*

Townsend, A. R., A. J. McMichael, N. P. Carter, J. A. Huddleston, and G. G. Brownlee. 1984. Cytotoxic T cell recognition of the influenza nucleoprotein and hemagglutinin expressed in transfected mouse L cells. *Cell 39:13.*

Townsend, A. R., and J. J. Skehel. 1984. The influenza A virus nucleoprotein gene controls the induction of both subtype specific and cross-reactive cytotoxic T cells. *J Exp Med 160:552.*

Townsend, A. R., J. Rothbard, F. M. Gotch, G. Bahadur, D. Wraith, and A. J. McMichael. 1986. The epitopes of influenza nucleoprotein recognized by cytotoxic T lymphocytes can be defined with short synthetic peptides. *Cell 44:959.*

Townsend, A., and H. Bodmer. 1989. Antigen recognition by class I-restricted T lymphocytes. *Annu Rev Immunol 7:601.*

Tozeren, A., K. L. Sung, and S. Chien. 1989. Theoretical and experimental studies on cross-bridge migration during cell disaggregation. *Biophys J 55:479.*

Trambas, C. M., and G. M. Griffiths. 2003. Delivering the kiss of death. *Nat Immunol 4:399.*

Trapani, J. A., B. S. Kwon, C. A. Kozak, C. Chintamaneni, J. D. Young, and B. Dupont. 1990. Genomic organization of the mouse pore-forming protein (perforin) gene and localization to chromosome 10. Similarities to and differences from C9. *J Exp Med 171:545.*

Trapani, J. A., M. J. Smyth, V. A. Apostolidis, M. Dawson, and K. A. Browne. 1994. Granule serine proteases are normal nuclear constituents of natural killer cells. *J Biol Chem 269:18359.*

Trapani, J. A. 1998. Dual mechanisms of apoptosis induction by cytotoxic lymphocytes. *Int Rev Cytol 182:111.*

Trapani, J. A., D. A. Jans, P. J. Jans, M. J. Smyth, K. A. Browne, and V. R. Sutton. 1998. Efficient nuclear targeting of granzyme B and the nuclear consequences of apoptosis induced by granzyme B and perforin are caspase-dependent, but cell death is caspase-independent. *J Biol Chem 273:27934.*

Trapani, J. A., P. Jans, M. J. Smyth, C. J. Froelich, E. A. Williams, V. R. Sutton, and D. A. Jans. 1998. Perforin-dependent nuclear entry of granzyme B precedes apoptosis, and is not a consequence of nuclear membrane dysfunction. *Cell Death Differ 5:488.*

Trapani, J. A., J. Davis, V. R. Sutton, and M. J. Smyth. 2000. Proapoptotic functions of cytotoxic lymphocyte granule constituents in vitro and in vivo. *Curr Opin Immunol 12:323.*

Trapani, J. A., V. Sutton, K. Y. Thia, Y. Li, C. J. Froelich, D. Jans, M. Sandrin, and K. Browne. 2003. A clathrin/dynamin- and mannose-6-phosphate receptor-independent pathway for granzyme B-induced cell death. *J Cell Biol 160:155.*

Traversari, C., P. van der Bruggen, I. F. Luescher, C. Lurquin, P. Chomez, A. Van Pel, E. De Plaen, A. Amar-Costesec, and T. Boon. 1992. A nonapeptide encoded by human gene MAGE-1 is recognized on HLA-A1 by cytolytic T lymphocytes directed against tumor antigen MZ2-E. *J Exp Med 176:1453.*

Traversari, C., P. van der Bruggen, B. Van den Eynde, P. Hainaut, C. Lemoine, N. Ohta, L. Old, and T. Boon. 1992. Transfection and expression of a gene coding for a human melanoma antigen recognized by autologous cytolytic T lymphocytes. *Immunogenetics 35:145.*

Trenn, G., H. Takayama, and M. V. Sitkovsky. 1987. Exocytosis of cytolytic granules may not be required for target cell lysis by cytotoxic T-lymphocytes. *Nature 330:72.*

Trimble, L. A., and J. Lieberman. 1998. Circulating CD8 T lymphocytes in human immunodeficiency virus-infected individuals have impaired function and downmodulate CD3 zeta, the signaling chain of the T-cell receptor complex. *Blood 91:585.*

Trinchieri, G., and D. Santoli. 1978. Anti-viral activity induced by culturing lymphocytes with tumor-derived or virus-transformed cells. Enhancement of human natural killer cell activity by interferon and antagonistic inhibition of susceptibility of target cells to lysis. *J Exp Med 147:1314.*

Trinchieri, G., D. Granato, and B. Perussia. 1981. Interferon-induced resistance of fibroblasts to cytolysis mediated by natural killer cells: specificity and mechanism. *J Immunol 126:335.*

Trinchieri, G. 1989. Biology of natural killer cells. *Adv Immunol 47:187.*

Trinchieri, G. 1994. Interleukin-12: a cytokine produced by antigen-presenting cells with immunoregulatory functions in the generation of T-helper cells type 1 and cytotoxic lymphocytes. *Blood 84:4008.*

Troye-Blomberg, M., S. Worku, P. Tangteerawatana, R. Jamshaid, K. Soderstrom, G. Elghazali, L. Moretta, M. Hammarstrom, and L.

Mincheva-Nilsson. 1999. Human gamma delta T cells that inhibit the in vitro growth of the asexual blood stages of the Plasmodium falciparum parasite express cytolytic and proinflammatory molecules. *Scand J Immunol 50:642.*

Truneh, A., F. Albert, P. Golstein, and A. M. Schmitt-Verhulst. 1985a. Early steps of lymphocyte activation bypassed by synergy between calcium ionophores and phorbol ester. *Nature 313:318.*

Truneh, A., F. Albert, P. Golstein, and A. M. Schmitt-Verhulst. 1985b. Calcium ionophore plus phorbol ester can substitute for antigen in the induction of cytolytic T lymphocytes from specifically primed precursors. *J Immunol 135:2262.*

Tschopp, J., D. Masson, and K. Stanley. 1986. Structural/functional similarity between proteins involved in complement- and cytotoxic T lymphocyte-mediated cytolysis. *Nature 322:831.*

Tschopp, J., S. Schafer, D. Masson, M. C. Peitsch, and C. Heusser. 1989. Phosphorylcholine acts as a Ca2+-dependent receptor molecule for lymphocyte perforin. *Nature 337:272.*

Tseng, C. T., E. Miskovsky, M. Houghton, and G. R. Klimpel. 2001. Characterization of liver T-cell receptor gammadelta T cells obtained from individuals chronically infected with hepatitis C virus (HCV): evidence for these T cells playing a role in the liver pathology associated with HCV infections. *Hepatology 33:1312.*

Tsotsiashvilli, M., R. Levi, R. Arnon, and G. Berke. 1998. Activation of influenza-specific memory cytotoxic T lymphocytes by Concanavalin A stimulation. *Immunol Lett 60:89.*

Tsuchida, M., Y. Matsumoto, H. Hirahara, H. Hanawa, K. Tomiyama, and T. Abo. 1993. Preferential distribution of V beta 8.2-positive T cells in the central nervous system of rats with myelin basic protein-induced autoimmune encephalomyelitis. *Eur J Immunol 23:2399.*

Tsuchida, T., K. C. Parker, R. V. Turner, H. F. McFarland, J. E. Coligan, and W. E. Biddison. 1994. Autoreactive CD8+ T-cell responses to human myelin protein-derived peptides. *Proc Natl Acad Sci U S A 91:10859.*

Tsukaguchi, K., K. N. Balaji, and W. H. Boom. 1995. CD4+ alpha beta T cell and gamma delta T cell responses to Mycobacterium tuberculosis. Similarities and differences in Ag recognition, cytotoxic effector function, and cytokine production. *J Immunol 154:1786.*

Tsunetsugu-Yokota, Y. 2002. Selective expansion of perforin-positive T cells by immature dendritic cells infected with live Bacillus Calmette-Guerin mycobacteris. *J Leukoc Biol 72:115.*

Turka, L. A., P. S. Linsley, H. Lin, W. Brady, J. M. Leiden, R. Q. Wei, M. L. Gibson, X. G. Zheng, S. Myrdal, D. Gordon, and et al. 1992. T-cell activation by the CD28 ligand B7 is required for cardiac allograft rejection in vivo. *Proc Natl Acad Sci U S A 89:11102.*

Tyler, D. S., S. D. Stanley, C. A. Nastala, A. A. Austin, J. A. Bartlett, K. C. Stine, H. K. Lyerly, D. P. Bolognesi, and K. J. Weinhold. 1990. Alterations in antibody-dependent cellular cytotoxicity during the course of HIV-1 infection. Humoral and cellular defects. *J Immunol 144:3375.*

Uchida, A., and H. Fukata. 1993. Role of NK cell cytotoxic factor against fresh human tumors. *Nat Immun 12:267.*

Uellner, R., M. J. Zvelebil, J. Hopkins, J. Jones, L. K. MacDougall, B. P. Morgan, E. Podack, M. D. Waterfield, and G. M. Griffiths. 1997. Perforin is activated by a proteolytic cleavage during biosynthesis which reveals a phospholipid-binding C2 domain. *Embo J 16:7287.*

Uhr, J. W. 1966. Delayed hypersensitivity. *Physiol Rev 46:359.*

Uhrberg, M., N. M. Valiante, N. T. Young, L. L. Lanier, J. H. Phillips, and P. Parham. 2001. The repertoire of killer cell Ig-like receptor and CD94:NKG2A receptors in T cells: clones sharing identical alpha beta TCR rearrangement express highly diverse killer cell Ig-like receptor patterns. *J Immunol 166:3923.*

Ullberg, M., and M. Jondal. 1981. Recycling and target binding capacity of human natural killer cells. *J Exp Med 153:615.*

Urban, J. L., and H. Schreiber. 1992. Tumor antigens. *Annu Rev Immunol 10:617.*

Valentin, A., M. Rosati, D. J. Patenaude, A. Hatzakis, L. G. Kostrikis, M. Lazanas, K. M. Wyvill, R. Yarchoan, and G. N. Pavlakis. 2002. Persistent HIV-1 infection of natural killer cells in patients receiving highly active antiretroviral therapy. *Proc Natl Acad Sci U S A 99:7015.*

Valitutti, S., M. Dessing, and A. Lanzavecchia. 1993. Role of cAMP in regulating cytotoxic T lymphocyte adhesion and motility. *Eur J Immunol 23:790.*

van Binnendijk, R. S., M. C. Poelen, P. de Vries, H. O. Voorma, A. D. Osterhaus, and F. G. Uytdehaag. 1989. Measles virus-specific human T cell clones. Characterization of specificity and function of CD4+ helper/cytotoxic and CD8+ cytotoxic T cell clones. *J Immunol 142:2847.*

van den Broek, M. F., D. Kagi, R. M. Zinkernagel, and H. Hengartner. 1995. Perforin dependence of natural killer cell-mediated tumor control in vivo. *Eur J Immunol 25:3514.*

van den Broek, M. E., D. Kagi, F. Ossendorp, R. Toes, S. Vamvakas, W. K. Lutz, C. J. Melief, R. M. Zinkernagel, and H. Hengartner. 1996. Decreased tumor surveillance in perforin-deficient mice. *J Exp Med 184:1781.*

van der Bruggen, P., C. Traversari, P. Chomez, C. Lurquin, E. De Plaen, B. Van den Eynde, A. Knuth, and T. Boon. 1991. A gene encoding an antigen recognized by cytolytic T lymphocytes on a human melanoma. *Science 254:1643.*

van der Merwe, P. A. 2002. Formation and function of the immunological synapse. *Curr Opin Immunol 14:293.*

Van der van Merwe P. A. 2002. Formation and function of the immunological synapse. Curr Opin Immunol. 14:293-8]

Van Epps, H. L., M. Terajima, J. Mustonen, T. P. Arstila, E. A. Corey, A. Vaheri, and F. A. Ennis. 2002. Long-lived memory T lymphocyte responses after hantavirus infection. *J Exp Med 196:579.*

Van Parijs, L., Y. Refaeli, J. D. Lord, B. H. Nelson, A. K. Abbas, and D. Baltimore. 1999. Uncoupling IL-2 signals that regulate T cell proliferation, survival, and Fas-mediated activation-induced cell death. *Immunity 11:281.*

Van Pel, A., F. Vessiere, and T. Boon. 1983. Protection against two spontaneous mouse leukemias conferred by immunogenic variants obtained by mutagenesis. *J Exp Med 157:1992.*

van Ravenswaay Claasen, H. H., R. J. van de Griend, D. Mezzanzanica, R. L. Bolhuis, S. O. Warnaar, and G. J. Fleuren. 1993. Analysis of production, purification, and cytolytic potential of bi-specific antibodies reactive with ovarian-carcinoma-associated antigens and the T-cell antigen CD3. *Int J Cancer 55:128.*

van Seventer, G. A., K. C. Kuijpers, R. A. van Lier, E. R. de Groot, L. A. Aarden, and C. J. Melief. 1987. Mechanism of inhibition and induction of cytolytic activity in cytotoxic T lymphocytes by CD3 monoclonal antibodies. *J Immunol 139:2545.*

van Stipdonk, M. J., E. E. Lemmens, and S. P. Schoenberger. 2001. Naive CTLs require a single brief period of antigenic stimulation for clonal expansion and differentiation. *Nat Immunol 2:423.*

Van Stipdonk, M. J., G. Hardenberg, M. S. Bijker, E. E. Lemmens, N. M. Droin, D. R. Green, and S. P. Schoenberger. 2003. Dynamic programming of CD8(+) T lymphocyte responses. *Nat Immunol 4:361.*

Van Wauwe, J., and J. Goossens. 1981. Mitogenic actions of Orthoclone OKT3 on human peripheral blood lymphocytes: effects of monocytes and serum components. *Int J Immunopharmacol 3:203.*

Vance, R. E., J. R. Kraft, J. D. Altman, P. E. Jensen, and D. H. Raulet. 1998. Mouse CD94/NKG2A is a natural killer cell receptor for the nonclassical major histocompatibility complex (MHC) class I molecule Qa-1(b). *J Exp Med 188:1841.*

Vanky, F., S. Argov, and E. Klein. 1981. Tumor biopsy cells participating in systems in which cytotoxicity of lymphocytes is generated. Autologous and allogeneic studies. *Int J Cancer 27:273.*

Vass, K., and H. Lassmann. 1990. Intrathecal application of interferon gamma. Progressive appearance of MHC antigens within the rat nervous system. *Am J Pathol 137:789.*

Veiga-Fernandes, H., U. Walter, C. Bourgeois, A. McLean, and B. Rocha. 2000. Response of naive and memory CD8+ T cells to antigen stimulation in vivo. *Nat Immunol 1:47.*

Vella, A. T., S. Dow, T. A. Potter, J. Kappler, and P. Marrack. 1998. Cytokine-induced survival of activated T cells in vitro and in vivo. *Proc Natl Acad Sci U S A 95:3810.*

Vely, F., M. Peyrat, C. Couedel, J. Morcet, F. Halary, F. Davodeau, F. Romagne, E. Scotet, X. Saulquin, E. Houssaint, N. Schleinitz, A. Moretta, E. Vivier, and M. Bonneville. 2001. Regulation of inhibitory and activating killer-cell Ig-like receptor expression occurs in T cells after termination of TCR rearrangements. *J Immunol 166:2487.*

Vergelli, M., B. Hemmer, P. A. Muraro, L. Tranquill, W. E. Biddison, A. Sarin, H. F. McFarland, and R. Martin. 1997. Human autoreactive CD4+ T cell clones use perforin- or Fas/Fas ligand-mediated pathways for target cell lysis. *J Immunol 158:2756.*

Verret, C. R., A. A. Firmenich, D. M. Kranz, and H. N. Eisen. 1987. Resistance of cytotoxic T lymphocytes to the lytic effects of their toxic granules. *J Exp Med 166:1536.*

Vignaux, F., E. Vivier, B. Malissen, V. Depraetere, S. Nagata, and P. Golstein. 1995. TCR/CD3 coupling to Fas-based cytotoxicity. *J Exp Med 181:781.*

Vilches, C., and P. Parham. 2002. KIR: Diverse, Rapidly Evolving Receptors of Innate and Adaptive Immunity. *Annu Rev Immunol 20:217.*

Vivier, E., E. Tomasello, and P. Paul. 2002. Lymphocyte activation via NKG2D: towards a new paradigm in immune recognition? *Curr Opin Immunol 14:306.*

von Boehmer, H., H. Hengartner, M. Nabholz, W. Lernhardt, M. H. Schreier, and W. Haas. 1979. Fine specificity of a continuously growing killer cell clone specific for H-Y antigen. *Eur J Immunol 9:592.*

von Boehmer, H., P. Kisielow, W. Leiserson, and W. Haas. 1984. Lyt-2- T cell-independent functions of Lyt-2+ cells stimulated with antigen or concanavalin A. *J Immunol 133:59.*

von Herrath, M. G., M. Yokoyama, J. Dockter, M. B. Oldstone, and J. L. Whitton. 1996. CD4-deficient mice have reduced levels of memory cytotoxic T lymphocytes after immunization and show diminished resistance to subsequent virus challenge. *J Virol 70:1072.*

Vose, B. M., F. Vanky, M. Fopp, and E. Klein. 1978. In vitro generation of cytotoxicity against autologous human tumour biopsy cells. *Int J Cancer 21:588.*

Vose, B. M., and G. D. Bonnard. 1982. Human tumour antigens defined by cytotoxicity and proliferative responses of cultured lymphoid cells. *Nature 296:359.*

Vujanovic, N. L. 2001. Role of TNF family ligands in antitumor activity of natural killer cells. *Int Rev Immunol 20:415.*

Vyas, Y. M., K. M. Mehta, M. Morgan, H. Maniar, L. Butros, S. Jung, J. K. Burkhardt, and B. Dupont. 2001. Spatial organization of signal transduction molecules in the NK cell immune synapses during MHC class I-regulated noncytolytic and cytolytic interactions. *J Immunol 167:4358.*

Vyse, T. J., and J. A. Todd. 1996. Genetic analysis of autoimmune disease. *Cell 85:311.*

Wagner, F. H., and M. Rollinghoff. 1973. In vitro induction of tumor-specific immunity: I. Parameters of activation and cytotoxic reactivity of mouse lymphoid cells immunized in vitro against syngeneic and allogeneic plasma cell tumors. *J Exp Med 138:1.*

Wagner, H., and M. Rollinghoff. 1974. T cell-mediated cytotoxicity: discrimination between antigen recognition, lethal hit and cytolysis phase. *Eur J Immunol 4:745.*

Wagner, H., D. Gotze, L. Ptschelinzew, and M. Rollinghoff. 1975. Induction of cytotoxic T lymphocytes against I-region-coded determinants: in vitro evidence for a third histocompatibility locus in the mouse. *J Exp Med 142:1477.*

Wagner, H., A. Starzinski-Powitz, H. Jung, and M. Rollinghoff. 1977. Induction of I region-restricted hapten-specific cytotoxic T lymphocytes. *J Immunol 119:1365.*

Waksman, B. 1963. The pattern of rejection in rat skin homografts and its relation to the vascular network. *Lab Invest 12:46.*

Walczak, H., R. E. Miller, K. Ariail, B. Gliniak, T. S. Griffith, M. Kubin, W. Chin, J. Jones, A. Woodward, T. Le, C. Smith, P. Smolak, R. G. Goodwin, C. T. Rauch, J. C. Schuh, and D. H. Lynch. 1999. Tumoricidal activity of tumor necrosis factor-related apoptosis-inducing ligand in vivo. *Nat Med 5:157.*

Walden, P. R., and H. N. Eisen. 1990. Cognate peptides induce self-destruction of CD8+ cytolytic T lymphocytes. *Proc Natl Acad Sci U S A 87:9015.*

Wallach, D., E. E. Varfolomeev, N. L. Malinin, Y. V. Goltsev, A. V. Kovalenko, and M. P. Boldin. 1999. Tumor necrosis factor receptor and Fas signaling mechanisms. *Annu Rev Immunol 17:331.*

Walsh, C. M., M. Matloubian, C. C. Liu, R. Ueda, C. G. Kurahara, J. L. Christensen, M. T. Huang, J. D. Young, R. Ahmed, and W. R. Clark. 1994a. Immune function in mice lacking the perforin gene. *Proc Natl Acad Sci U S A 91:10854.*

Walsh, C. M., A. A. Glass, V. Chiu, and W. R. Clark. 1994b. The role of the Fas lytic pathway in a perforin-less CTL hybridoma. *J Immunol 153:2506.*

Walsh, C. M., F. Hayashi, D. C. Saffran, S. T. Ju, G. Berke, and W. R. Clark. 1996. Cell-mediated cytotoxicity results from, but may not be critical for, primary allograft rejection. *J Immunol 156:1436.*

Walter, E. A., P. D. Greenberg, M. J. Gilbert, R. J. Finch, K. S. Watanabe, E. D. Thomas, and S. R. Riddell. 1995. Reconstitution of cellular immunity against cytomegalovirus in recipients of allogeneic bone marrow by transfer of T-cell clones from the donor. *N Engl J Med 333:1038.*

Walter, E. A., P. D. Greenberg, M. J. Gilbert, R. J. Finch, K. S. Watanabe, E. D. Thomas, and S. R. Riddell. 1995. Reconstitution of cellular immunity against cytomegalovirus in recipients of allogeneic bone marrow by transfer of T-cell clones from the donor. *N Engl J Med 333:1038.*

Walunas, T. L., D. J. Lenschow, C. Y. Bakker, P. S. Linsley, G. J. Freeman, J. M. Green, C. B. Thompson, and J. A. Bluestone. 1994. CTLA-4 can function as a negative regulator of T cell activation. *Immunity 1:405.*

Wang, Z. E., S. L. Reiner, F. Hatam, F. P. Heinzel, J. Bouvier, C. W. Turck, and R. M. Locksley. 1993. Targeted activation of CD8 cells and infection of beta 2-microglobulin-deficient mice fail to confirm a primary protective role for CD8 cells in experimental leishmaniasis. *J Immunol 151:2077.*

Wang, B., A. Gonzalez, C. Benoist, and D. Mathis. 1996. The role of CD8+ T cells in the initiation of insulin-dependent diabetes mellitus. *Eur J Immunol 26:1762.*

Wang, F., E. Bade, C. Kuniyoshi, L. Spears, G. Jeffery, V. Marty, S. Groshen, and J. Weber. 1999. Phase I trial of a MART-1 peptide vaccine with incomplete Freund's adjuvant for resected high-risk melanoma. *Clin Cancer Res 5:2756.*

Ware, C. F., P. D. Crowe, M. H. Grayson, M. J. Androlewicz, and J. L. Browning. 1992. Expression of surface lymphotoxin and tumor necrosis factor on activated T, B, and natural killer cells. *J Immunol 149:3881.*

Ware, C. F., S. VanArsdale, and T. L. VanArsdale. 1996. Apoptosis mediated by the TNF-related cytokine and receptor families. *J Cell Biochem 60:47.*

Watanabe-Fukunaga, R., C. I. Brannan, N. G. Copeland, N. A. Jenkins, and S. Nagata. 1992. Lymphoproliferation disorder in mice explained by defects in Fas antigen that mediates apoptosis. *Nature 356:314.*

Waterfield, J. D., E. M. Waterfield, and G. Moller. 1975. Lymphocyte-mediated cytotoxicity against tumor cells. I. Con A activated cytotoxic effector cells exhibit immunological specificity. *Cell Immunol 17:392.*

Watson, J. 1979. Continuous proliferation of murine antigen-specific helper T lymphocytes in culture. *J Exp Med 150:1510.*

Waugh, S. M., J. L. Harris, R. Fletterick, and C. S. Craik. 2000. The structure of the pro-apoptotic protease granzyme B reveals the molecular determinants of its specificity. *Nat Struct Biol 7:762.*

Weaver, J., and e. al. 1955. The growth of cells in vivo in diffusion chambers. II. The role of cells in the destruction of homografts. *J Nat'l Canc Inst 15:1737.*

Webb, S., and J. Sprent. 1987. Downregulation of T cell responses by antibodies to the T cell receptor. *J Exp Med 165:584.*

Weber, W. E., W. A. Buurman, M. M. Vandermeeren, and J. C. Raus. 1985. Activation through CD3 molecule leads to clonal expansion of all human peripheral blood T lymphocytes: functional analysis of clonally expanded cells. *J Immunol 135:2337.*

Weber, J. 2002. Peptide vaccines for cancer. *Cancer Invest 20:208.*

Wei, W., and R. R. Lindquist. 1981. Alloimmune cytolytic T lymphocyte activity: triggering and expression of killing mechanisms in cytolytic T lymphocytes. *J Immunol 126:513.*

Wei, S., D. L. Gilvary, B. C. Corliss, S. Sebti, J. Sun, D. B. Straus, P. J. Leibson, J. A. Trapani, A. D. Hamilton, M. J. Weber, and J. Y. Djeu. 2000. Direct tumor lysis by NK cells uses a Ras-independent mitogen-activated protein kinase signal pathway. *J Immunol 165:3811.*

Weiss, W. R., S. Mellouk, R. A. Houghten, M. Sedegah, S. Kumar, M. F. Good, J. A. Berzofsky, L. H. Miller, and S. L. Hoffman. 1990. Cytotoxic T cells recognize a peptide from the circumsporozoite protein on malaria-infected hepatocytes. *J Exp Med 171:763.*

Welsh, R. M., Jr. 1978. Cytotoxic cells induced during lymphocytic choriomeningitis virus infection of mice. I. Characterization of natural killer cell induction. *J Exp Med 148:163.*

Welsh, R. M., K. Karre, M. Hansson, L. A. Kunkel, and R. W. Kiessling. 1981. Interferon-mediated protection of normal and tumor target cells against lysis by mouse natural killer cells. *J Immunol 126:219.*

Welsh R.M., K. Bahl and X. Z. Wang 2004. Apoptosis and loss of virus-specific CD8+ T-cell memory. Curr Opin Immunol. 16:271-6.

Weninger, W., M. A. Crowley, N. Manjunath, and U. H. von Andrian. 2001. Migratory properties of naive, effector, and memory CD8(+) T cells. *J Exp Med 194:953.*

West, W. H., G. B. Cannon, H. D. Kay, G. D. Bonnard, and R. B. Herberman. 1977. Natural cytotoxic reactivity of human lymphocytes against a myeloid cell line: characterization of effector cells. *J Immunol 118:355.*

West, W. H., K. W. Tauer, J. R. Yannelli, G. D. Marshall, D. W. Orr, G. B. Thurman, and R. K. Oldham. 1987. Constant-infusion recombinant interleukin-2 in adoptive immunotherapy of advanced cancer. *N Engl J Med 316:898.*

White, B. 1996. Immunopathogenesis of systemic sclerosis. *Rheum Dis Clin North Am 22:695.*

White, D. W., and J. T. Harty. 1998. Perforin-deficient CD8+ T cells provide immunity to Listeria monocytogenes by a mechanism that is independent of CD95 and IFN-gamma but requires TNF-alpha. *J Immunol 160:898.*

Whitmire, J. K., R. A. Flavell, I. S. Grewal, C. P. Larsen, T. C. Pearson, and R. Ahmed. 1999. CD40-CD40 ligand costimulation is required for generating antiviral CD4 T cell responses but is dispensable for CD8 T cell responses. *J Immunol 163:3194.*

Wicker, L. S., E. H. Leiter, J. A. Todd, R. J. Renjilian, E. Peterson, P. A. Fischer, P. L. Podolin, M. Zijlstra, R. Jaenisch, and L. B. Peterson. 1994. Beta 2-microglobulin-deficient NOD mice do not develop insulitis or diabetes. *Diabetes 43:500.*

Wigzell, H. 1965. Quantitative titrations of mouse H-2 antibodies using Cr51-labeled target cells. *Transplantation 85:423.*

Wigzell, H. 1969. Specific fractionation of immunocompetent cells. *Transplant Rev 5:76.*

Williams, G. T., and C. A. Smith. 1993. Molecular regulation of apoptosis: genetic controls on cell death. *Cell 74:777.*

Williams, N. S., and V. H. Engelhard. 1996. Identification of a population of CD4+ CTL that utilizes a perforin- rather than a Fas ligand-dependent cytotoxic mechanism. *J Immunol 156:153.*

Wilson, D. B. 1963. The reaction of immunologically activated lymphoid cells against homologous target tissue cells in vitro. *J Cell Comp Physiol 62:273.*

Wilson, J. L., L. C. Heffler, J. Charo, A. Scheynius, M. T. Bejarano, and H. G. Ljunggren. 1999. Targeting of human dendritic cells by autologous NK cells. *J Immunol 163:6365.*

Wingren, C., M. P. Crowley, M. Degano, Y. Chien, and I. A. Wilson. 2000. Crystal structure of a gammadelta T cell receptor ligand T22: a truncated MHC-like fold. *Science 287:310.*

Winkler-Pickett, R. T., H. A. Young, A. Kuta, and J. R. Ortaldo. 1991. Analysis of rat natural killer cytotoxic factor (NKCF): mechanism of action and relationship to other cytotoxic/cytostatic factors. *Cell Immunol 135:42.*

Wirthmueller, U., T. Kurosaki, M. S. Murakami, and J. V. Ravetch. 1992. Signal transduction by Fc gamma RIII (CD16) is mediated through the gamma chain. *J Exp Med 175:1381.*

Wizel, B., M. Palmieri, C. Mendoza, B. Arana, J. Sidney, A. Sette, and R. Tarleton. 1998. Human infection with Trypanosoma cruzi induces parasite antigen-specific cytotoxic T lymphocyte responses. *J Clin Invest 102:1062.*

Wong, F. S., I. Visintin, L. Wen, R. A. Flavell, and C. A. Janeway, Jr. 1996. CD8 T cell clones from young nonobese diabetic (NOD) islets can transfer rapid onset of diabetes in NOD mice in the absence of CD4 cells. *J Exp Med 183:67.*

Wong, P., and E. G. Pamer. 2003. Feedback regulation of pathogen-specific T cell priming. *Immunity 18:499.*

Woods, G., K. Kitagami, and A. Ochi. 1989. Evidence for an involvement of T4+ cytotoxic T cells in tumor immunity. *Cell Immunol 118:126.*

Wright, S. C., and B. Bonavida. 1981. Selective lysis of NK-sensitive target cells by a soluble mediator released from murine spleen cells and human peripheral blood lymphocytes. *J Immunol 126:1516.*

Wright, S. C., and B. Bonavida. 1982. Studies on the mechanism of natural killer (NK) cell-mediated cytotoxicity (CMC). I. Release of cytotoxic factors specific for NK-sensitive target cells (NKCF) during co-culture of NK effector cells with NK target cells. *J Immunol 129:433.*

Wright, S. C., and B. Bonavida. 1983. Studies on the mechanism of natural killer cell-mediated cytotoxicity. IV. Interferon-induced inhibition of NK target cell susceptibility to lysis is due to a defect in their ability to stimulate release of natural killer cytotoxic factors (NKCF). *J Immunol 130:2965.*

Wright, S. C., and B. Bonavida. 1983. YAC-1 variant clones selected for resistance to natural killer cytotoxic factors are also resistant to natural killer cell-mediated cytotoxicity. *Proc Natl Acad Sci U S A 80:1688.*

Wright, S. C., M. L. Weitzen, R. Kahle, G. A. Granger, and B. Bonavida. 1983. Studies on the mechanism of natural killer cytotoxicity. II. coculture of human PBL with NK-sensitive or resistant cell lines stimulates release of natural killer cytotoxic factors (NKCF) selectively cytotoxic to NK-sensitive target cells. *J Immunol 130:2479.*

Wright, S. C., and B. Bonavida. 1987. Studies on the mechanism of natural killer cell-mediated cytotoxicity. VII. functional comparison of human natural killer cytotoxic factors with recombinant lymphotoxin and tumor necrosis factor. *J Immunol 138:1791.*

Wrightsman, R. A., K. A. Luhrs, D. Fouts, and J. E. Manning. 2002. Paraflagellar rod protein-specific CD8+ cytotoxic T lymphocytes target Trypanosoma cruzi-infected host cells. *Parasite Immunol 24:401.*

Wu, Y. J., W. T. Tian, R. M. Snider, C. Rittershaus, P. Rogers, L. LaManna, and S. H. Ip. 1988. Signal transduction of gamma/delta T cell antigen receptor with a novel mitogenic anti-delta antibody. *J Immunol 141:1476.*

Wu, Y., and Y. Liu. 1994. Viral induction of co-stimulatory activity on antigen-presenting cells bypasses the need for CD4+ T-cell help in CD8+ T-cell responses. *Curr Biol 4:499.*

Wu, Z., E. R. Podack, J. M. McKenzie, K. J. Olsen, and M. Zakarija. 1994. Perforin expression by thyroid-infiltrating T cells in autoimmune thyroid disease. *Clin Exp Immunol 98:470.*

Wu, J., Y. Song, A. B. Bakker, S. Bauer, T. Spies, L. L. Lanier, and J. H. Phillips. 1999. An activating immunoreceptor complex formed by NKG2D and DAP10. *Science 285:730.*

Wu, J., V. Groh, and T. Spies. 2002. T cell antigen receptor engagement and specificity in the recognition of stress-inducible MHC class I-related chains by human epithelial gamma delta T cells. *J Immunol 169:1236.*

Wucherpfennig, K. W., J. Newcombe, H. Li, C. Keddy, M. L. Cuzner, and D. A. Hafler. 1992. Gamma delta T-cell receptor repertoire in acute multiple sclerosis lesions. *Proc Natl Acad Sci U S A 89:4588.*

Wyllie, A. H., R. G. Morris, A. L. Smith, and D. Dunlop. 1984. Chromatin cleavage in apoptosis: association with condensed chromatin morphology and dependence on macromolecular synthesis. *J Pathol 142:67.*

Yamashita, K., K. Yui, M. Ueda, and A. Yano. 1998. CTL-mediated lysis of Toxoplasma gondii-infected target cells does not lead to death of intracellular parasites. *Infect Immun 66:4651.*

Yang, Y., Q. Su, I. S. Grewal, R. Schilz, R. A. Flavell, and J. M. Wilson. 1996. Transient subversion of CD40 ligand function diminishes immune responses to adenovirus vectors in mouse liver and lung tissues. *J Virol 70:6370.*

Yang, Y., and J. M. Wilson. 1996. CD40 ligand-dependent T cell activation: requirement of B7-CD28 signaling through CD40. *Science 273:1862.*

Yannelli, J. R., J. A. Sullivan, G. L. Mandell, and V. H. Engelhard. 1986. Reorientation and fusion of cytotoxic T lymphocyte granules after interaction with target cells as determined by high resolution cinemicrography. *J Immunol 136:377.*

Yasukawa, M., A. Inatsuki, and Y. Kobayashi. 1988. Helper activity in antigen-specific antibody production mediated by CD4+ human cytotoxic T cell clones directed against herpes simplex virus. *J Immunol 140:3419.*

Yasukawa, M., Y. Yakushijin, H. Hasegawa, M. Miyake, Y. Hitsumoto, S. Kimura, N. Takeuchi, and S. Fujita. 1993. Expression of perforin and membrane-bound lymphotoxin (tumor necrosis factor-beta) in virus-specific CD4+ human cytotoxic T-cell clones. *Blood 81:1527.*

Yasukawa, M., H. Ohminami, J. Arai, Y. Kasahara, Y. Ishida, and S. Fujita. 2000. Granule exocytosis, and not the fas/fas ligand system, is the main pathway of cytotoxicity mediated by alloantigen-specific CD4(+) as well as CD8(+) cytotoxic T lymphocytes in humans. *Blood 95:2352.*

Yee, C., S. R. Riddell, and P. D. Greenberg. 1997. Prospects for adoptive T cell therapy. *Curr Opin Immunol 9:702.*

Yee, C., J. A. Thompson, P. Roche, D. R. Byrd, P. P. Lee, M. Piepkorn, K. Kenyon, M. M. Davis, S. R. Riddell, and P. D. Greenberg. 2000. Melanocyte destruction after antigen-specific immunotherapy of melanoma: direct evidence of t cell-mediated vitiligo. *J Exp Med 192:1637.*

Yokoyama, W. M., and B. F. Plougastel. 2003. Immune functions encoded by the natural killer gene complex. *Nat Rev Immunol 3:304.*

Yoon, J. W., C. S. Yoon, H. W. Lim, Q. Q. Huang, Y. Kang, K. H. Pyun, K. Hirasawa, R. S. Sherwin, and H. S. Jun. 1999. Control of autoimmune diabetes in NOD mice by GAD expression or suppression in beta cells. *Science 284:1183.*

Yoon, J. W., and H. S. Jun. 2001. Cellular and molecular pathogenic mechanisms of insulin-dependent diabetes mellitus. *Ann N Y Acad Sci 928:200.*

Yoshimoto, T., A. Bendelac, J. Hu-Li, and W. E. Paul. 1995. Defective IgE production by SJL mice is linked to the absence of CD4+, NK1.1+ T cells that promptly produce interleukin 4. *Proc Natl Acad Sci U S A 92:11931.*

Young, W. W., Jr., S. I. Hakomori, J. M. Durdik, and C. S. Henney. 1980. Identification of ganglio-N-tetraosylceramide as a new cell surface marker for murine natural killer (NK) cells. *J Immunol 124:199.*

Young, J. D., Z. A. Cohn, and E. R. Podack. 1986. The ninth component of complement and the pore-forming protein (perforin 1) from cytotoxic T cells: structural, immunological, and functional similarities. *Science 233:184.*

Young, H. A., and J. R. Ortaldo. 1987. One-signal requirement for interferon-gamma production by human large granular lymphocytes. *J Immunol 139:724.*

Young, L. H., L. S. Klavinskis, M. B. Oldstone, and J. D. Young. 1989. In vivo expression of perforin by CD8+ lymphocytes during an acute viral infection. *J Exp Med 169:2159.*

Young, L. H., S. V. Joag, L. M. Zheng, C. P. Lee, Y. S. Lee, and J. D. Young. 1990. Perforin-mediated myocardial damage in acute myocarditis. *Lancet 336:1019.*

Young, N. T., M. Uhrberg, J. H. Phillips, L. L. Lanier, and P. Parham. 2001. Differential expression of leukocyte receptor complex-encoded Ig-like receptors correlates with the transition from effector to memory CTL. *J Immunol 166:3933.*

Yu, Y., M. Hagihara, K. Ando, B. Gansuvd, H. Matsuzawa, T. Tsuchiya, Y. Ueda, H. Inoue, T. Hotta, and S. Kato. 2001. Enhancement of human cord blood CD34+ cell-derived NK cell cytotoxicity by dendritic cells. *J Immunol 166:1590.*

Yunis, E. J., and D. B. Amos. 1971. Three closely linked genetic systems relevant to transplantation. *Proc Natl Acad Sci U S A 68:3031.*

Yurovsky, V. V., P. A. Sutton, D. H. Schulze, F. M. Wigley, R. A. Wise, R. F. Howard, and B. White. 1994. Expansion of selected V delta 1+ gamma delta T cells in systemic sclerosis patients. *J Immunol 153:881.*

Zagury, D., J. Bernard, N. Thierness, M. Feldman, and G. Berke. 1975. Isolation and characterization of individual functionally reactive cytotoxic T lymphocytes. Conjugation, killing and recycling at the single cell level. *Eur J Immunol 5:818.*

Zagury, D., J. Bernard, P. Jeannesson, N. Thiernesse, and J. C. Cerottini. 1979. Studies on the mechanism of T cell-mediated lysis at the single effector cell level. I. Kinetic analysis of lethal hits and target cell lysis in multicellular conjugates. *J Immunol 123:1604.*

Zajac, A. J., D. G. Quinn, P. L. Cohen, and J. A. Frelinger. 1996. Fas-dependent CD4+ cytotoxic T-cell-mediated pathogenesis during virus infection. *Proc Natl Acad Sci U S A 93:14730.*

Zajac, A. J., J. N. Blattman, K. Murali-Krishna, D. J. Sourdive, M. Suresh, J. D. Altman, and R. Ahmed. 1998. Viral immune evasion due to persistence of activated T cells without effector function. *J Exp Med 188:2205.*

Zakai, N., R. G. Kulka, and A. Loyter. 1977. Membrane ultrastructural changes during calcium phosphate-induced fusion of human erythrocyte ghosts. *Proc Natl Acad Sci U S A 74:2417.*

Zaks, T. Z., D. B. Chappell, S. A. Rosenberg, and N. P. Restifo. 1999. Fas-mediated suicide of tumor-reactive T cells following activation by specific tumor: selective rescue by caspase inhibition. *J Immunol 162:3273.*

Zalman, L. S., M. A. Brothers, and H. J. Muller-Eberhard. 1988. Self-protection of cytotoxic lymphocytes: a soluble form of homologous restriction factor in cytoplasmic granules. *Proc Natl Acad Sci U S A 85:4827.*

Zamai, L., M. Ahmad, I. M. Bennett, L. Azzoni, E. S. Alnemri, and B. Perussia. 1998. Natural killer (NK) cell-mediated cytotoxicity: differential

use of TRAIL and Fas ligand by immature and mature primary human NK cells. *J Exp Med 188:2375.*
Zand, M. S., Y. Li, W. Hancock, X. C. Li, P. Roy-Chaudhury, X. X. Zheng, and T. B. Strom. 2000. Interleukin-2 and interferon-gamma double knockout mice reject heterotopic cardiac allografts. *Transplantation 70:1378.*
Zanovello, P., V. Cerundolo, V. Bronte, M. Giunta, M. Panozzo, G. Biasi, and D. Collavo. 1989. Resistance of lymphokine-activated T lymphocytes to cell-mediated cytotoxicity. *Cell Immunol 122:450.*
Zanovello, P., V. Bronte, A. Rosato, P. Pizzo, and F. Di Virgilio. 1990. Responses of mouse lymphocytes to extracellular ATP. II. Extracellular ATP causes cell type-dependent lysis and DNA fragmentation. *J Immunol 145:1545.*
Zarling, J. M., R. C. Nowinski, and F. H. Bach. 1975. Lysis of leukemia cells by spleen cells of normal mice. *Proc Natl Acad Sci U S A 72:2780.*
Zeine, R., R. Pon, U. Ladiwala, J. P. Antel, L. G. Filion, and M. S. Freedman. 1998. Mechanism of gammadelta T cell-induced human oligodendrocyte cytotoxicity: relevance to multiple sclerosis. *J Neuroimmunol 87:49.*
Zhang, X., S. Sun, I. Hwang, D. F. Tough, and J. Sprent. 1998. Potent and selective stimulation of memory-phenotype CD8+ T cells in vivo by IL-15. *Immunity 8:591.*
Zhang, D., P. Shankar, Z. Xu, B. Harnisch, G. Chen, C. Lange, S. J. Lee, H. Valdez, M. M. Lederman, and J. Lieberman. 2002. Most antiviral CD8 T cells during chronic viral infection do not express high levels of perforin and are not directly cytotoxic. *Blood.*
Zheng, L. M., A. Zychlinsky, C. C. Liu, D. M. Ojcius, and J. D. Young. 1991. Extracellular ATP as a trigger for apoptosis or programmed cell death. *J Cell Biol 112:279.*
Zheng, L., G. Fisher, R. E. Miller, J. Peschon, D. H. Lynch, and M. J. Lenardo. 1995. Induction of apoptosis in mature T cells by tumour necrosis factor. *Nature 377:348.*
Zhou, S., R. Ou, L. Huang, and D. Moskophidis. 2002. Critical role for perforin-, Fas/FasL-, and TNFR1-mediated cytotoxic pathways in down-regulation of antigen-specific T cells during persistent viral infection. *J Virol 76:829.*
Zinkernagel, R. M., and P. C. Doherty. 1974. Restriction of in vitro T cell-mediated cytotoxicity in lymphocytic choriomeningitis within a syngeneic or semiallogeneic system. *Nature 248:701.*
Zinkernagel, R. M., and P. C. Doherty. 1975. H-2 compatability requirement for T-cell-mediated lysis of target cells infected with lymphocytic choriomeningitis virus. Different cytotoxic T-cell specificities are associated with structures coded for in H-2K or H-2D. *J Exp Med 141:1427.*
Zinkernagel, R. M., and P. C. Doherty. 1977. Major transplantation antigens, viruses, and specificity of surveillance T cells. *Contemp Top Immunobiol 7:179.*
Zinkernagel, R. M., and P. C. Doherty. 1979. MHC-restricted cytotoxic T cells: studies on the biological role of polymorphic major transplantation

antigens determining T-cell restriction-specificity, function, and responsiveness. *Adv Immunol 27:51.*

Zinkernagel, R. M., E. Haenseler, T. Leist, A. Cerny, H. Hengartner, and A. Althage. 1986. T cell-mediated hepatitis in mice infected with lymphocytic choriomeningitis virus. Liver cell destruction by H-2 class I-restricted virus-specific cytotoxic T cells as a physiological correlate of the 51Cr-release assay? *J Exp Med 164:1075.*

Zinkernagel, R. M., and P. C. Doherty. 1997. The discovery of MHC restriction. *Immunol Today 18:14.*

Zinkernagel, R. M. 2002. On cross-priming of MHC class I-specific CTL: rule or exception? *Eur J Immunol 32:2385.*

Zitvogel, L., J. I. Mayordomo, T. Tjandrawan, A. B. DeLeo, M. R. Clarke, M. T. Lotze, and W. J. Storkus. 1996. Therapy of murine tumors with tumor peptide-pulsed dendritic cells: dependence on T cells, B7 costimulation, and T helper cell 1-associated cytokines. *J Exp Med 183:87.*

Zitvogel, L. 2002. Dendritic and natural killer cells cooperate in the control/switch of innate immunity. *J Exp Med 195:F9.*

Zocchi, M. R., M. Ferrarini, and C. Rugarli. 1990. Selective lysis of the autologous tumor by delta TCS1+ gamma/delta+ tumor-infiltrating lymphocytes from human lung carcinomas. *Eur J Immunol 20:2685.*

Zychlinsky, A., L. M. Zheng, C. C. Liu, and J. D. Young. 1991. Cytolytic lymphocytes induce both apoptosis and necrosis in target cells. *J Immunol 146:393.*

INDEX